CAPACITORS

Capacitors

R. P. Deshpande

B. Tech. Hon. Elect. (IIT, Bombay)
Fellow, The Institution of Engineers (India)

Mc Graw Hill Education

New York Chicago San Francisco
Athens London Madrid
Mexico City Milan New Delhi
Singapore Sydney Toronto

1 2 3 4 5 6 7 8 9 0 QVS/QVS 1 2 0 9 8 7 6 5 4

ISBN 978-0-07-184856-5
MHID 0-07-184856-8

The sponsoring editor for this book was Michael McCabe and the production supervisor was Pamela A. Pelton. The art director for the cover was Jeff Weeks.

This book was previously published as *Capacitors: Technology and Trends* by McGraw-Hill Education (India) Private Limited, New Delhi, copyright © 2013.

To

All electrical and electronic engineering professionals

ABOUT THE AUTHOR

R. P. Deshpande has more than 40 years of experience in the capacitor technology sector. He has worked as a consultant with capacitor manufacturing organizations, both growth-oriented start-ups and established companies, to help them achieve leading industry positions. Throughout his career, Mr. Deshpande has pioneered the development of many capacitor products, technologies, processes, and related applications. His recent research has focused on ultracapacitors, and his work on the use of capacitors for energy storage and alternative energy is widely recognized. Mr. Deshpande is a Fellow of the Institution of Engineers (India) and an electrical engineering graduate of the Indian Institute of Technology, Mumbai.

CONTENTS

FOREWORD

The role of capacitors to help improve grid efficiency and reliability can hardly be overstated. By producing reactive power, capacitors compensate for the reactive power consumption of electrical motors, transformers, etc. This results in more stable power grids with increased transmission capacity and reduced losses. Though an elementary electrical device, its applications in emerging fields, such as smart-grids and renewable energy, have underscored capacitor's versatility and business value.

The book "Capacitors" is an authoritative account about capacitors and related developments. Mr. Deshpande's deep theoretical knowledge and an envious industry experience provide for ample academic treatment about the subject matter and, an evolutionary perspective of capacitor technology.

The book covers capacitors for both electrical and electronic fields in sufficient details, including latest development trends. Possibility of using ultracapacitors for grid stability and power quality through active power management are very interesting. The new breed ultracapacitors are seen to have opened many new avenues for capacitor applications.

Importantly for electrical or power engineering students, as also for electronic engineers, the book brings forward capacitors' diverse business applications and this, in my opinion, could prove inspirational for many to choose it as a career option. As a reference guide, this work is equally valuable to young and mid-career professionals and as also to entrepreneurs connected with capacitors.

I recommend this book to anyone who wants to learn about the exciting field of capacitors, and urge colleges and universities to make it available through their libraries.

J.S.S. Rao
Principal Director
National Power Training Institute
Ministry of Power, Government of India

FOREWORD

Capacitor, a basic component in electrical and electronic engineering, engulfs varied functions in electronic circuits, automobiles traction, mobiles, motors and power supplies, to mention a few. The author, in this book, brings together up to date information on most present-day capacitors. This book is the outcome of vast experience of the author in the capacitor industry.

The strength of this book lies in the comprehensive technical data and latest manufacturing practices relevant to the users in the capacitor industry. The richness of the book lies in its wide coverage of the subject matter, starting with a brief history of capacitors, their types, as also suitability for different applications. It dwells at length on basics like dielectrics, electrodes, construction and manufacture of capacitors and describes the latest trends and developments taking place in capacitor fields. All types of dielectrics and capacitors including XY safety capacitors are covered. Several applications of capacitors in power electronic circuits, automotive industry, power factor management, harmonic control are discussed in separate chapters.

Electrochemical capacitors, which have come up as a new class by themselves in addition to hitherto known electrostatic and electrolytic capacitors, hold huge potential. Of these, ultracapacitors (also known as supercapacitors), already in use in many parts of the world, but not yet properly known, hold a promising future in mobiles, computers and large power grid systems. In fact, the new capacitors dealt in the book are yet to find a place in study material or text books.

The book also gives useful data on practical aspects on the best ways to select appropriate products from alternatives available in the market, or for making the most economical choice for a given application. It also guides on proper testing methods for a capacitor user to ensure best performance of their product. Causes of field failures and guidelines to prevent capacitor failures presented in this book will be useful to users in designing long life products.

Overall, capacitor users in most electrical and electronic industries will find the book as a ready reference for most of their design work and for ensuring proper selection and testing of capacitors.

I am happy to get an opportunity to read this work on capacitors as laid out by the author in this book. The book will serve as a useful study material for students, as well as a source of information for researchers.

Dr. M.L. Kothari
Ex-Professor
Department of Electrical Engineering,
IIT Delhi

PREFACE

With the advent of new technology, the application of capacitors has exploded. Though a fundamental electrical component, their utility seems limitless. Today, these are used extensively in many modern appliances and gadgets and still newer applications are emerging in diverse industries.

In the auto sector, introduction of energy saving systems—such as auto ignition, CDI units, regenerative braking—has spurred the demand for capacitors. A modern car may use as many as 1700 capacitors for various functions and its accessories. Mobile phones, computers, entertainment industry and other modern day systems are consuming unprecedented quantities of capacitors. Spaceships and satellite communication equipment have capacitors working reliably for decades under extreme conditions.

As expected then, to keep up with these diverse demands, capacitor technology itself has undergone considerable changes over the past few decades. The evolution has been rapid.

Capacitors have become smaller in size, and eco-friendly technologies have emerged. Besides changes in dielectrics, impregnants and electrodes, metallized films have undergone transition, and improved materials with better properties are making entry in new generation capacitors. Range and variety of capacitors has extended many fold. For instance, tantalum polymer and Niobium capacitors are replacing conventional Tantalum capacitors in many applications. Technology for manufacturing has transformed with most production made now on high speed machines, with faster output and precision quality products. Additionally, chip technologies, SMDs, coupled with electronically controlled variable capacitors are now commonplace.

Through this book, I wish to share some of my experience and knowledge areas for the benefit of readers. The book provides a brief history of capacitors, as also the state of technology as of today. It starts with a brief history of capacitors. Capacitor basics, including its physics, like dipole moments, and mechanism of current flow in capacitors are explained. Mathematical equations and a good part of theoretical treatment have been purposely kept to bare minimum.

Properties of modern day dielectrics are discussed in the next chapter, including films, oils, glass, ceramics and others. Various types of dielectrics, like paper, and plastic film capacitors (both foil type and metallized construction), mica, glass, ceramic and others are in common usage. Metal oxides serve as the dielectric in electrolytic capacitors. Plastic films, mainly polypropylene and polyester, form a major chunk of capacitors today, while hazy PP film is used for oil filled A.C. capacitors. Others like PC, PS, PTFE, and others are used for many applications. Chapter 3 describes all these films, including new ones like PPS, PEN replacing present ones. Metallized films are most common due their cost and self-healing nature, and form the subject matter of fourth chapter. Self-healing process of these capacitors is fully detailed, so also merits and demerits of metallized and non-metallized films.

Capacitors are classified on the basis of dielectrics used, and another classification is based on their usage. Both these classifications are discussed along with their peculiarities and applications. Characteristics of various dielectrics are discussed in respect of their critical properties.

In the traditional electricity sector, new methods of power factor control have emerged and, along with power factor correction, harmonic control and power quality through electronic circuits is becoming common. Use of electronics in this process has put more stringent demands on the system as also on capacitors. Furthermore, with utility companies and industries stressing on efficient energy utilization, capacitors are increasingly playing a significant role at LV and HV supply sides. IGBT, commutation and snubber capacitors, used in these and other applications are gaining popularity.

With utility companies and industries stressing on efficient energy utilization, capacitors are increasingly playing a significant role at LV and HV supply sides. The definition of power factor has changed, and new methods of power factor control have emerged. Power capacitor switching is a typical operation, which could turn dangerous for the switchgear and the system and is a separate subject by itself. Special capacitor duty contactors have been developed, and one needs to take due care in fuse ratings and capacitor selection along the length of power system. Three chapters are dedicated to power factor correction, power capacitors, their effect on power quality, and capacitor switching. Harmonics have become common with increasing use of electronics in process and motor speed control. Harmonics generation and their mitigation in power systems are detailed in a separate chapter.

Capacitors for fans, motors, household appliances and lighting consume huge quantities of capacitors for either running or for power factor improvement. Capacitor type energy saving fan regulators have driven out conventional resistance type regulators; and CFL are replacing incandescent lamps and tube lights. Capacitors for these applications are covered in separate chapters, and recommended ratings of capacitors for both are tabulated in appendix. Manufacturing process of film / foil and metallized capacitors is outlined in a separate chapter. In the auto sector, advent of energy saving systems—such as auto ignition, CDI units, regenerative braking—has spurred the demand for capacitors. One chapter is devoted to discussion on auto ignition systems, both conventional and the new generation Capacitor Discharge Ignition (CDI) units.

Power electronics make extensive use of snubber, commutation and damping capacitors, as also DC link capacitors in PWM and grid control circuits. These capacitors could be inductive or non-inductive types, and among these are metallized / non-metallized film, vacuum capacitors, glass dielectric, feed through type, trimmers, variable and voltage dependent capacitors to mention a few. Electromagnetic RF interference (EMI) is causing concern in most circuits today with the advent of chopper circuits and PMW motors, electronic control circuits, electronic regulators and dimmers. Further, XY safety capacitors are an integral part of large number of equipment for the safety of instruments and personnel. These come in wide variety of sizes, configurations and ratings. Energy storage capacitors are widely used in pulsed circuits and energy storage devices for space and military applications. All these capacitors are covered in detail from Chapters 15 through 19.

Electrolytic capacitors are used in large numbers in many applications requiring large values, mostly in DC applications (except AC motor start use) from small PCB mounted to large energy storage for ships and military applications. With variety of voltage and value ranges, Aluminium and Tantalum capacitors of various construction and energy storage capacity are widely used in most electronic and electrical applications. No rectifiers are complete without their stabilizing effect. Chapter 10 deals with electrolytic capacitors of all types. New generation Niobium and Tantalum polymer capacitors have made entry in electrolytic field, and find a place in this chapter.

Ceramic capacitors are widely preferred for their small size and cost, as well as their ideal suitability in various applications. From highly stable COG to cheapest class II varieties, and from coupling, isolation and RF interference, they are found everywhere. Mica's thermal, electrical, and chemical properties make for excellent capacitors. Mica capacitors have extreme stability over temperature and frequency ranges (even in GHz), very low drift and desirable properties even for very high voltages. Ceramic and mica capacitors are dealt in Chapters 11 and 12 in detail.

Electrochemical capacitors, developed over the past few decades, are gaining momentum. Farad is no more unthinkably large a unit and capacitors of thousands of Farads are available. A class of these capacitors, known as ultracapacitors (or supercapacitors), are changing the way energy was being stored, and are supplementing or even replacing batteries in some applications. Ultracapacitors are adding to reliability and stability of grid systems, and helping energy efficient vehicles and automobiles with new alternative technologies. Mobile phones and computers are already using them in large numbers. A separate chapter is devoted to development of these capacitors, their principle, construction and applications.

The book concludes with a discussion on capacitor failure mechanism and their mitigation techniques. It becomes necessary to ensure that when a capacitor fails at the end of its life, it does not cause explosion or harm the surrounding environment. Special constructional features are incorporated in some fail-safe capacitors. Steps towards proper utilization of capacitors and due care can substantially increase capacitor life and also reduce the failure rate.

Wherever possible, on-going developments and research topics are highlighted. Effort is made to include the latest developments in respective fields. Useful tables for capacitor applications and a list of Indian Standard Specifications and their IEC Standard equivalents are given at the end of the book for ready reference. I hope this book serves as a good reference on capacitors. The contents herein could be useful for engineering students, researchers, and industry practitioners alike.

R. P. Deshpande

ACKNOWLEDGEMENTS

I have been fortunate to receive valuable suggestions and guidelines from two experts in academic field with large experience and knowledge. Dr M. U. Deshpande, ex-advisor to AICTE, Director, Distance Education Council, New Delhi, ex-principal of VNIT and VRCE, Nagpur has notable contribution in several spheres of education and research to his credit. Dr M. L. Kothari, Ex-Professor, I.I.T., Delhi, has vast lifetime academic and professional experience and is currently Emeritus Fellow with I.I.T., Delhi.

Particular mention must be made of the tireless support from my wife Aruna and son Abhijeet in editing and updating at various stages. They, along with daughter-in-law Navita were the inspiration to write this book, and have worked with me to complete the work. My friends and well-wishers have helped me by giving their precious time, data and personal views to improve the book. I am deeply obliged to them all.

I want to thank the entire team at McGraw Hill Education (India), who have worked meticulously towards this publication.

R. P. Deshpande

CAPACITORS

1

INTRODUCTION TO CAPACITORS

1.1 HISTORY OF CAPACITOR DEVELOPMENT

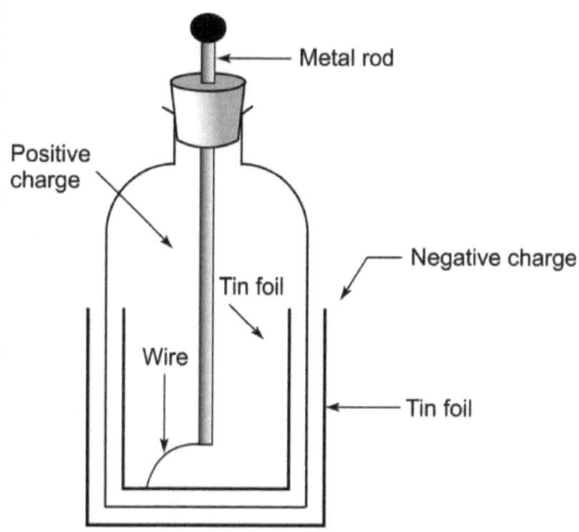

Fig. 1.1 *Leyden jar.*

The Leyden jar (Fig. 1.1) was originally invented in 1745 by Pieter van Musschenbroek, physics and mathematics professor at the University of Leiden, the Netherlands. It was a device used to build and store static electricity. A Leyden jar consists of a glass jar with an outer and inner metal coating covering the bottom and sides nearly to the neck. A brass rod with an external knob passes through a wooden stopper and is connected to the inner coating by a loose chain. When an electrical charge is applied to the external knob (by means of a static generator), positive and negative charges accumulate on the two metal coatings respectively, but are unable to discharge due to the glass between them. The charges will hold each other in equilibrium until a discharge path is provided. If the inner and outer layers of foil are shorted by a conductor, the opposite charges cause a spark that discharges the jar. Leyden jars were first used to store electricity in experiments, and later as condensers in early wireless equipment.

Records indicate that a German scientist named Ewald Georg von Kleist also invented the capacitor in November 1745. Now both these scientists are credited with the invention of capacitors as they worked around the same time independently to develop the first capacitor. Benjamin Franklin worked with the Leyden jar in his experiments with electricity and soon found that a flat piece of glass worked as well as the jar model, prompting him to develop the flat capacitor, or the Franklin square. Years later, English chemist Michael Faraday pioneered the first practical applications for the capacitor in trying to store unused electrons from his experiments. The capacitance unit Farad was coined in his honour.

Capacitors were originally known as condensers, a term that is still occasionally used today (e.g. auto condensers). The term was first used for this purpose by Alessandro Volta in 1782, with reference to the device's ability to store a higher density of electric charge than a normal isolated conductor and their value in μF (micro farads) was "capacity". In the late 1950s it was decided to harmonize the nomenclature of most components, and the term "capacitor" was coined, to fall in line with "resistor" and "inductor", and similarly, its value came to be called "capacitance".

Leyden jars or more powerful devices employing flat glass plates alternating with foil conductors were used exclusively up to about 1900, when the invention of wireless radio created a demand for standard capacitors, and the development of higher frequency applications required capacitors with lower inductance. A more compact construction began to be used with a flexible dielectric sheet such as oiled paper sandwiched between sheets of metal foil, rolled or folded into a small package.

There are a large number and types of capacitors in use today: Paper, plastic, ceramic, vacuum, glass and many more. We have fixed and variable capacitors and trimmers. Then a number of types of electrolytic capacitors are available. To add to the list, electrochemical capacitors have been developed over the past few years. There are a large variety of capacitors to suit different applications and functions.

The history of modern day A.C. capacitors can be viewed from the developments since the 1930s. It will be interesting to see the evolution of present day capacitors as shown in Table 1.1 below:

Table 1.1 *AC Capacitor Development*

	Paper/ Foil + Tr. Oil	*Paper/Foil + PCB**	*Paper/PP/Foil + PCB*	*PP/Paper /foil + Non-PCB oil*	*Met. PP + oil or resin*	*MKV*	*Met. PP dry/ oil*
Year	< 1940	1940–42	1965–77	1977–70	1977	1979	1984>
V/Micron	12–15	18–20	53	53	44	72	65
Loss Angle	0.2%	0.3%	0.15%	0.15%	0.02%	0.02%	.02%
BDV V/μm,	55	78	N. A.	N. A.	152	185	200
Corona Voltage AC	>750 V	>750 V	>750 V	>750 V	250–350 V	>750 V	300–500 V
Capacitor element Vol. (cc)/μF	18	11.6	12	12	5	4.5	3
Can Size- 4 μF 440 V~	45 × 55	40 × 55	40 × 55	40 × 55	32 × 55 35 × 55	32 × 55 35 × 55	30 × 55

Polychlorinated biphenyls (PCBs), a class of chemicals used as capacitor fluid, were banned in the 1980s due to environmental hazards, and their place was taken by vegetable oils and synthetic fluids. The development of dry capacitors did away with fluids altogether for some types of capacitors.

1.2 BASIC PRINCIPLES

The capacitor is one of the three basic passive circuit components of any electronic or electrical circuit. Resistance in a circuit gives rise to ohmic or watt losses, and its current is in phase with the applied voltage waveform. Inductance or a capacitance gives rise to currents out of phase with voltage by 90° in AC circuits, and is the cause of transient currents in many circuits. Inductance is an electromagnetic activity, a basic principle behind all transformers, motors, pumps, electromagnets, chokes etc. It resists a change in current, and stores energy when carrying a current. An ideal inductor is a short circuit path to a steady DC current. In AC circuits, its current lags behind the voltage by 90°.

A capacitor on the other hand, works in electric field. Its properties are exactly opposite to those of inductance, i.e. it resists a change in voltage. It stores energy when a steady voltage is applied. It gets charged to the applied voltage and keeps the energy as well as the voltage even after removal of external voltage. This factor makes handling of capacitors quite dangerous at times, and caution must be exercised when working with them. A capacitor offers an open circuit to the flow of DC current in steady state. Current in an ideal capacitor leads the voltage by 90° in AC circuits.

A capacitor is defined as two conductors (or sets of interconnected conductors) separated by a dielectric. The conductors may be plates, foil, solid shapes, or even wires. The separator can be air, vacuum, solids, an oxide layer on metal (as in electrolytic capacitors), flat thin paper or film, placed or wound on the conductors. A pair of cables near each other will have a capacitance, however small. A capacitor is also formed on two sides of a PCB by coating metal on opposite sides of a given area.

Figure 1.2 shows the general construction of a parallel plate capacitor. The plates shown may be metal foil, or more commonly for many AC applications, the conductive surface of a metallized film. This set of electrodes and dielectrics is very thin and typically long and narrow, which is rolled up and encapsulated. In some cases, the capacitor is made flat, with interleaved plates and dielectric. This allows maximum capacitance for a given volume.

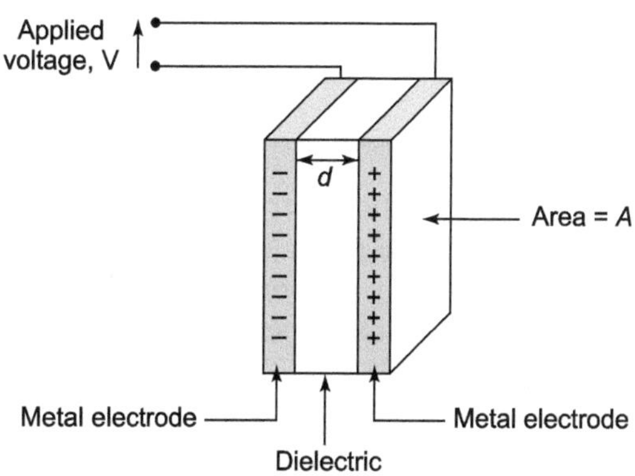

Fig. 1.2 *Construction of a capacitor.*

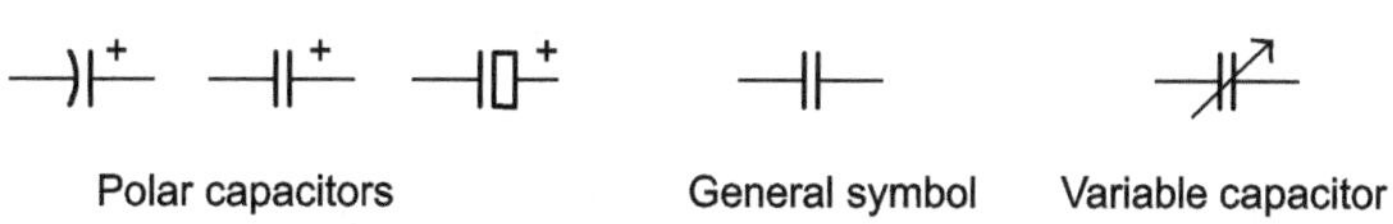

Fig. 1.3 *Symbols for capacitors*

Figure 1.3 shows the general symbols used for capacitors.

1.3 ELECTRIC FIELD

It will be useful to dwell upon the electric field for a better understanding of capacitors. In physics, an electric field surrounds electrically charged particles and time-varying magnetic fields. The electric field strength at a point is the force per unit charge exerted on a positive charge placed at that point. This electric field exerts a force on other electrically charged objects.

The English chemist, Michael Faraday, introduced the concept of an electric field. All charged objects create an electric field that extends outward into the space that surrounds it. The charge alters that space, causing any other charged object that enters the space to be affected by this field. The strength of the electric field is dependent upon how charged the object creating the field is and upon the distance of separation from the charged objects.

Electric Field Strength = Force/Charge, or $E = F/Q$

A uniform electric field may be created by charging two plates. Increasing the voltage between them will increase the field strength, and moving the plates further apart will decrease the field strength. A simple equation for field strength appears as follows:

$$E = -V/d$$

Where V is the voltage between the plates, and d is the distance between them. The minus sign in the equation shows that the force that a positive charge will experience in the field is away from the positively charged plate.

An electric field being defined as a force per charge, its unit is force units divided by charge units, viz. newton/coulomb or N/C in metric units. The electric field is defined as the force per unit charge that would be experienced by a stationary point charge at a given location in the field.

$$E = F/q$$

Where F is the electric force experienced by the particle, q is its charge, and E is the electric field wherein the particle is located. Coulomb's law states that the electric force between two charges is directly proportional to the product of their charges and inversely proportional to the square of the distance between their centres.

The electric field is a vector field with SI units of newtons per coulomb (NC^{-1}) or, equivalently, volts per meter (Vm^{-1}). The SI base units of the electric field are $kg \cdot m \cdot s^{-3} \cdot A^{-1}$. The strength or magnitude of the field at a given point is defined as the force that would be exerted on a positive test charge of 1 coulomb placed at that point; the direction of the field is given by the direction of that force. Electric fields contain electrical energy proportional to the square of the field amplitude. The electric field is to charge as gravitational acceleration is to mass and force density is to volume.

As is clear from the definition, the direction of the electric field is the same as the direction of the force it would exert on a positively charged particle, and opposite to the direction of the force on a negatively charged particle. Since like charges repel and opposites attract, the electric field tends to point away from positive charges and towards negative charges. The fields created by a charged sheet and a spherical body are depicted in Fig. 1.4.

Electric flux Φ is a measure of the number of electric field lines passing through an area. To calculate the flux through a particular surface, multiply the surface area by the component of the electric field perpendicular to the surface. If the electric field is parallel to the surface, no field lines

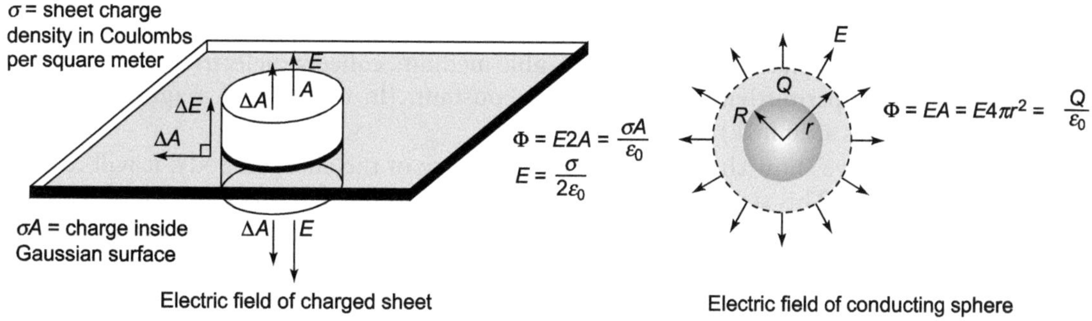

Fig. 1.4 *Electric fields of a sheet and sphere.*

pass through the surface and the flux will be zero. The maximum flux occurs when the field is perpendicular to the surface.

Oppositely charged parallel conducting plates may be treated like infinite planes (neglecting fringing), and Gauss' law used to calculate the electric field between the plates. Presuming the plates to be at equilibrium with zero electric field inside the conductors, then the result from a charged conducting surface can be used:

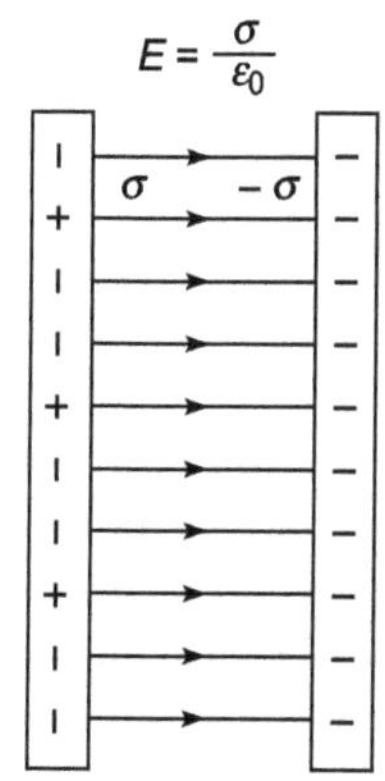

Fig. 1.5 *Field across charged parallel plates.*

The energy density of the electric field is given by

$$u = \varepsilon (E)^2/2$$

where ε is the permittivity of the medium in which the field exists, and $\boldsymbol{E}$ is the electric field vector. The total energy stored in the electric field in a given volume V is therefore

$$u = \varepsilon\int (E)^2\, dV/2$$

where dV is the differential volume element.

1.4 PERMITTIVITY

In the equations describing electric and magnetic fields and their propagation, three constants are normally used. One is the speed of light C, and the other two are the electric permittivity of free space ε_0 and the magnetic permeability of free space, μ_0. The magnetic permeability of free space is taken to have the exact value

$$\mu_0 = 4\pi \times 10^{-7}\text{N/A}^2$$

This contains the force unit N for newton and the unit A is the ampere, the unit of electric current. The magnetic permeability and the electric permittivity are related by the equation

$C = 1/\sqrt{(\mu_0\varepsilon_0)}$ where C is the speed of light, $\mathbf{C = 2.99792458 \times 10^8 \approx 3 \times 10^8}$ **m/s**

This gives a value of permittivity of free space (or vacuum):

$$\varepsilon_0 = 8.854187817 \times 10^{-12}\ \text{F/m}$$

The vacuum permittivity μ_0 (also called permittivity of free space) is the ratio D/E (D being the flux density of electric field) in free space. It also appears in the Coulomb force constant $1/4\,\pi\,\varepsilon_0$ where C_0 is the speed of light in space, and μ_0 is the permeability of vacuum.

These expressions contain the units *F* for Farad, the unit of capacitance, and C for Coulomb, the unit of electric charge. In the presence of polarizable or magnetic media, the effective constants will have different values. In the case of a polarizable medium, called a dielectric, the comparison is stated as a relative permittivity or a dielectric constant. (In the case of magnetic field, the relative permeability is specified.)

The examples above show that whatever the geometry of the charged body, it will carry an associated electric field. This statement can be extrapolated to say that an electrified wire, or any charged substance inside or near another body or equipment can affect it by its electric field, and will exert a capacitive effect. This has a very important bearing on various sensitive equipment and measurements.

In electromagnetism, permittivity is the measure of resistance encountered when forming an electric field in a medium. In other words, permittivity is a measure of how an electric field affects, and is affected by, a dielectric medium. Permittivity is determined by the ability of a material to polarize in response to the field, and thereby reduce the total electric field inside the material. Thus, permittivity relates to a material's ability to transmit (or 'permit') an electric field. It is directly related to electric susceptibility, which is a measure of how easily a dielectric polarizes in response to an electric field. In SI units, permittivity μ is measured in farads per meter (F/m); electric susceptibility χ is dimensionless. Permittivity of a dielectric is

$$\varepsilon = \varepsilon_r \varepsilon_0$$

where ε_r is the relative permittivity of the material. The electric displacement field $\boldsymbol{D}$ represents how an electric field E influences the organization of electrical charges in a given medium, including charge migration and electric dipole reorientation. Its relation to permittivity in the very simple case of linear, homogeneous, isotropic materials with 'instantaneous' response to changes in electric field is

$$D = \varepsilon E$$

where the permittivity ε is a scalar. In general, permittivity is not a constant, as it can vary with the position in the medium, the frequency of the field applied, humidity, temperature and other parameters. In a nonlinear medium, the permittivity can depend on the strength of the electric field. Permittivity as a function of frequency can take on real or complex values.

Materials with a large amount of loss inhibit the propagation of electromagnetic waves. Dielectrics are associated with lossless or low-loss materials. A *perfect dielectric* is a material that has no conductivity, thus exhibiting only a displacement current. Therefore it stores and returns electrical energy as if it were an ideal capacitor.

The **electric susceptibility** χ_e (Latin: *susceptibilis* 'receptiveness') of a dielectric material is a measure of how easily it polarizes in response to an electric field. This, in turn, determines the electric permittivity of the material and thus influences many other phenomena in that medium, from the capacitance of capacitors to the speed of light. χ_e is defined as the constant of proportionality (which may be a tensor) relating an electric field E to the induced dielectric polarization density P such that

$$P = \varepsilon_0 \chi_e E$$

where ε_0 is the electric permittivity of free space. In general, a material cannot polarize instantaneously in response to an applied field, and is time dependent. The susceptibility of a medium is related to its relative permittivity ε by

$$\chi_e = \varepsilon_r - 1 \quad \text{(so vacuum has zero susceptibility)}.$$

1.5 CAPACITANCE

A capacitor is a passive component that stores the electric charge. It is essentially a pair of conductors which contains movable electric charge separated by a dielectric (or insulator). A potential difference must be present to create a voltage between these conductors. As the energy is stored, a mechanical force is produced between the conductors. This is most common between flat and narrowly separated conductors.

The concept of capacitance was first conceived in 1861 by James Clerk Maxwell. He invented the concept of displacement current, as the rate of change within the electromagnetic field. Maxwell understood the concepts of flow of electricity through insulators. He also understood how electromotive force would produce a state of polarization.

An ideal capacitor has a constant capacitance C, defined as the ratio of charge $\pm Q$ on each conductor to the voltage V between them. It will not have any resistive or inductive properties, hence would not dissipate any power. A real world capacitor consumes a small amount of power whenever current flows through it, due to ohmic losses. In addition, under continuous *AC* there are dielectric losses, which are minor at supply frequencies but can get significant at higher frequencies, depending on the type of capacitor.

Capacitance is a measure of the ability of a capacitor to store electric charge. Capacitance is also a measure of the amount of electrical energy stored (or separated) for a given electric potential. If a capacitor is charged to a voltage V, it holds a charge $+Q$ at one plate and $-Q$ on the other; 1 coulomb is a unit of charge equal to 6.28×10^{18} electrons. A capacitor of 1 Farad will accept a charge of 1 coulomb to change its potential by 1 volt.

$$C = Q/V \quad \textbf{So 1 Farad = 1 Coulomb/ Volt.}$$

Sometimes charge build-up affects the capacitor mechanically, causing its capacitance to vary. In this case, capacitance is defined in terms of incremental changes:

$$C = dq/dv$$

A coulomb is 1 ampere second. If 51 mA current for 1 second causes the voltage to change by 1 volt, the capacitance will be 51 mF. It has base SI representation of $s^4 \cdot A^2 \cdot m^{-2} kg^{-1}$. This gives rise to following dimensional equalities:

$$\mathbf{F = A.s/V = J/V^2 = C/V = C^2/J = C^2/\,N.m = s^2C^2/\,m^2.kg = s^4A^2/m^2.kg = s/\Omega}$$

C represents electric charge in SI system of units. Farad was considered too big a unit for practical use, and much smaller fractions, microfarad (μF), nanofarad (nF), picofarad (pF), and even ppF have been in circulation all along. Over the past two decades, with the development of electrochemical capacitors, unit of Farad (F) has come into use, along with kilofarad (kF). The capacitance of a pair of plates is determined by the formula

$$C = \varepsilon_0 \varepsilon . A/d$$

Where C = capacitance (Farads), ε_0 = permittivity of vacuum, 8.85×10^{-12} F/m, A = area (m^2) and d = dielectric thickness (m), and ε_r is called the relative permittivity of the dielectric material, and denotes energy stored in a material by an applied voltage, relative to that stored in vacuum for a given geometry. This is also denoted by **k**, called the dielectric constant of the material. The above formula then becomes

$$C = 8.85 \times 10^{-12}\, kA/d$$

So, for example, a pair of plates of 0.01 m² area, separated by 1 μm, and having an insulation with a dielectric constant of 3 (e.g. polyester), will have a capacitance of about 260 nF. These plates might typically be a metallized layer of 10 mm width, and having a length of 1 m.

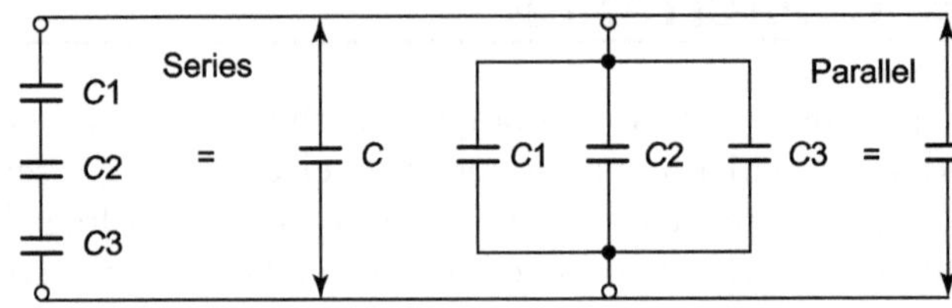

Fig. 1.6 *Series and parallel connection of capacitors.*

If a number of capacitors are connected in series or parallel connections, as shown in Fig. 1.6 above, the resultant capacitance is given as follows:

Capacitors in series: $\frac{1}{C} = \frac{1}{C1} + \frac{1}{C2} + \frac{1}{C3}$

Capacitors in parallel: $C = C1 + C2 + C3$

(It may be noted that these relations are exactly opposite to those applicable for resistors or inductors.)

1.6 CHARGING A CAPACITOR

When a DC supply is connected to a series resistor and capacitor, the initial current is high as the supply source transports charge from one plate of the capacitor to the other. The charging current asymptotically approaches zero as the capacitor becomes charged up to the supply voltage (Fig. 1.7). Charging the capacitor stores energy in the electric field between the capacitor plates. The rate of charging is typically described in terms of a time constant *RC*, *R* being the resistance connected in series with capacitor, inclusive of equivalent series resistance of capacitor (see Fig. 1.8).

Note that the charging proceeds exponentially with time, and current goes on reducing, from a maximum at time zero to almost zero after 5 time constants. The voltage followsa reverse curve, starting from zero and reaching near - full voltage at time 5T. A capacitor acts like a short circuit at the instant of applied voltage, and an open circuit when the full voltage is reached. Table 1.2 shows the progress of charge voltage on a capacitor in terms of multiples of time constant RC for the charging circuit shown in Fig.1.7.

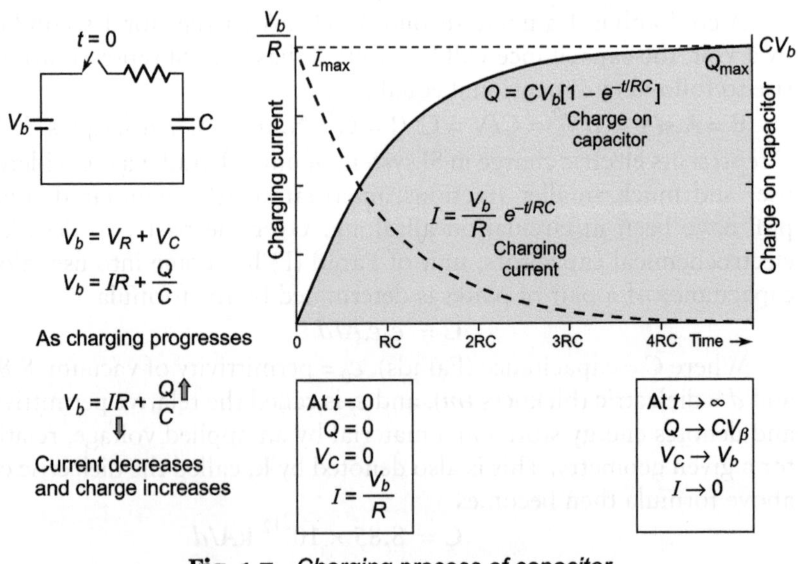

Fig. 1.7 *Charging process of capacitor.*

Table 1.2 *Charging Voltage Variation with Time Constant*

Time	0	1 RC	2 RC	3 RC	4 RC	5 RC
Voltage	0	63%	86%	95%	98%	99%

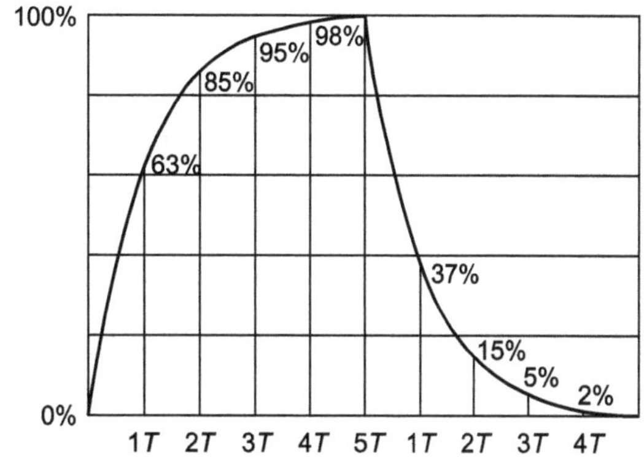

Fig. 1.8 *Charge–discharge cycle. (T = RC = Time constant)*

Stored energy in capacitor

Capacitors return their stored energy fully to the circuit. The energy stored by a capacitor is much smaller than that stored by a battery, so capacitors are not a practical source of energy for most purposes.

As opposite charges accumulate on the plates of a capacitor due to the separation of charge, a voltage develops across the capacitor due to the electric field of these charges. Ever-increasing work must be done against this increasing electric field as more charge is separated. The energy (in joules) stored in a capacitor is equal to the amount of work required to establish the voltage across the capacitor. The energy stored is given by:

Stored Energy $E = ½CV^2 = ½Q^2/C = ½VQ$

where V = voltage across the capacitor.

1.7 CURRENT IN A CAPACITOR

The impedance of a capacitor can be calculated based on the current through it and the voltage across its terminals. A real capacitor is made from two conductors separated by a dielectric. Process of current flow from one conductor to the other, when it has an insulating dielectric between them, is a fundamental question. Real current does not really flow through a capacitor, though it appears so when the voltage across the capacitor changes. If the voltage across the capacitor in Fig. 1.9 were to increase, some positive charge would have to be added to the top conductor and some negative charge would have to be added to the bottom conductor. Adding negative charge to the bottom conductor is the same as pushing positive charge out; it is as though positive charges were added to the top terminal and positive charges were pushed out of the bottom terminal.

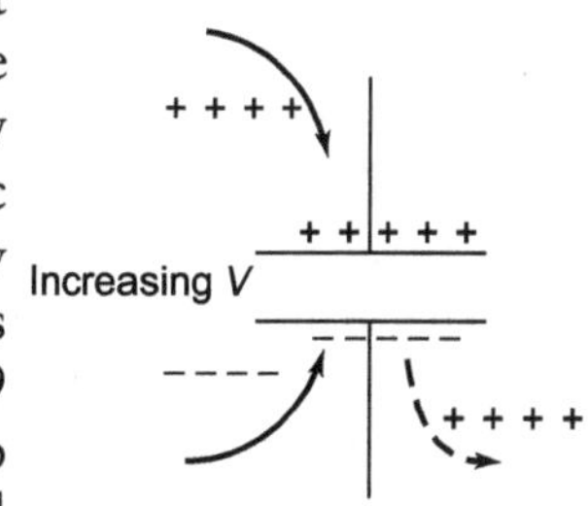

Fig. 1.9 *Increasing voltage across a capacitor.*

Assuming that the width of the plates is much greater than their separation d, the electric field near the centre of the device will be uniform with the magnitude $E = \rho/\varepsilon$. The voltage is defined as the line integral of the electric field between the plates.

$$V = \int_0^d E dz = \int_0^d \frac{\rho}{\varepsilon} dz = \frac{\rho d}{\varepsilon} = \frac{Qd}{\varepsilon A}$$

Solving this for $C = Q/V$ reveals that capacitance increases with area and decreases with separation

$$C = \varepsilon A/d$$

The capacitance is therefore greatest in devices made from materials with a high permittivity. A capacitor opposes a change in voltage, and its current depends on rate of change of voltage.

$$I = CdV/dt$$

Current I is in amperes, voltage V in volts and time t is in seconds. This relationship reaffirms that the only way current flows through a capacitor is when the voltage across it changes. If the voltage is constant, the current through a capacitor is zero. The current through it doubles only if the rate of change of the voltage across it doubles. No current flows in a capacitor once a steady state DC voltage is established across it. However, a current will flow with every change of voltage, or under surge conditions. If a steady AC voltage of an ideal sinusoidal waveform is applied, it can be derived that the current through a capacitor is

$$I = 2\pi fCV, \text{ or } I = \omega CV,$$

where $\omega = 2\pi f$ is the rotational angular frequency of applied voltage. The current in a capacitor is directly proportional to the frequency of applied voltage.

The real capacitor can be modelled as an ideal capacitor in series with an ideal inductor and an ideal resistor in series. The ideal inductor represents the capacitor's Equivalent Series Inductance (ESL) and the ideal resistor, the capacitor's Equivalent Series Resistance (ESR). The ideal capacitor impedance is infinite at DC and decreases as the applied voltage increases in frequency. Eventually, the ideal capacitor's impedance reaches zero. The real capacitor's impedance never reaches zero because of ESR and ESL.

Non-ideal behaviour

Real life capacitors deviate from the ideal capacitor in a number of ways. Some of these deviations, like leakage current and parasitic effects are linear or near linear, and can be dealt with by adding resistive and inductive components in the equivalent circuit of the capacitor. In other cases, such as with breakdown voltage, the effect is nonlinear and normal linear network analysis cannot be used. There is yet another group, where capacitance may be linearly related to some other variable like temperature or pressure. These will be excluded from the present discussion for simplicity.

[The symbol $j = \sqrt{-1}$ is used in electrical engineering to denote the vectorial reactance component of impedance, 90° out of phase with the resistive (active) component. +j indicates inductive, while –j indicates capacitive reactance.]

$X_c = V/I = 1/\omega C$ is called reactance of a capacitor. Capacitive reactance (X_c) is a measure of an ideal capacitor's opposition to *AC* (alternating current). Like resistance, it is measured in ohms, but reactance is more complex because its value depends on the frequency (*f*) of the electrical signal passing through the capacitor as well as on the capacitance, *C*.

$$VAR = I \times V = 2\pi fCV^2$$

The resistive component of capacitor is too small, so the above formula holds true almost all the time. It follows that a capacitor acts as an open circuit for *DC* voltage, and a short circuit path for very high frequencies. If its equivalent series resistance is R then the watt loss in capacitor becomes $W = I^2R$.

ESR and Q

A capacitor's quality factor (*Q*) is numerically equal to the ratio of its net reactance ($X_C - X_L$) to its equivalent series resistance. It can be seen that the capacitor's Q varies inversely to its ESR and directly with the net reactance.

$$Q = (X_c - X_L)/ESR$$

Equivalent Series Resistance (ESR) is the summation of all losses resulting from the dielectric and metal elements of a capacitor and is typically expressed as milliohms. ESR is a key parameter to consider when utilizing capacitors in RF bypass applications. In electronic circuits, a capacitor's ESR should be known at all frequencies within the passband, especially at frequencies above the capacitor's series resonant frequency. The ESR will increase for increasing frequencies and may become the dominant loss factor. It will largely determine the attenuation at the parallel resonant frequency.

Impedance: (Z) The magnitude of a capacitor's impedance, accounting for ESR and ESL, is equal to:

$$Z = \sqrt{[(\text{ESR})^2 + (X_L - X_C)^2]}$$

If inductance ESL is negligible, absolute value of $Z_c = \sqrt{(\text{ESR}^2 + X_C^2)}$.

(Usually ESR << X_c, and Z_c is the reactance of ideal capacitor.) Z_c is a vector quantity, and j shows vector quantity at right angles to the resistive component R. The minus sign is for leading power factor.

Hence $I = \omega CV$ (presuming $R << X_c$).

The apparent power drawn by capacitor, called reactive power, is calculated as

$Z_c = R - j(1/\omega C)$, where R is the equivalent series resistance of a capacitor.

Usually, R is too small and negligible, and the above equation reduces to

$$Z_c = V_c/I_c = -jX_c = -j(1/\omega C).$$

DC current in capacitors

The dielectric between the plates is an insulator and blocks the flow of electrons. A steady current through a capacitor deposits electrons on one plate and removes the same quantity of electrons from the other plate. This process is commonly called 'charging' the capacitor. The current I through the capacitor is the rate at which charge Q is forced through the capacitor (dQ/dt). This can be expressed mathematically as:

$$I = dQ/dt = CdV/dt$$

For circuits with a constant DC voltage source, and consisting of only resistors and capacitors, the voltage across the capacitor cannot exceed the voltage of the source. Equilibrium is reached where the voltage across the capacitor is constant and the current through the capacitor is zero. For this reason, it is commonly said that capacitors block DC.

AC current in capacitors

The current through a capacitor due to an AC source reverses direction every half cycle. Except for the instant that the current changes direction, the capacitor current is non-zero at all times during a cycle. For this reason, it is commonly said that capacitors 'pass' AC.

The voltage across a capacitor is proportional to the integral of the current, with sine waves in AC or signal circuits. This results in a phase difference of 90°, the current leading the voltage phase angle, as shown in

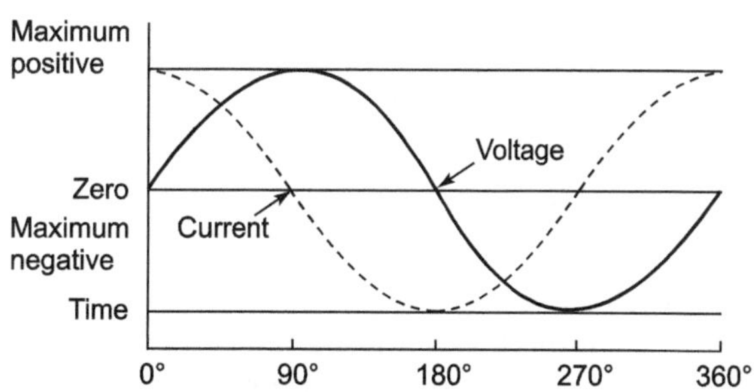

Fig. 1.10 *Current and voltage waveforms for a capacitor.*

Fig. 1.10. The amplitude of the voltage (v) depends on the amplitude of the current (I) divided by the product of the frequency (f) of the current with the capacitances, C.

1.8 EQUIVALENT CIRCUIT OF CAPACITOR

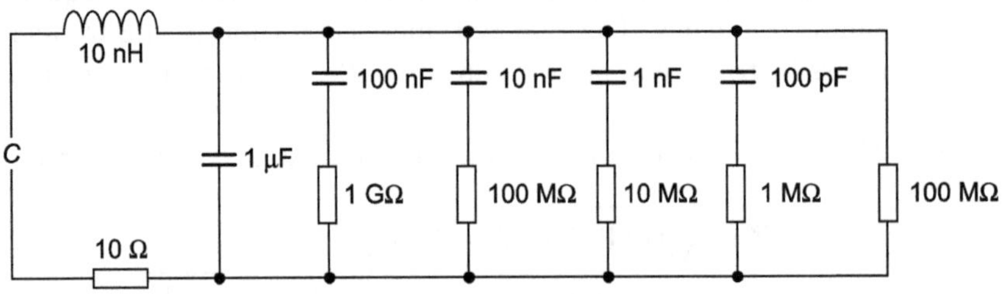

Fig. 1.11 *Lumped equivalent circuit of 1 µF capacitor.*

The generalized equivalent circuit of a capacitor is shown in Fig. 1.11. The nominal capacitance is the value of C, with ESR and ESL (equivalent series resistance and inductance) in series. The parasitic capacitances (C_2 – C_n) and their series resistances represent the dielectric loss (resistance) and dielectric absorption. These are infinite, with ever diminishing capacitance and increasing series resistance. The values used are for simulation purposes. It must be understood that dielectric loss, residual charge, series resistance and inductance are quite normal, and appear in all components.

If a meter is connected between the capacitor terminal and its metal case (or a metal foil around the outer surface), it measures a capacitance value, although small. This is because the case serves as one capacitor electrode while the main capacitor element is the second electrode.

1.9 DIELECTRICS

A dielectric is an electrical insulator that can be polarized by an applied electric field. In an electric field, electric charges do not flow through the dielectric material, as in a conductor, but shift only slightly from their average equilibrium positions (dielectric polarization). Positive charges are displaced toward the field and negative charges in the opposite direction. This creates an internal electric field which reduces the overall field within the dielectric itself. If a dielectric is composed of weakly bonded molecules, those molecules further reorient to align with the field.

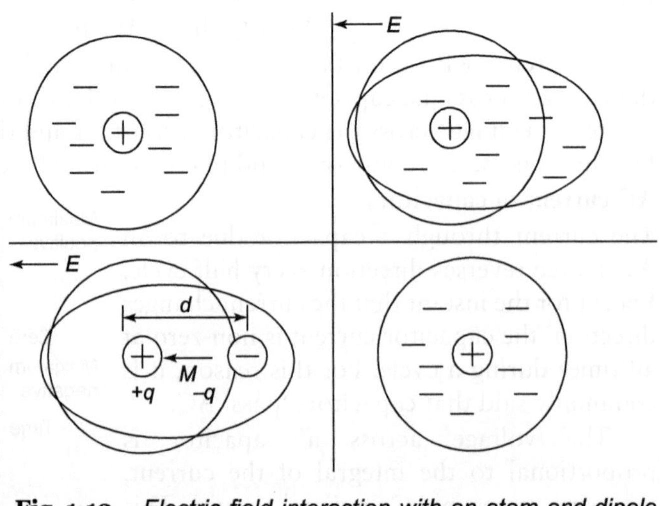

Fig. 1.12 *Electric field interaction with an atom and dipole action.*

Although 'insulator' implies low electrical conduction, the term 'dielectric' is used to describe materials with a high polarizability, which is expressed by a number called the dielectric constant. A common example of a dielectric is the electrically insulating material between the metallic plates of a capacitor. The study of dielectrics is concerned with the storage and dissipation of electric and magnetic energy in materials. The term 'dielectric' was coined by William Whewell in response to a request from Michael Faraday.

Atomic model of a dipole

A dielectric material is made up of atoms. Each atom effectively consists of a cloud of negative charge (electrons) bound to and surrounding a positive point charge at its centre. An electric field distorts the cloud as in Fig. 1.12. The cloud thus displaced can be represented by a dipole and its dipole moment M, a vector quantity. The relationship between the electric field and the dipole moment gives rise to the behaviour of the dielectric. (Note that the dipole moment is shown to be pointing in the same direction as the electric field, which is mostly the case.) When the electric field is removed the atom returns to its original state. The time required to do so is the so-called relaxation time, an exponential decay.

1.10 DIELECTRIC MATERIALS

The dielectric of a capacitor and its thickness is determined by factors such as the type of capacitor, its voltage rating, application, operating temperatures, duty cycle and life expectancy. An important property of a dielectric is its ability to support an electrostatic field while dissipating minimal energy in the form of heat. The lower the dielectric loss (the proportion of energy lost as heat), the more effective is a dielectric material. Another consideration is the dielectric constant, the extent to which a substance concentrates the electrostatic flux. Substances with low dielectric constant include a perfect vacuum, dry air and most pure, dry gases such as helium and nitrogen. Materials with moderate dielectric constants include ceramics, distilled water, paper, mica, polyethylene and glass. Metal oxides, in general, and some ceramics have high dielectric constants.

The prime asset of high-dielectric-constant substances is the fact that they make possible the manufacture of high-value capacitors with small physical volume. But these materials generally do not withstand electrostatic fields as intense as low-dielectric-constant substances such as air.

(a) Permittivity and Dielectric Constant

Permittivity (ε) may be visualized as space factor, or an indicator of capacitance packing in a given space. This property of dielectrics has been dealt with in detail in Sec. 1.3. It is known that most dielectrics have permittivity higher than vacuum. The higher the permittivity, the smaller is the space requirement for a given capacitance value. It is customary to mention these values in comparison with this reference, and as a multiplier of permittivity of vacuum (ε_0). This is called relative permittivity, or Dielectric Constant, and is represented by letter ε_r or **k**, which appears in the capacitance formula. Values for **k** differ widely, from 1 (vacuum) to several hundred or a few thousands. While permittivity is relatively constant, it tends to vary with temperature, and sometimes with frequency.

(b) Dielectric Strength

Voltage applied on a capacitor may be increased gradually, and its energy stored will go up in squared ratio to voltage. If the voltage across a dielectric material becomes too great – that is, if the electrostatic field becomes too intense – the material will suddenly begin to conduct current. This phenomenon is called dielectric breakdown. In components that use gases or liquids as the dielectric medium, this condition reverses itself if the voltage decreases below a critical point. But in components containing solid dielectrics, dielectric breakdown usually results in permanent damage. The operating voltage stress of capacitor dielectric is kept much below this limit. The life of a capacitor is much longer and operational safety higher at safer voltage levels. Various dielectrics have different breakdown strengths, and have to be considered for this property for a given application.

Dielectric strength is dependent on temperature, frequency, shape of electrodes or clear insulating margins towards winding ends. The strength of base material like kraft paper may be substantially increased by use of impregnants.

The working voltage stress on a capacitor dielectric is among the highest compared to most insulation materials used anywhere. A modern metallized capacitor for a motor may use a working stress in excess of 60 KV AC per mm. (The dielectric will have breakdown strength of over 400 KV DC per mm. By comparison, vacuum or dry air has breakdown strength of 3 KV DC per mm.) Even higher stresses may be used in future capacitors.

In high current applications, there will be significant voltage across the capacitors. They must withstand the voltage and current that they will be subjected to. This is an area where dielectric loss may cause the capacitors to heat up with sustained high power, and the devices used need to be stable with time and temperature.

The size of capacitors varies widely from miniscule (the size of sand granules) to large ones requiring special enclosures or rooms. They come in cylindrical shape, disc type, in boxes, with or without mounting studs or brackets. Case materials may be steel, aluminium, or a variety of plastics. Terminations can be of wires, solder tags, and slip type terminals, screws and bolts, with or without special bushings. They may also be Surface Mounting (SMD) type, which are soldered directly to PCBs in electronic circuits, without any wire leads.

Capacitors may be used anywhere from – 40°C to +100°C or even higher temperatures, and may be used on mountaintops or under the sea. Space ships also use them under extreme conditions of reliability in zero gravity. From highly salty conditions on seashores to deserts, a capacitor has to be found for any weather.

1.11 DIELECTRIC ABSORPTION

This is the phenomenon that allows a capacitor to recover some of its original charge due to dielectric absorption. If a capacitor is charged fully and then discharged, the voltage is seen to rise again, as shown in Fig. 1.13, even though it was obviously zero for the duration of the short.

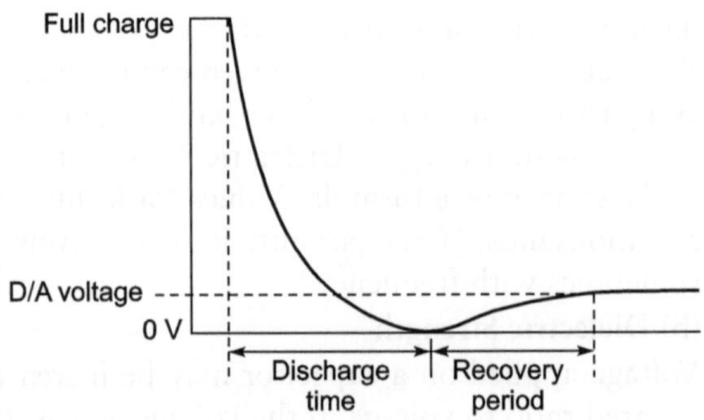

Fig. 1.13 *Dielectric absorption phenomenon.*

The dipole orientation of capacitor dielectric materials reverts to its natural neutral position when its external voltage is removed, and the capacitor is discharged. However, all dipole energy is not discharged instantly (due to inertia), but the dipoles get reoriented after removal of external discharge path. Since dipoles do not lose their energy fully, when the external discharge path is removed within a short time, a process of dipole reorientation begins by virtue of their residual inertial energy, resulting in a small voltage appearing across the capacitor terminals. However, if capacitor terminals are shorted for a sufficiently long time, the dipoles get enough time to lose all their energy and fall back to their natural positions, and the voltage does not reappear. This inertia is quite large in electrolytic and electrochemical capacitors: so large that these capacitors have to be kept shorted for a long time to neutralize their charge fully. In fact, electrochemical capacitors take anywhere from fifteen minutes to a few hours to reach stable voltage zero, i.e. to have no adsorption effect apparent.

1.12 GENERAL CONSTRUCTION OF A CAPACITOR

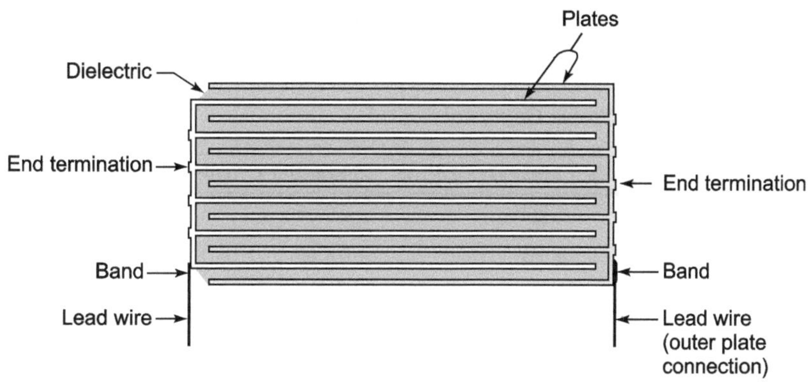

Fig. 1.14 *Multilayer capacitor construction, or cross-section of extended electrode wound element.*

Figure 1.14 shows the general construction of a multilayer capacitor, and is also representative of the cross-section of a traditional extended foil capacitor. With some capacitors, one end is marked with a band or is otherwise indicated as the outer foil. This can be useful for sensitive circuits, where the outer foil (or plate) ends may be connected to earth (ground/chassis) to shield the capacitor against interference. This is usually only needed in very high impedance circuits, or where there is considerable external noise.

The way the ends of the foil are joined in these capacitors prevents the signal from having to traverse the length of the plates. This is referred to as 'non-inductive capacitor', as practically all turns or layers have been shorted, and only the inductance of terminal wires and the width between ends are significant for inductance.

General notes on soldering of capacitor leads

Permissible heat exposure loads on capacitors are determined by the upper category temperature T_{max}. Long exposure to temperatures above this type-related temperature limit can lead to changes in the plastic dielectric and thus change a capacitor's electrical characteristics irreversibly. High temperatures are encountered during soldering, but these are only applied briefly and not allowed to affect the plastic base.

Apart from being dependent on the solder bath temperature and the soldering time, the thermal load is also affected by the initial (pre-heating) and the post-soldering (cooling) temperatures. Shadowing by neighbouring components or subsequent heating due to their effect has a similar result. Since the soldering heat is transmitted into the components mainly via the leads, the thermal resistance of the terminals is the deciding factor for the heat transmitted, especially for smaller capacitor sizes. Thus a poor thermal conductivity is desirable from this aspect, however, this is contrary to good electrical conductivity required in order to achieve low dissipation factors, since the electrical conductivity is generally proportional to the thermal conductivity.

1.13 CODES FOR VALUES OF ELECTRONIC CAPACITORS

Electronic capacitors which are small in size carry their values, tolerances and voltages in the form of colour codes or digital codes. Table 1.3 below gives the standard colour codes in use for capacitors.

Table 1.3 *Colour Codes for Electronic Capacitors*

Colour	*Digit A*	*Digit B*	*Multiplier D*	*Tolerance T > 10 pF*	*Tolerance T < 10 pF*	*Temp. coeff. TC*	*Working voltage V*
Black	0	0	× 1	± 20%	± 2.0 pF		
Brown	1	1	× 10	± 1%	± 0.1 pF	-33×10^{-6}	
Red	2	2	× 100	± 2%	± 0.25 pF	-75×10^{-6}	250
Orange	3	3	× 1000	± 3%		-150×10^{-6}	
Yellow	4	4	× 10k	+ 100%, – 0%		-220×10^{-6}	400
Green	5	5	× 100k	± 5%	± 0.5 pF	-330×10^{-6}	100
Blue	6	6	× 1m			-470×10^{-6}	630
Violet	7	7				-750×10^{-6}	
Grey	8	8	× 0.01	+ 80%, – 20%			
White	9	9	× 0.1	± 10%			

For memory, these may be remembered as follows:

B. B. R O Y of Great Britain has a Very Good Wife

0 1 2 3 4 5 6 7 8 9

Capacitor on the left is 160 KPF ± 5% = 160 nF ± 5%

(Brown –1, Blue –6, Orange –3 zeros gives 160 K and Green = ± 5%)

Capacitor on right is rated 470 KPF 250 V = 470 nF ± 10%

(Yellow –4, Violet –7, Orange –4 zeros, White = ±10%, Red –250V)

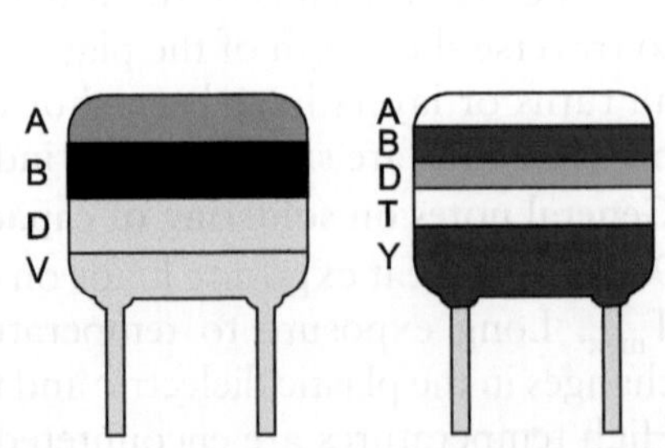

Fig. 1.15 *Colour coding on capacitors.*

It may be noted that all colour coding is in picofarads, and conversion may be used to read in nF and μF. Another way to codify the capacitance value is via digital code (Fig. 1.16) The first two digits are two significant figures, and the third shows the number of zeroes added. The value is then read in pF.

In digital notation, capacitor tolerance is indicated by the following codes:

Table.1.4 *Digital Notation for Capacitor Marking*

	Letter	*B*	*C*	*D*	*F*	*G*	*J*	*K*	*M*	*Z*
Tolerance	C <10 pF ±pF	0.1	0.25	0.5	1	2				+80
	C >10 pF ±%			0.5	1	2	5	10	20	–20

In Fig. 1.16 Capacitor value is 47 pF × 1000 (3 zeros) = 47,000 pF, or .047 μF

Hence capacitor rating is .047 μF ± 5%

An electronic DC rated plastic film capacitor may also be used for AC voltage with suitable derating. The AC voltage depends upon the nature of dielectric viz. polypropylene (PP) or polyester (PET), and whether the construction is metallized or non-metallized.

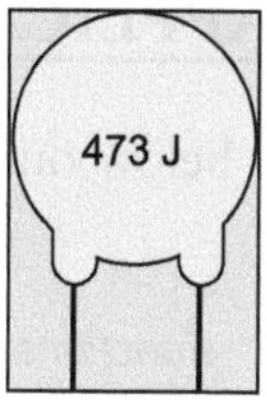

Fig. 1.16 *A ceramic disc type capacitor with code 473J printed.*

1.14 CAPACITOR CHARACTERISTICS

Each type of capacitor is distinguished by its own unique characteristics. Even within a given type, the choice will depend upon the specific dielectric selected for desired end use. As temperatures change, capacitance values also change. This is termed as the temperature coefficient of capacitance (TCC).

For ceramic capacitors, the voltage applied to the capacitor also affects the capacitance value (the electric field strength across the dielectric changes the effective dielectric constant K of the material). The capacitor cost and size need to be considered, as well as its packaging type.

End-of-life reliability issues may also be important. Each capacitor type has its own set of characteristics that will make it the most logical choice for a given application.

Capacitor – physical sizes

- Smallest discreet capacitors can be just the size of sugar granules.
- Large Power Factor Capacitor banks need more space to house them.
- SMD capacitors are among the smallest variety.
- Integrated circuits have capacitors etched in semiconductor circuits.

1.15 CAPACITOR PERFORMANCE REQUIREMENTS

- Long working life.
- Wide range of working temperature, (typically – 25°C to +85°C).
- No damages to surroundings in case of failure.
- High stability of the capacitance value during the working life, in particular for series connected KVAR capacitors.

- The life of a capacitor depends mainly on its working conditions. Temperature and voltage are particularly important in this respect.
- It may be noted that each 10°C rise of the working temperature causes a 50% reduction of the capacitor life. Usually the following formula is adopted:

$$L(t_x) = L(t_n) \cdot 2^{\frac{-(t_x - t_n)}{10}}$$

Where $L(t_x)$ is the capacitor life at temperature t_x;
$L(t_n)$ is the capacitor life at the maximum rated temperature.

1.16 APPLICATIONS IN INDUSTRY

Table 1.5 below outlines typical functions performed by capacitors in electrical and electronics circuits:

Table 1.5 *Functions and Applications of Capacitors*

Function	*Representative application*
Energy storage	Photoflash, Timers, Vehicles
Resonant circuits	Oscillators, Tuning
Smoothing	Power Supplies
H.F. filters	DC Supplies, R.F. Suppression
Phase shifting	Motors, Fans
Measurement and sensors	Vacuum, electrical and mechanical parameters
Capacitive switching	Touch Control
Transient suppression	Power Supplies
Peak voltage generation	Auto Industry
Commutation	Traction, series motors
Isolators and couplers	DC electronic circuits
Motor starting and running	AC capacitor run/capacitor start

This is only representative and capacitors find place in wide range of applications in the electrical and electronics industry. Practically every electronic circuit and most electrical systems use capacitors in some form, performing a variety of functions. Radios, TVs, cars, planes, mobile phones, washing machines, pumps, motors, and generation and distribution systems, traction, commutation…the list is endless.

Electrolytic capacitors are excellent for power supply, and in most places where high values of capacitance are needed. They are unsuitable for filters, because they have wide tolerance, should be biased, and may change capacitance depending on applied frequency. Bipolar electrolytics are excellent where high values are needed, and no polarizing voltage is available. Because of wide tolerance, they too are unsuitable for filter circuits.

1.17 DEFINITIONS, RATINGS AND SERVICE CONDITIONS

It will be useful to know common engineering terminology related to capacitors, and to understand the ratings and service conditions in which a capacitor has to perform:

- **Rated voltage:** The maximum DC or AC RMS voltage of specified frequency, which may continuously be applied to a capacitor at any ambient temperature below the rated temperature.
- **Capacitance:** The capacitive part of the equivalent circuit composed of capacitance, series resistance and inductance.
- **Capacitance tolerance:** The permissible deviation of the capacitance at an ambient temperature of 23 ± 2°C and a RH of 50–55%. Preferred tolerances are 5%, 10% and 15%. Closer tolerances are permitted and required for certain applications. Asymmetric tolerances are permitted (e.g. –1% to +3%, +3% to +7%, etc.)
- **Insulation Resistance (IR):** The insulation resistance at a specified applied DC voltage divided by the leakage current after a well-defined minimum time. The IR may be measured between terminals, or between terminals shorted and the capacitor case.
- **Time constant:** The product of IR between terminals and the capacitance value. This has a time dimension, and is expressed in seconds, or as $M\Omega \times \mu F$.
- **Dissipation factor:** The dissipation factor, loss factor or tan δ (tan delta, or tangent of loss angle) is the power loss of capacitor divided by the reactive power of the capacitor at a sinusoidal voltage of specified frequency.
- **Equivalent Series Resistance (ESR):** Resistive part of equivalent circuit composed of series resistance and inductance.
- **Equivalent Series Inductance (ESI):** The inductance part of the above circuit.
- **Temperature coefficient:** The rate of capacitance change with temperature, measured over a specified temperature range – normally measured in parts per million (ppm) per kelvin.
- **Capacitance drift:** Maximum irreversible variation of capacitance observed at room temperature during or after the completion of specified number of temperature cycles – expressed as a percentage of capacitance related to a reference temperature (usually 20°C).
- **Climatic category:** Climatic category code (e.g. –10/85/21) indicates the minimum and maximum temperature of operation, while the third figure specifies maximum number of days of exposure to steady state damp heat that the capacitor will withstand.

  The minimum temperature is usually chosen from – 40°C, – 25°C, – 10°C and 0°C. The maximum preferred temperatures are 55°C, 70°C, 85°C and 100°C. Damp heat severity can be between 4 and 56 days. The preferred severity is 21 days.
- **Pulse rise (dV/dt):** The maximum pulse rise (volt/µs) the capacitor can withstand with a pulse voltage equal to the rated voltage. The voltage pulse rise rate (calculated as per second) multiplied by the capacitance gives the peak current in amperes for the capacitor.
- **Self-healing:** The process by which electrical properties of metallized capacitor are rapidly restored to their original values after a local breakdown.
- **Storage temperature:** The temperature range with a RH maximum of 80% without condensation at which the initial characteristics can be guaranteed for at least two years.

- **Rapid change of temperature:** The rapid change of temperature is to determine the effect of succession of temperature changes, and consists of 5 cycles of 30 minutes at lower category temperature and 30 minutes at higher category temperature.
- **Temperature characteristics:** Variations of capacitance values with variation of temperature, expressed as a percentage of value at reference temperature (usually 20°C). This property is not always relevant, but applies where the capacitance changes considerably with temperature.
- **Damp heat cyclic:** This test determines the suitability for use and storage under high humidity when combined with cyclic temperature changes and generally producing condensation on capacitor surface. One cycle usually consists of 24 hours exposure to 55°C and 95% to 100% RH.
- **Dry heat:** This tests the suitability of capacitors to be used or stored at high temperature. The standard test is for 16 hours at upper category temperature.
- **Cold:** This test determines the ability of capacitors to be used at low temperature, and involves 2 hours at this temperature.
- **Resonant frequency:** In the equivalent circuit of capacitors at low frequencies, i.e. the frequencies below the resonant frequency, the ESL inductive component is lower than the capacitive component. When these two become equal at a frequency, the circuit is resistive, and this is the resonance frequency. Above this frequency, the impedance is inductive and rises with frequency. ESL depends on the construction and dimensions of the capacitor.
- **Rated duty cycle:** A rated cycle indicating the type of intermittent or starting duty for which a capacitor is suitable. It is specified by the duty cycle duration, in minutes or seconds and the percentage of time during which the capacitor is energized.
- **Duty cycle duration:** Total time of one energized and one unenergized interval during the intermittent operation.
- **Class of operation:** The minimum total life for which a motor capacitor has been designed at rated duty, voltage, temperature and frequency.

Table 1.6 *Class of Operation Based on Life Expectancy*

Class	*Life expectancy min. hours*	*Endurance test voltage*	*Endurance test duration*	*Permitted capacitance change*
A	30,000	1.25 Vn	6000 h	3%
		Or 1.35 Vn	3000 h	
B	10,000	1.25 Vn	2000 h	3%
		Or 1.35 Vn	1000 h	
C	3,000	1.25 Vn	600 h	3%
D	1,000	1.25 Vn	200 h	3%

The above tests are carried out at rated maximum service temperature of capacitors. For intermittent duty capacitors, energization time is the same as above, but voltage is reduced to 1.15 Vn, with maximum on time being 50% of cycle. These classes of operation are intended to represent a true failure rate not over 3% during the life of a capacitor. It is possible to have more than one class for a capacitor at different voltages.

Fan capacitors have been specified an endurance test duration of 500 hours at 1.25 Vn.

- **Class of safety protection:** The degree of safety protection identified by one of the three codes to be marked on capacitor:

  (P2) indicates that the capacitor has been designed to fail in the open circuit mode only, and is protected against fire or shock hazard.

  (P1) indicates that the capacitor may fail in the open-circuit or short-circuit mode, and is protected against fire or shock hazard.

  (P0) indicates that the capacitor has no specific failure protection.
- **Normal service conditions:** Usual operation conditions include (a) altitude not exceeding 2000 m above sea level, (b) designed for operation in lightly polluted atmosphere (high degree of pollution may adversely affect capacitor performance), (c) residual voltage at energization not exceeding 10% of rated voltage.
- **Robustness of terminations:** The terminations of capacitors, whether solder type, wire leads, screws/bolts or snap action type, must be able to withstand the torsion, soldering heat, torque, force or pull as the case may be.
- **Vibrations:** Capacitors are subjected to varying degrees of mechanical vibrations depending upon their usage, location and service conditions including mounting system.
- **Leakage:** Capacitors containing liquid or semi-solid/gel type impregnants should not show any leakage of impregnant during thermal cycling or prolonged exposure at highest rated service temperature.
- **Exposure to salty atmosphere:** Capacitors exposed to sea weather or corrosive atmospheres can get chemically attacked and subsequently the inner contents can get damaged, putting the capacitor out of service. Such capacitors need to have enough protection provided to make them withstand these conditions during their lifetime.
- **Discharge devices:** Capacitors connected permanently to motor windings do not need discharge devices. Most capacitors in other applications like lighting capacitors and power capacitors need to have discharge devices connected across their terminals, which will reduce the voltage across terminals from peak level to a safe level below 50 V within one minute after switching off. This allows the capacitor to be handled safely whenever required, e.g. for replacement or maintenance. The discharge device is mostly in the form of a resistor connected across terminals of capacitors, but it may also be in other forms like inductors.
- **Sealing:** Capacitors containing oil, waxes, jellies etc. should be constructed in a manner so as not to allow any seepage of these impregnants at the highest continuous rated service temperature, or during exposure to extreme cyclic variations of temperatures as may be experienced in service.

All capacitors have to be designed and constructed taking these factors into account. Relevant testing procedures are designed or specified to ensure trouble-free performance of capacitors under the relevant specified conditions.

In addition, certain safety precautions have to be built into capacitors. These include creepage distances over surfaces between live parts of different polarity, those between live parts and case, distances over air gaps between these points, as also distances from earthing points.

2

PROPERTIES OF DIELECTRICS

2.1 CAPACITOR DIELECTRICS

Dielectrics are materials that define most of the parameters of capacitors, such as capacitance, loss factor, voltage capacities etc. All dielectrics are insulators. An insulator used between plates or electrodes of capacitor as a supporter of electrostatic field is called a dielectric. If the flow of current in oppositely charged polarities in an electric field is kept at a minimum, and lines of electrostatic field are not impeded or obstructed, an electrostatic field can store energy. An important property of a dielectric is to cause minimum heat while supporting the electric field. The lower the dielectric loss, the more effective is the dielectric.

In an electric field, atoms of materials undergo orientation of their charge cloud, or its dipoles from their neutral position (Fig. 2.1). This absorbs certain energy, and the energy is stored so long as the field exists, or even after the field is removed. Once the dipoles are oriented and the external field removed, they do not revert on their own, but must be neutralized externally by some means. No insulator being ideal, there is a leakage current between opposite poles, although this is quite small or miniscule.

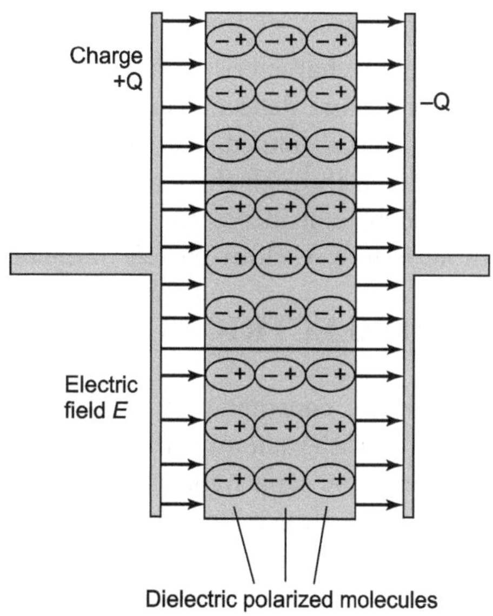

Fig. 2.1 *Polarized molecules in electric field.*

The electrons within dielectric molecules are influenced by the electric field, causing the molecules to rotate slightly from their equilibrium positions. The air gap is shown for clarity; in a real capacitor, the dielectric is in direct contact with the plates. An electric field E is created in the region between the plates that is proportional to the amount of charge that has been moved from one plate to the other. This electric field creates a potential difference $V = E \times d$ between the plates of the capacitor.

Most dielectrics are solids, although vacuum and dry air are excellent dielectrics. Some oils and chemicals like transformer oil, petroleum jelly, askarels (polychlorinated biphenyls), wax, resins, Phenyl Xylyl Ethane (PXE) Dioctyl Pthalate (DOP), castor oil and some other vegetable oils have been used in composite dielectrics with solids. Some gases also find application in capacitors. In electrolytic capacitors, an oxide layer is formed on the electrode material as a result of chemical reaction with the electrolyte, and this layer acts as a dielectric. Liquids, when used as insulation in high voltage equipment, have the advantage that any puncture path is self-healing. They have excellent dielectric properties and high breakdown strength, but these are affected by the presence of impurities and dissolved gases. They are generally used at 50–60 KV/mm, but when used by themselves, can have working stresses of up to 200 KV/mm. Askarels, a class of polychlorinated biphenyls, were the most used liquid impregnants for paper capacitors, but were banned in 1980s due to environmental pollution hazards. Other substitutes like PXE, silicone oils and other synthetic oils, as also certain vegetable origin oils like castor oil and rapeseed oil were developed to take their place.

In composite dielectrics, both components should be inherently stable, and any impurities or pollutants should be scrupulously kept away. They should not react with each other in the presence of an electric field, or thermal and mechanical stress conditions encountered in service or during test conditions. They should have nearly equal dielectric constants to ensure a uniform electric field. Before the onset of askarels, i.e. prior to the 1950s, transformer oil (dielectric constant 2.2) was being used with kraft paper (dielectric constant 6.0), and voltage stress was limited to 12–13 KV/mm. Use of askarels (dielectric constant 5.6–6.1) allowed stresses to increase to 17 KV/mm, and subsequently, with improvement in kraft paper, they were further increased to 22 KV/mm.

The main characteristics of a dielectric are its dielectric strength, dielectric constant **k** (relative permittivity) and dielectric loss factor. The dielectric strength is a measure of a material as an insulator, and is defined as the maximum voltage that can be applied to an insulator to produce a breakdown through the material, measured in volts per unit thickness. It is one of the main factors to be considered for a capacitor to arrive at electric field stress over its working life without failure. Various aspects of dielectrics need to be understood.

2.2 DIELECTRIC CONSTANT

The dielectric constant may be defined as the ratio of capacitance with a dielectric, to the capacitance with vacuum as the dielectric in the same geometry. The higher the dielectric constant, the higher is the capacitance in a given geometry, or smaller capacitor volume for a given value. Vacuum is an ideal material with $k = 1$, and various materials have values of k ranging from nearly 2 for plastics to 5.8 – 6 for kraft tissue papers, and up to several hundreds or thousands for ceramics and certain oxides. Substances with low k include vacuum, dry air and most pure gases like helium and nitrogen. Moderate k materials include polypropylene, distilled water, glass, ceramics and paper. Metal oxides in general have high dielectric constants. Table 2.1 below scans through properties of some dielectrics.

Table 2.1 *Dielectric Constants (Relative Permittivity) of Common Dielectrics*

Dielectric material	*Dielectric constant*
Air	1.0059
Vacuum	1.000
Polypropylene	2.25–2.3
Polyester	3.2
Pure cellulose or paper	6.0
Ceramic (COG)	45
Barium titanate	1000–30,000
Tantalum pentoxide	27
Alumina	10
Glass (Silicon)	42

A comprehensive list of dielectric properties of a number of materials is given in Appendix A. Dielectrics may be pure materials or composites, with or without liquid/pastes/waxes/jellies. The electrode material can also be from a host of metals or electrolytes, and can take many shapes. Metals like aluminium, tin, silver, etc. in plate, foil, cylindrical or other formed shapes are in use. The electrodes in modern day capacitors also are in the form of a conducting metallized surface on base dielectric film. Integrated circuits have entire capacitors etched in.

The main asset of high k materials is their small volume but they have low electric strength and high losses compared with electrostatic capacitors. They are not able to stand a high electrostatic field unlike those with lower k. Each dielectric has its own utility in a typical capacitance and voltage range.

2.3 EQUIVALENT SERIES RESISTANCE (ESR)

Theoretically, an ideal capacitor has an infinite insulation resistance between terminals (as also from the terminal to the case). A real capacitor essentially has both finite series and insulation resistance components.

Series resistance arises out of electrode resistance, wire leads, solder, end connections etc. Insulation resistance (IR) is formed due to dielectric insulation resistance, electrolytes, oils or resins in capacitors, resistance between leads through external atmosphere, between the winding and the case, the case and terminals, resistance of capacitor seals, paints, etc. Atmospheric moisture and chemicals add their own effects. An additional component is AC resistance when the capacitor is used in AC circuits, due to dipole movements in dielectric material. A high frequency current passing through the capacitor also increases its losses.

The resistive components give rise to equivalent series resistance (ESR), which represents a lumped resistance depiction in series with the capacitor. The overall resistive effect gives rise to the loss factor/dissipation factor/loss angle/tangent of loss angle (tan delta)/power factor of capacitors. (All these terms are used synonymously.) The ESR is the value of resistance equal to the total effect of a large and complex set of energy loss mechanisms (Fig. 2.2) occurring under a particular set of measurement or operating conditions.

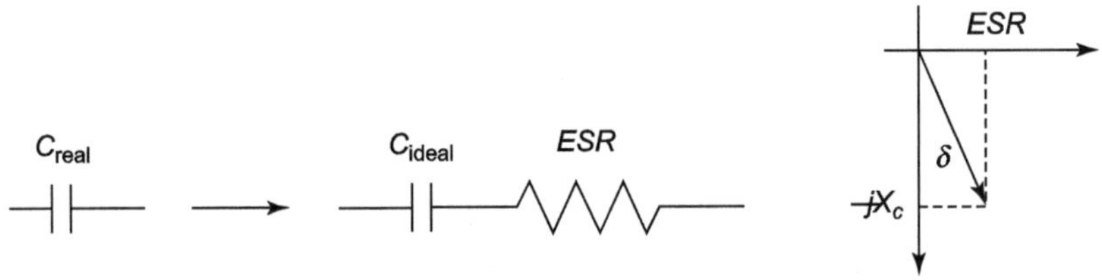

Fig. 2.2 *ESR and impedance vector of a practical capacitor.*

Q or quality factor of a capacitor is important in tuned circuits because they are more damped and have a broader tuning point as Q goes down. $Q = X_c/\text{ESR}$, where $X_c = 1/\omega C = 1/2\pi fC$ and ESR is the sum of in-phase AC resistance. It includes the resistance of the dielectric, plate material, electrolytic solution (for electrolytic capacitors) and terminal leads at a particular frequency. This resistance often is the cause of failures in capacitor circuits due to heat build-up. As the plate area increases, the ESR will go down for the same plate thickness.

2.4 DISSIPATION FACTOR, LOSS FACTOR, TAN DELTA OF CAPACITOR

The dissipation factor (DF) is given as D, or more commonly as tan delta (or tan δ). It may be noted that the term D is exactly opposite to Q (reactance/resistance) the quality factor, of an inductor, and the relationship is $Q = 1/D$.

Hence, $$DF = 1/Q = \text{ESR}/X_c$$

The power factor or loss factor of capacitor is the ratio of real power to apparent power. **Loss factor** = $I^2 \times \text{ESR}/I^2X_c = \omega C \times \text{ESR} = 1/Q$

The loss factor is thus seen to be directly proportional to the frequency of voltage. However, it is often also a function of ambient temperature and biasing voltage as also the voltage stress on dielectric. A good capacitor design also accounts for the resistance introduced by electrode thickness or its shape. In general, electrolytic capacitors have much higher loss factor compared with other types. Loss factor is given as an absolute per unit quantity, or a percentage of capacitor VA (volt × ampere). For example, it could be written as 0.05, or 5 per cent.

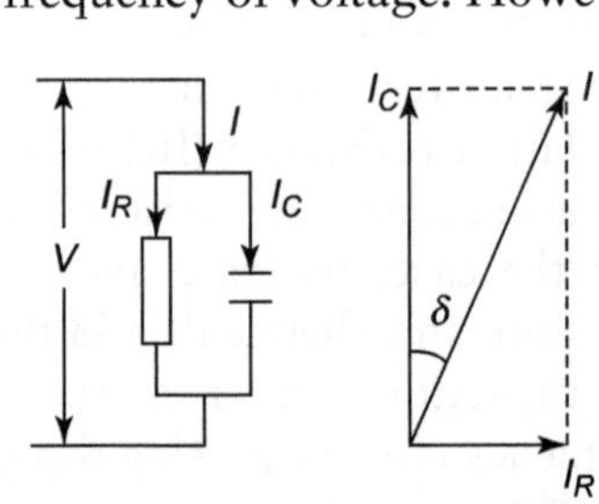

Fig. 2.3 *Current components and loss angle of a capacitor.*

The presence of a resistive component in capacitors gives rise to two components of AC currents – one 90° leading the voltage, and the other in phase with voltage (resistive component), as shown in Fig. 2.3. However, the resistive component in most cases is so small that the total current leads the voltage by almost 90°, deviating only by a small fraction of a degree. Hence the power factor of a capacitor is a very small figure. For this reason, instead of customary power factor angle, its complementary angle is mentioned, the loss angle, whose sine is equal to the cosine of conventional power factor. At this small angle, its sine is equal to its tangent value, and when the angle is expressed in radians, the tangent is equal to the angle itself. Hence the loss factor is also given as tangent of loss angle, or tan δ (δ being the value of angle in radians).

The loss tangent at any given frequency is the ratio of real and imaginary parts of its impedance. This is an important parameter, and can be vital to the performance of a capacitor as also the circuit where it is used. In electronic circuits it can give unnecessary resistance, while in an *AC* circuit it may cause unacceptable heat loss. If we consider a large capacitor of 36 μF working at 440 V 50 Hz AC, with a loss tangent of 0.1 per cent, the watt loss will be roughly 2.2 watts, while the capacitor generates nearly 2200 VA. For a capacitor unit of 25 KVAR in a power system, the corresponding loss will be 25 watts. Such high losses will cause the capacitors to fail under thermal runaway.

All dielectrics (except vacuum) have two types of losses. One is a conduction loss, representing the flow of actual charge through the dielectric. The other is a dielectric loss due to movement or rotation of atoms or molecules in an alternating electric field. Dielectric losses in water are the reason for food and drink getting hot in a microwave oven.

Most capacitors lose a fraction of the energy when an alternating current is applied. In other words, the dielectric is less than perfect. The simplest model for a capacitor with a lossy dielectric is also represented by a capacitor with a perfect dielectric in parallel with a resistor (Fig. 2.3) giving the power dissipation. The current now leads the voltage by a very little less than 90°, where the difference δ (Greek letter delta) is termed the dielectric loss angle.

The dielectric constant and losses depend on temperature as well as the frequency of applied voltage. This is because the dipole movements cannot keep pace with change in the electric field direction as frequency rises. The dipoles also face higher resistance because of frequency of movements, causing higher heat losses.

The dielectric loss factor represents watt loss in a dielectric material through conduction, slow polarization currents and other dissipative phenomena. Polarization currents occur due to frequent reorientations of dipoles in every half cycle of external *AC* voltage, as they try to adjust to the rapidly changing electric field. This oscillation of dipoles involves energy consumption depending upon material properties, and adds to watt loss of capacitor. The loss tangent is a parameter of a dielectric material that quantifies its inherent dissipation of energy. The term refers to the angle in a complex plane between the resistive component of an electric current and its reactive (lossless) component. Note that the ESR is not simply the resistance that would be measured across a capacitor by an ohmmeter, but is a derived quantity representing the loss due to both the dielectric's conduction electrons and the bound dipole relaxation phenomena. Table 2.2 below gives the dissipation factor (tan δ) and ESR of 1 μF capacitor with different dielectrics.

Table 2.2 *Dissipation Factor and ESR for Common Dielectrics*

Dielectric	*D. F. (maximum)*	*ESR for 1 μF capacitor*
Aluminium electrolytic (50 V)	10% at 120 Hz	132.6 Ω
Tantalum	4% at 120 Hz	53 Ω
Polyester	1% at 1 kHz	1.59 Ω
Polypropylene	0.1% at 1 kHz	0.159 Ω

Most electrolytic capacitors tend to have a high loss factor due to the nature of electrolytes and the dielectric layer characteristics. Their use in *AC* is therefore very limited, the most common application being motor start capacitors, where a capacitor remains in circuit for just under three seconds, and is switched off once the motor attains speed. The heat generation can be imagined from the fact that their loss factor can be between 1.8 to 7 per cent, and capacitor values ranges between 40 μF and 300 μF. A capacitor of 200 μF having tan δ of 2% and an applied voltage of 230 V AC will generate 66 W of heating load.

Heat generated in a capacitor must be dissipated from its surface to the environment. This factor has to be built into the design, and sometimes, special measures are necessary for this purpose. Most common methods include use of large containers, providing fins, water-cooled bodies, forced air ventilation etc. It is common practice to check the heat dissipation surface during design stage of common AC capacitors. Power capacitors on supply systems at substations have to be well ventilated, particularly when used indoors.

2.5 EQUIVALENT SERIES INDUCTANCE (ESI) IN CAPACITORS

In a wound capacitor, the coil structure leads to the formation of an inductance between leads of capacitors. In such a capacitor, if terminals are brought out from the ends of each electrode foil, an inductance can be measured between these terminal pairs. Further, the coil structure also causes formation of inductance between the two electrodes. In certain applications like railway carrier line communication networks superimposed on main lines, the inductance of a capacitor winding used to be specified (in μH) between two terminals of each electrode, and played a part in the circuit.

Modern metallized capacitors have their ends sprayed to collect current all through the length and thus have all turns shorted at both electrode ends, making them essentially non-inductive. Extended electrode foil construction in many capacitors also makes them non-inductive.

Leads of capacitors are also a source of inductance. A wire, however small, will present an inductance. If measurements are made at a distance from capacitor ends, they could be slightly different from the one at the ends. The wire length, their shape and gauge affect the value of this inductance.

In small capacitors and at higher frequencies, these inductances could be a nuisance. They may even cause resonance problems at certain frequencies. They need to be compensated/accounted for when dealing with high frequencies or higher order harmonics, and when used for filters.

The equivalent series inductance (ESI) of a capacitor is caused by the inductance of the electrodes and leads. It sets the limiting factor of how well (or fast) a capacitor can decouple noise off a power bus. The ESL (L) of a capacitor also sets the resonance point of a capacitor. Because the inductance appears in series with the capacitor, they form a tank circuit. At resonance frequency F_r, inductive reactance $X_l = \omega L$ becomes equal to capacitive reactance

$$X_c = 1/2\pi fC = X_L = 2\pi fL$$

The resonance frequency is then $F_r = \left(1/2\pi\sqrt{LC}\right)$

2.6 COMMON DIELECTRIC MATERIALS

Several materials of different types, varieties and compositions are in use for different types and grades of capacitors. Electrostatic, electrolytic and electrochemical capacitors use dielectrics of different characteristics, and demands on their dielectric layers are different.

2.6.1 Plastic Films

Plastic film capacitors are commonly used in modern high-performance electrostatic capacitor applications. Strips of plastic film are wound or stacked between electrodes to serve as dielectrics. Polyester and polypropylene dielectrics are the mainstay of film capacitors. They are the ones that most film capacitor makers use, although use of polyphenylene sulphide (PPS) is on the rise, replacing polycarbonate film because of its superior properties, along with polyethylene naphthalate (PEN) for a number of applications.

[A] Polyester (PET)

Polyester is probably the most popular of film capacitors. Polyester is a generic term for a class of polymers. The one used in polyester capacitors is polyethylene terephthalate film (DuPont's Mylar). This is commonly known as PET. Low cost, small size and versatility make it a good choice for most applications. It has higher dissipation factor, hence it is best used in DC or relatively low-frequency or low-current pulse and AC power applications. Poor temperature drift, dielectric absorption, and leakage currents limit its usage to non-critical circuit applications. Polyester capacitors can typically be found in values from 0.01 μF through at least 10 μF and beyond. Polyester has a high temperature drift. PET capacitors are available to 125°C. Good heat resistance allows polyester capacitors to be made in surface-mount styles. PET motor run capacitors and SCR snubbers are replacing older electrolytic capacitors. PET capacitors are the mainstay of most *DC* capacitors, and voltages up to a maximum of 16 KV DV are available.

[B] Polypropylene (PP)

Polypropylene (PP) capacitors are available in a wide range of sizes and voltages, and are used in a wide variety of circuits. Bi-axially oriented polypropylene film (BOPP) has a very low dissipation factor over its entire temperature range and over a wide frequency spectrum. Dielectric strength is considerably higher than PET or PC film. This makes polypropylene capacitors popular for high-frequency, high-current applications like switching power supplies. Large capacitors made from *PP* film, film–oil, and paper–oil–film types are found in power-line applications like power-factor correction. These films can also be used for high voltage capacitors.

Low leakage and low dielectric absorption make small polypropylene capacitors suitable for integrators and sample-and-hold circuits. Moisture absorption is negligible. Only their higher temperature drift makes them inferior to polystyrene. Polypropylene has limited heat resistance (to 105°C), and is not found in surface mount capacitors.

[C] Polystyrene (PS)

Polystyrene (*PS*), (often called 'Styroflex') has long been the material of choice for critical analogue circuits. Low leakage, low dielectric absorption and a flat temperature curve (Fig. 2.4) makes these capacitors suitable for timing circuits, filters, integrators and sample-and-hold circuits. Moisture absorption is very low. Size, cost, availability and temperature range limitations make polystyrene unsuitable for most other applications. Heat resistance is limited to about 85°C. It can be damaged by soldering and by chlorinated cleaning solvents.

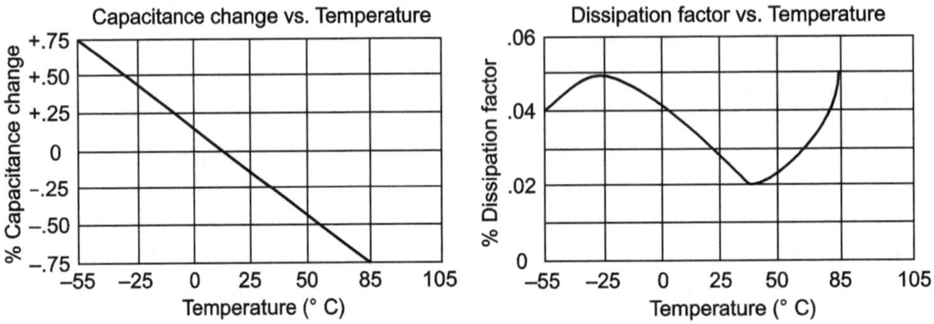

Fig. 2.4 *Capacitance and D of PS in relation to temperature.*

Because of the poor heat resistance, polystyrene has largely been replaced by polypropylene and C0G ceramics, and the capacitor-grade film is no longer being made. PS's other electrical properties are mostly very similar to those of *PP*. Polystyrene has mostly been discontinued from capacitors due to better materials being available.

[D] Polycarbonate (PC)

Polycarbonate capacitors have been used in a wide variety of applications because of their superior performance. Typically they are used in applications where precision capacitors are needed (<5%). They are generally used in electronics circuits such as filters, as well as for timing and precision coupling applications.

Polycarbonate capacitors can also be used for AC applications. They are sometimes found in switching power supplies. Although dissipation factor is low, current must be restricted to prevent them from overheating, although they can tolerate temperature better than many other types of capacitor.

Polycarbonate has a fairly low temperature drift (lower than most films), dissipation factor and dielectric absorption.The polycarbonate dielectric material is very stable, has a high temperature tolerance and can operate over a temperature range of typically –55°C to +125°C without de-rating. Additionally the insulation resistance and dissipation factors are good and the dielectric constant of 3.0 means that polycarbonate capacitors are a reasonable size for their capacitance.

It can be used in timing circuits, although ceramics are a better choice for small sizes. It is suitable for some pulse applications, and for some precision analogue applications, especially for good temperature stability and relatively high temperature rating. The excellent stability of PC capacitors can be seen from Fig. 2.5, which shows hardly any variation in value and dissipation factor over the full working range. Moisture absorption is high compared to most other film dielectrics, a problem for some critical applications. It has good heat resistance, to 125°C, but not good enough for in surface-mount packages. It was also being used for automotive applications. This material is becoming obsolete as a dielectric and is being replaced by PPS.

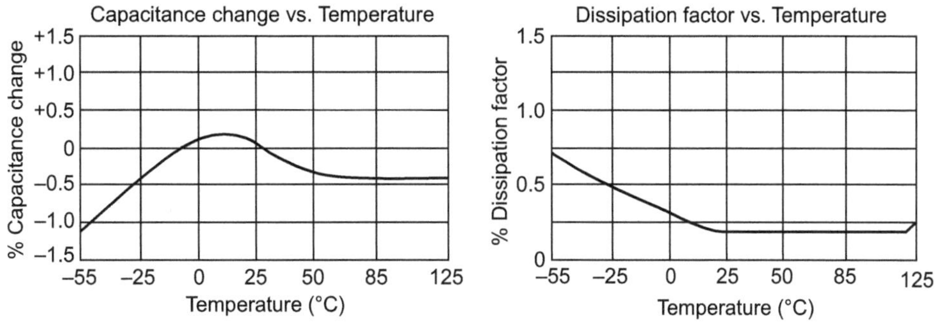

Fig. 2.5 *Temperature stability of polycarbonate film.*

While polycarbonate capacitors have been widely used within many electronics circuits and found favour with many electronics design engineers, they are not as widely used these days. The Bayer Corporation, the major manufacturer of polycarbonate, discontinued its production around 2000. There are still some smaller sources of the dielectric material and some capacitors are still made. However, there are fewer suppliers today, and the film is on the way out from capacitor materials.

Table 2.3 *Comparative Properties of PP, PET, PC and PS (*Source: Yunstar Electronic Co., Ltd*)*

Property	*Polypropylene*	*Polyester (Mylar)*	*Polycarbonate*	*Polystyrene*
Dielectric Constant	2.3	3.2	3.0	2.5
Density	0.905	1.395	1.20	0.95
Area factor in^2/ lb/ mil	30.600	19.800	23.100	26.500
Temperature rating (for capacitors) °C	105	125	125	85
Dissipation factor % at 10^6 Hz	0.02	1.6	1.0	0.01
Flatness of electrical properties	Excellent	Fair	Good	Excellent
Dielectric strength V/mil at 1 mil	7.000	7.000	4.000	5.000
Tensile strength lb/in^2	28.000	30.000	8.000	5.000
Moisture sensitivity	None	Small	Moderate	None

[E] Polyphenylene sulphide (PPS)

Like PC, PPS film is available from only one source of supply. Toray (Tokyo, Japan), a film-manufacturing company, makes PPS film trademarked as Torelina. In general, since PPS and PC have the same dielectric constant, the sizes of PPS and PC capacitors will be the same. PPS has properties closest to PC, and has replaced it in most applications. It is flame resistant, and has good resistance to moisture and solvents. Its cost is comparable to PC, while its dielectric constant is same as PC, and the capacitor size also remains the same.

The ESR performance at room temperature over a frequency range from 100 to 100,000 Hz for polyphenylene sulphide is superior to polycarbonate. A PPS device may run hotter than a PC device without any problem in some AC applications. PPS, unlike PC, can operate without degradation at capacitor temperatures exceeding the +125°C in DC applications.

Economics is a major consideration in today's competitive business environment. While the cost of a capacitor depends on many factors, the cost of PPS dielectric film is comparable to the historical cost of PC dielectric film. Furthermore, the supply of PPS has never been interrupted — unlike PC — since the material was first made available as a capacitor dielectric film some 20 years ago.

Dielectric strength of PPS is higher at 400 V compared with 300 V for PC, so film thicknesses could be lower, reducing capacitor size. Capacitance stability is better, with almost a flat characteristic from –55 to 150°C. Capacitance variation is about only 7 per cent over a wide range of –55 to 85°C. The ESR and loss factor are also superior up to about 100°C. This has become popular for military applications due to the temperature tolerance and relatively smaller sizes. Short-term exposure to 250°C does not degrade the capacitors.

[F] Polyethylene naphthalate (PEN)

This is a new film (Teonex from DuPont) dielectric, with electric strength 25% greater than PET. Its working temperature extends to 160–180°C, has high mechanical strength, good thermal conductivity and lower thermal shrinkage. This permits use of thinner films, and a smaller capacitor size. Costs of materials and those of capacitors may go down once this is fully commercialized. High temperature characteristics make it ideal for SMD capacitors. Figure 2.6 brings out the superiority of PPN over PET even at higher temperatures.

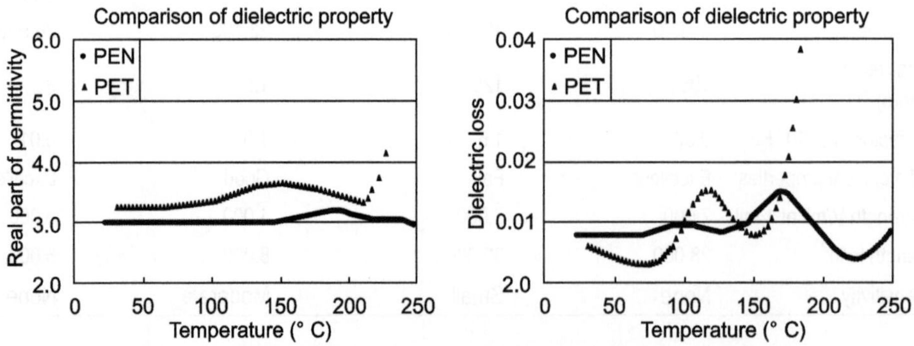

Fig. 2.6 *PEN and PET comparison (*Source: SKC Inc, Georgia*).*

PEN is especially suited for the manufacture of chip type capacitors for surface mounting and for increased service temperatures.

[G] Teflon

Teflon TFE is DuPont's trade name for poly-tetra-fluoro-ethylene or PTFE. It is also sold under a number of other trademarks such as Fluoroplast-4 and Fluon PTPE. PTFE has very low leakage, very low dielectric absorption (probably the lowest in both cases), a very low dissipation factor, a wide temperature range (up to 200°C for some), low temperature drift, negligible moisture absorption, and very good stability. It is about the best capacitor film for critical analogue applications. It is expensive, and few companies still make it. The film has exceptionally poor mechanical properties and inconsistent thickness, and manufacturers find it difficult to work with.

PTFE capacitors are available from 0.001 μF to at least 2 μF, but not normally used in SMD. However, a few Teflon SMD parts in very low values, <10 pF are being made. Teflon is the lowest loss solid dielectric, with operating temperatures up to 250°C, extremely high insulation resistance and good stability. It is used in stringent, mission-critical applications. Its disadvantages are large size (due to low dielectric constant), and higher cost than other film capacitors. The Teflon film trademark was coined by DuPont and registered in 1945; the first products were sold commercially under the trademark beginning in 1946.

Dielectric constants of all materials depend upon temperature, and the temperature dependence has to be accounted for when selecting a material for an application. Fig. 2.7 shows the relationship between capacitance and temperature of PP/foil, metallized PP (MPP), metallized PPS (MPPS), metallized PS (MPS), metallized polycarbonate (MPC) and polystyrene/foil (PS/FOIL) capacitors.

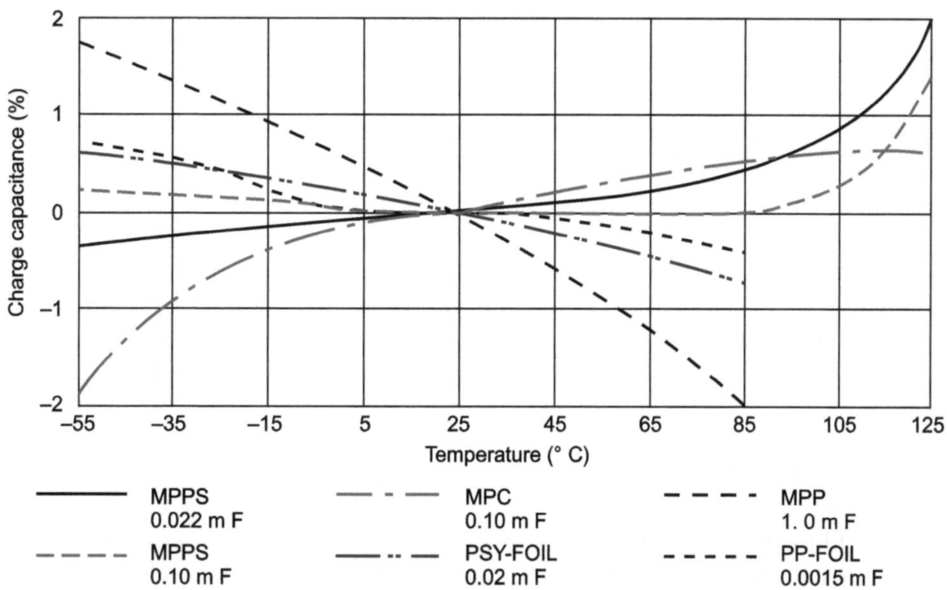

Fig. 2.7 *Effect of temperature on dielectric constant of different plastic films.*

It will be noted that MPPS capacitors undergo a very small capacitance change from –55°C to +85°C, with maximum variation under 0.5% from their value at 25°C, whereas MPP film capacitors show a reduction in capacitance value up to –2% from 25°C to 85°C. MPC capacitors are most stable between –15°C to 125°C, and practically do not undergo any change from –5 to +85°C.

The dissipation factor (tan δ) of most dielectrics goes up with temperature. It can be seen from Fig. 2.8 that 0.1 μF MPPS is the most stable in the whole spectrum up to 100 kHz. Frequency dependence also varies with the size or rating of capacitor for the same film. Width and volume, the aspect ratio (length of winding/winding diameter), the leads and joints in relation to capacitance value are factors which influence capacitor loss.

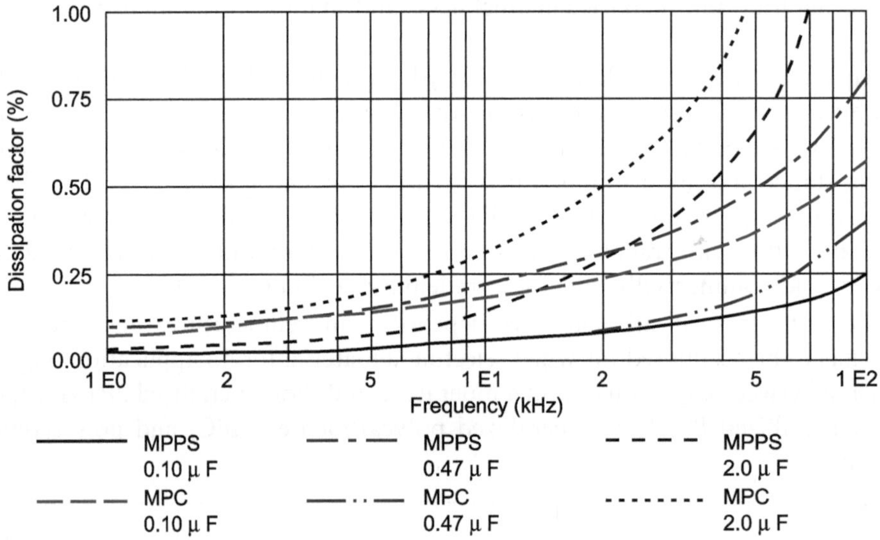

Fig. 2.8 *Relationship between frequency and dissipation factor for MPPS and MPC.*

2.6.2 Composite Dielectrics

Solid dielectrics like kraft paper are porous, and inherently have air and moisture in them. These inhibit the properties of the solid, and need to be removed in order to get full properties as a dielectric. Even film/foil systems will have some air/moisture between layers. Special films (hazy films) are made, and special processes (vacuum impregnation) are used to remove the air and moisture under high vacuum and temperatures and replace the spaces with a compatible liquid dielectric. The liquid should have dielectric constant as close to the solid as possible to ensure uniform electric field, and the working voltage stress is dependent solely on the weaker dielectric strength material, which is generally the solid.

Film or paper has to be of very high dielectric strength and must be ideally free from any impurities, conducting paths or pinholes, since each of these reduces the electric strength. The manufacture of films or paper involves rigorous dust and environment controls, and the base material and other items used have to be as pure as practically possible. However, any film or paper will always unavoidably have certain imperfections, which are inherent or introduced due to manufacturing limitations.

Hence, in capacitors using foil electrodes, paper or PP is generally used in a minimum of two layers, and the number of layers can go up, depending upon the reliability requirements and the voltages used. For example, if the desired thickness is 20 μm, two layers of 10 μm are used. The chance of weak spots of two layers coming into alignment is very remote and the dielectric can work to its capacity. So the composite film or paper dielectrics generally have two or more layers of base materials, impregnated with a suitable impregnant.

The impregnant may be chemicals, synthetic oils, resins, wax, petroleum jelly, vegetable oil (castor oil and rapeseed oil are common), nitrogen or some inert gas. A choice of impregnants is based on the desired capacitor, its properties, end use and environments of use. The impregnant has also to be in its purest form and must be processed to remove any impurities and moisture. Flash point and fire point, viscosity and gas contents are also considered. Generally oils are used in most power applications, but waxes and resins are used when it is desired to have a capacitor dielectric system in solid state at a working temperature.

Working voltage stresses of 50–60 KV/mm are common in impregnated film dielectric capacitors, and still higher in metallized capacitors. Impregnants are also used in solid systems to fill the voids in high voltage bushings, where the working stresses can go to 200 KV/mm.

[A] Mixed dielectrics

Often, the dielectrics are used in combination, e.g. PP with paper (with aluminium foil electrode). This allows drawing maximum benefits out of the composite dielectric system. These are impregnated with suitable dielectric oils. Sometimes one of the dielectrics may be a film or paper metallized on one side. In some other capacitors, paper or PP layer metallized on both sides serves as extended electrode, while the other plain film/paper works as the plain dielectric. This construction is common in power factor capacitors, commutation capacitors or some high frequency applications.

[B] Metallized paper or plastic films

With the development in materials and technology, the search for smaller sizes and better reliability in terms of short circuit failures in service, coupled with economy of design prompted the development of metallized electrodes. A metal (mostly zinc or aluminium, or an alloy of these) is vacuum-deposited on the film/paper in a thin layer, of the order of 0.03 μm or less in thickness. This does away with 5–8 μm aluminium foil, and the capacitor size is drastically reduced by that much.

The metallization layer is so thin that its current carrying capacity is limited, and current is collected all over the winding length, by spraying of metal on metallized layer ends, and brought out on either side of the winding. Any short circuit or surge current that may pass through a weak spot creates immense heat at the spot, and the metal around it evaporates. The dielectric is thus rid of the weak spot (which is isolated), and the capacitor continues to operate in the system.

Use of metallized electrode also does away with the necessity to use a minimum two layers of film/paper, as weak spots are isolated during manufacture by a process called 'Short clearing'. Failures at any weak spot in service are taken care of by self-healing. These advantages have seen the dominance of metallized film capacitors in both electrical and electronic fields in most wound capacitors.

In one type of capacitor construction, kraft capacitor tissue paper metallized on both sides without any side margins is used as electrodes, in place of foil. The capacitor undergoes self-healing on one side of an electrode in case of failure, the second side remaining unaffected. The decrease in capacitance is less than that in a normal metallized capacitor. The construction is known as MKV capacitor (it was also known as Bosch type capacitors). Metallized electrodes are extended on sides and sprayed with metal to form current collectors.

Electrical properties of the insulating system change due to age and continuous electrical stress. The principal contributor to the unexpected breakdown of the high voltage equipment is insulation failure. As compared to the magnetic, conducting and insulating materials which form the basis of any electrical equipment, the insulating material in a capacitor is more prone to service stresses like thermal stress, electrical stress, mechanical stress, environmental stress etc.

By measuring electrical properties such as capacitance and tan delta periodically, it is possible to ensure an operation free of unexpected breakdowns. The dissipation factor (tan delta) is one of the most powerful offline non-destructive diagnostic tools to monitor the condition of solid insulation of high voltage equipment.

2.6.3 Ceramics

Ceramics are a unique family of dielectrics with dielectric constants ranging from 6 to 10,000. These can be easily manufactured to desired physical and electrical characteristics by applying ceramic chemistry. For ceramic capacitors, voltage applied to the capacitor also affects its capacitance value (the electric field strength across the dielectric changes the effective dielectric constant **k** of the material).

Ceramic capacitors are available in three classes. Class I ceramics are used for resonant circuits and high-frequency bypass and coupling. These capacitors have a wider temperature range compared to Class II and Class III capacitors. Class II ceramics are used where miniaturization is required for bypassing at radio frequencies, filtering and inter-stage coupling. Class III ceramics are used where low-voltage coupling and bypassing in transistor circuits are necessary.

2.6.4 Mica

Mica, a mineral, is one of the oldest dielectric materials used in capacitor construction. There are several kinds of mica, with differing properties, but mica is in general very stable electrically, mechanically and chemically. There are many types of mica, but only six or so are common rock-forming minerals. Mica capacitors are normally made from muscovite mica, or potassium aluminum silicate, $KAl_2Si_3O_{10}(OH)_2$. It is thermally stable up to 500°C, and has a high dielectric strength. Phlogopite mica, or potassium magnesium silicate, $KMg_3Si_3AlO_{10}(OH)_2$, is softer than muscovite mica and has less desirable electrical characteristics, but it may be used up to 900°C. India is the biggest supplier of mica.

Mica has a dielectric constant in the range 5–7. Natural mica contains many other materials including iron, sodium, ferric oxide and lithium. Because of the variability in the composition of natural mica, mica destined for use in capacitors must be carefully inspected and classified, which adds to the manufacturing cost. Although there are several different forms of mica, they all have very similar properties. They are fundamentally very stable both mechanically and chemically, enabling capacitors manufactured with mica to exhibit similar properties.

The crystalline structure of mica has binding forces that are different in different planes. In one plane they are strong, but weak in the perpendicular plane. This gives it a layered structure and enables it to be spilt along the lines of the weak bond into very thin flat sheets. The sheets used in capacitor manufacture are from less than about 0.025 to 0.1 mm.

Mica is very stable and chemically inert. It does not react with oil, water, many acids, alkalis and solvents. As a result of this, ageing does not occur to any major degree, and the variations of water vapour in the atmosphere do not cause undue variations in the overall capacitor performance.

A desired application, size requirements, temperature and environmental conditions decide the choice of capacitor with a particular dielectric. Table 2.4 compares the properties of some of the most common types of dielectrics.

2.6.5 Glass

Glass is the preferred choice for long-life capacitors, which are highly stable over an entire range of frequencies and temperatures. These are characterized by loss angle as low as 0.0039 even at 1 GHz, and dielectric constant variation from 6.4 at 1 MHz to 6.13 at 1 GHz. Volume resistivity is of the order of 10^{14} ohm-cm or higher and surface resistivity of 10^{13} ohm-cm. Electric strength is also quite high at over 8 KV/mm.

Glass capacitors are used where ultimate performance is required for RF circuits. They offer very high levels of performance, although their cost is high when compared to many other forms of capacitor. Typically a glass capacitor will have a relatively low capacitance value. The values of glass capacitors may range between a fraction of pF up to 3,000 pF. As such these capacitors are used mainly in radio frequency circuit design.

On account of their costs, glass dielectric capacitors are reserved only for the most demanding RF requirements. These are usually low volume products where cost is not as important as in high volume products. These capacitors are made by only a few manufacturers and the capacitors may not be available ex-stock. Their properties offer real advantages in many applications over all other forms of capacitor. Their combination of robustness and high tolerance sets them above all other capacitors. When size and cost are not critical, a glass capacitor may solve a problem in a circuit that may otherwise not work properly in the particular environment in which it may need to operate.

Table 2.4 *Comparison of Some Common Dielectrics*

Property	*Film*	*Ceramic disc*	*Tantalum*	*Aluminium electrolytic*	*Multilayer ceramic*
Voltage range (DC)	50–2000	16–15000	4–50	1–600	63–4000
Capacitance range for DC	0.001 – 10 μF	1 pF – 0.1 μF	0.1 – 1500 μF	1 μF – 1 F	1 pF – 100 µF
Size	Large	Small	Small	Medium	Small
Capacitance	Medium	Low	High	High	Low
Polarity character	Non-polar	Non-polar	Polar	Polar	Non-polar
Thermal stability	Excellent	Medium	Poor	Good	Poor for high k
Max. operating temperature	105°C	125°C	125–150°C	85–120°C	125–150°C
Voltage de-rating	Not needed	Not needed	50–60% for better reliability	80% or V_n for double life	Not needed
Typical failure mode	Open circuit Drift	Drift/Short	Drift/short	Open circuit/Drift	Drift/Short
Cost	High	Low	High	Low	Low

3

POLYPROPYLENE AND POLYESTER FILM

3.1 MAJOR PLASTIC FILMS IN CAPACITORS

Plastic film capacitors are mostly used in high-performance applications. Polycarbonate, polyester and polypropylene have been the 'big three' of film capacitors. Most capacitor manufacturers have been using these all along, although use of polyethylene naphthalate (PEN) and polyphenylene sulphide (PPS) is on the rise, and polycarbonate is going out of favour in recent time. Therefore it is pertinent to study the characteristics of capacitors using these materials.

3.1.1 Classification of Film Capacitors

Short identification codes for the type of construction, describing the dielectric and the basic technology applied, are defined in DIN 41 379.

The last character in the short code indicates the type of dielectric:

T = Polyethylene terephthalate (PET)

P = Polypropylene (PP)

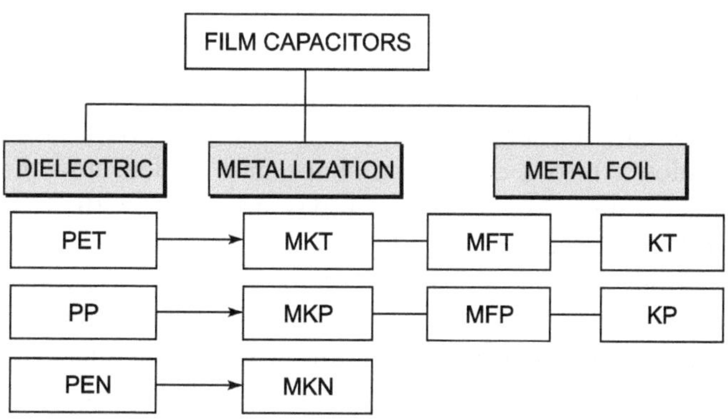

Fig. 3.1 *Main types of plastic film capacitors.*

N = Polyethylene naphthalate (PEN)

An M (= metallization) is prefixed for the identification code of capacitors with metallized films, while MF denotes one metallized film and one foil electrode with plain film. Figure 3.1 indicates the types and nomenclature of capacitors using these films.

3.2 POLYPROPYLENE FILM

Polypropylene has virtually taken over the capacitor scenario in most power frequency AC applications in the past few decades. Its low loss factor, ability to operate up to 100°C, and ease of use have been the deciding factors in this trend. In the 1980s, a search was on for an improved dielectric which could bear more stress and reduce capacitor size. PP was found to have superior properties in this respect, and in combination with paper, gave better electric strength. Paper capacitors were phased out after the ban on synthetic impregnating oil polychlorinated biphenyls (PCBs), due to ecological factors, and PP was seen as an alternative to be used with paper in mixed dielectric power capacitors.

The very low dissipation factor of PP has made it the only viable material for most high power AC applications. Self-healing properties of MPP in metallized capacitors are critical for reliable high voltage operation. When an arc-through of the dielectric occurs, PP tends to leave more carbon at the site than polyester. The arc is more quickly extinguished if the gas pressure at the failure site is as high as possible. However, oil impregnation helps improve and make up for this deficiency. Oil acts as a coolant and limits the spark, by conducting the heat away. Physical sizes of PP capacitors are comparatively higher due to its low dielectric constant. Polypropylene motor starter, motor run, and SCR snubber capacitors are replacing older electrolytic and paper–oil types all of which have much higher dissipation factors.

Polypropylene has virtually replaced paper–oil type capacitors which have much higher dissipation factors. PP is usable in high frequency applications. Larger values and higher voltages, up to 72 μF (and even higher) and 440 V capacitors are common in motor run applications. Higher voltage ratings have become common in power factor correction capacitors and certain traction capacitors. Its low tan delta and higher service temperature, coupled with the ease of manufacturing film has made it the favoured dielectric among designers. The capacitance variation over a temperature range is also within 3%, making it stable over the entire usable range. The DC high voltage tolerance of over 650 V per micron reached over the years has helped reduce capacitor dimensions.

Polypropylene is manufactured in two processes – blowing (bubble) and stenter. Initial developments were with the blowing process, which had a limitation of non-uniformity in thickness. The film was blown into a cylindrical shape and cooled, and then drawn, slit and rolled on machines to get a film. Most PP films today are made with stenter process on machines, which gives it uniform thickness and good mechanical properties. They have good tensile strength in both machining as well as cross direction. The dielectric constant of 2.22 to 2.25 agrees with most liquid impregnants which were introduced as replacements for PCBs. This was beneficial to ensure a uniform electric field throughout the composite dielectric. In paper/PP/foil capacitors, a major part of the applied voltage was shared by PP and oil, while paper helped proper impregnation of the dielectric system.

3.2.1 Hazy Film

Improvement in capacitor voltage stress necessitated the replacement of paper, which led to the development of hazy film. The film is made rough on one side, or on both sides, which allows the oil to penetrate the capacitor winding volume, and the need for paper has been overcome. The PP/oil dielectric has very high dielectric strength and matching dielectric constants. This gives the capacitor high uniform electric field stress and reduces the capacitor size. The loss factor and capacitor losses are also considerably lower. Hazy films from 6 to 17.5 μ are available today. The thickness of hazy film is not uniform because of serrations and wavy surface, and actual thickness is lower than measured micrometer thickness. The space factor depicts the correction factor to be taken into account while calculating actual film volume and thickness. It takes into account the effect of air spaces inside the overall film volume. When using hazy film, the space factor needs to be taken into consideration. A space factor of 7–10% is common.

In hazy film, the film surface develops depressions, with a hills and valley structure, which allows the air trapped inside the winding to get a chance to get out through these paths, and the oil fills up these spaces to create a void-free uniform dielectric system. In some designs, the electrode foil is also shaped on the machine additionally, to intentionally create spaces for this purpose. Given the nature of film and foil, it is very difficult, sometimes impossible, to completely evacuate the air and moisture trapped in the winding. Hence these extra measures are needed in the manufacturing process, and special provisions are made in winding machines for this purpose. The foil passes over forming rollers to get the serrations or shapes during its flow from take-off spools to winding cores on the machine.

3.2.2 Interaction Between PP Film and Oil

In film/foil capacitors the oil plays an important role. Proper knowledge of this interaction is necessary for perfect impregnation. Capacitor impregnation involves two processes: (1) Penetration of oil between layers of PP film and aluminium foil, and (2) Swelling of film due to penetration of oil in the interior of the film. The film partially dissolves into oil and absorbs it. This causes swelling of the film, in turn creating voids in the film, which are filled with oil, increasing its dielectric strength. Absorption of the oil by the film and dissolution of the film into the oil continue until saturation is reached. The time for the entire process depends on temperature.

To ensure full penetration in the winding and complete impregnation, the important parameters are space factor, viscosity of oil, capacitor temperature, oil temperature and vacuum during impregnation.

3.3 POLYESTER FILMS (PET)

PP is being used for both AC and DC applications, whereas PET has its limitations in AC due to its high loss factor. PET has a dielectric constant of 3.2 as against 2.2 for PP, so the size of PET capacitors is smaller in general. The density of PET is 1.35 as against 0.91 of PP, so the weight of PP and PET capacitors is the same for the same thickness. A PET capacitor is cheaper and in DC,

PET has the advantage of smaller size. PET also gives a higher working temperature capability, but when it comes to harmonics or high frequency applications, PP certainly has an edge due to its loss factor and stability over a wider range.

Polyester is probably the most popular dielectric for DC film capacitors. Polyester is a generic term for a class of polymers, the one used in polyester capacitors being polyethylene terephthalate, generally known as PET (Du Pont's trade name is Mylar). Low cost, small size and the ability to do many things well enough makes it a good choice for many noncritical applications. A high dissipation factor means it is best used in DC or relatively low-frequency/low-current pulse and AC power applications. Poor temperature drift, dielectric absorption and leakage limit its use to noncritical analogue circuit applications.

Polyester capacitors can typically be found in values from 0.01 µF to 10 µF and beyond. Polyester has a high temperature drift but can be layered with polypropylene to flatten the temperature curve (the two go in opposite directions). PET capacitors are available to 125°C. Good heat resistance allows capacitors to be made in surface-mount styles. Capacitor temperature should not exceed mounting heat resisting temperature during the soldering operation. Polyester has better heat resistance than polypropylene.

PET is tough material and easy to wind, and it does not undergo any stretch related elongation during winding. Very thin films are possible with PET, and films up to 1.5 microns have been made. PP gets stretched at high speeds or under tension, and its thickness can change due to this. Hence it is more difficult to wind, particularly for thin films. PP films are normally used up to 5-micron thicknesses, but nowadays 3.5-micron films have been developed.

3.4 CHOICE OF PP/PET FOR AC APPLICATIONS

All dielectrics have their advantages and shortcomings. For example, polypropylene behaves very uniformly and predictably over temperature and frequency changes allowing suggested operating limits to be calculated and plotted. It has very low dielectric losses. Its voltage tolerance per unit thickness is the highest of all capacitor films. However, polypropylene has a maximum service temperature limit of +105°C. It also has a low dielectric constant, resulting in a larger physical size for a given capacitance and voltage rating compared with most other film dielectrics. Table 3.1 compares the properties of these two film types.

Metallized polyester (and sometimes metallized polypropylene) is lately being used for motor start capacitors, where it is replacing conventional electrolytic capacitors. It draws benefit from the factors below:

- A comparable size of capacitor can be made.
- Its loss factor is extremely low compared with electrolytics.
- It can withstand the AC voltage for a much longer time.
- A smaller value capacitor may therefore be used.
- The motor can be redesigned to allow much higher voltage on capacitor (275 V in place of 230 V), saving on copper in motor, with consequent saving in cost.

Table 3.1 *Comparative Properties of BOPP and Polyester*

Properties	*Unit*	*PET film*	*BOPP film*
Thickness	µm	3.3 ~ 6.8	3.5 ~ 15
Density	G/cm^3	1.3 – 1.4	0.91 ± 0.01
Tensile strength	MD (N/mm^2)	255 ~ 275	≥130
	TD (N/mm^2)	~300	≥250
Elongation at break	MD (%)	110	≤180
	TD (%)	100	40~70
Heat shrinkage	MD (%)	2.2 (150°C, 30 min)	≤6 (≤6.0 µm)
			≤5 (>6.0 µm)
			(120°C, 15 min)
	TD (%)	0.8~1.0	1.0
Breakdown voltage (BDV)	V/µm DC	1900	≥700 (23°C)
Surface roughness	Ra (µm)	0.08~ 0.1	0.8
	R_{max} (µm)	1.39	1.09
Loss factor (20°C, 1KHz)	x 10^{-4}	50	≤ 3
Dielectric constant	-	3.2	2.25 ± 0.1
Insulation resistance	(20°C)	30,000	100,000
Water absorption	%	0.34	< 0.1
Number of pinholes (Insulation failure)	Nos./m^2	2	≤5 (6.5 µm)
			≤2 (7 µm to 11 µm)
			≤1 (>11 µm)

Capacitors made from PP and PET have their own advantages and limitations, as seen from Table 3.2 below.

Table 3.2 *Comparative Properties of PP & PET film capacitors*

Property	*Polyester*	*Polypropylene*
Dissipation factor typical, 25°C, 1kHz	0.5%	0.02%
Best tolerance	± 5%	± 1%
Max. service temperature	125 °C	105 °C
Insulation time constant (MΩ × µF)	25,000	100,000
Dielectric absorption	0.2%	0.05%
Corona Inception voltage AC RMS	275 V	315 V
Capacitance drift over 2 years at 40°C	3%	2%
Mounting heat resisting temp.	160°C	110°C

For metallized polypropylene film capacitors, increase in dissipation factor associated with increasing frequency is due to ohmic losses of lead wires and the metallization alloy deposited on the film. The dielectric losses are very small and do not change. This behaviour allows drawing performance curves that conservatively represent real behaviour in applications.

Figure 3.2 shows dependence of capacitance on temperature and frequency for PP capacitors. It may be noted that the value of PP capacitor drops by 2% from 25°C to 100°C. Overall insulation resistance, shown as time constant RC, drops considerably with frequency, whereas the loss tangent remains more or less unaffected up to over 50 kHz (Fig. 3.3). For metallized polypropylene film capacitors, the increase in dissipation factor associated with increasing frequency is due to the ohmic losses of lead wires and the metallization alloy deposited on the film. The dielectric losses are very small and do not change. This behaviour allows drawing performance curves that conservatively represent real behaviour in applications.

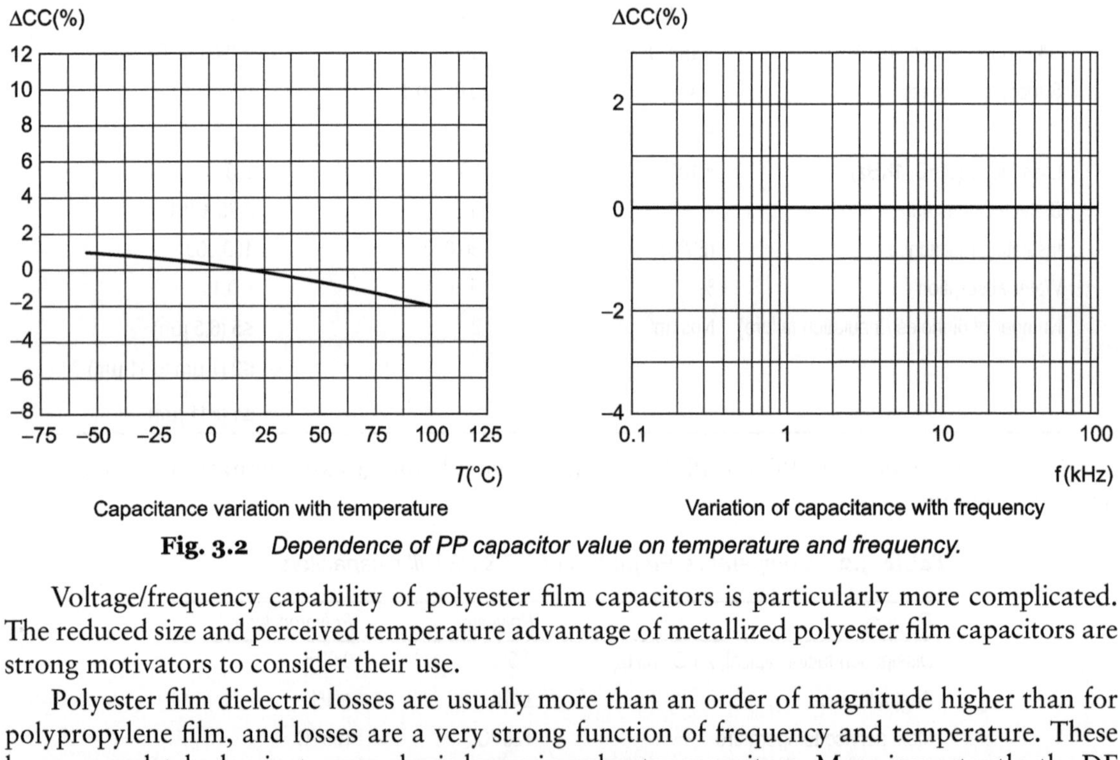

Fig. 3.2 *Dependence of PP capacitor value on temperature and frequency.*

Voltage/frequency capability of polyester film capacitors is particularly more complicated. The reduced size and perceived temperature advantage of metallized polyester film capacitors are strong motivators to consider their use.

Polyester film dielectric losses are usually more than an order of magnitude higher than for polypropylene film, and losses are a very strong function of frequency and temperature. These losses completely dominate over ohmic losses in polyester capacitors. More importantly, the DF behaviour with temperature and frequency cannot be modelled easily.

If the variations of PET capacitors shown in Fig. 3.4 are compared with those of PP capacitors, the PET shows a large variation of capacitance with both temperature as well as frequency. These capacitors therefore do not have a very good stability of value in these respects.

The typical temperature characteristics in Fig. 3.5 show a loss factor vs. temperature graph for polyester film at 1 kHz. Above +50°C the DF starts to climb. For large enough AC voltages the increasing losses with temperature can result in thermal runaway. It can also be seen that

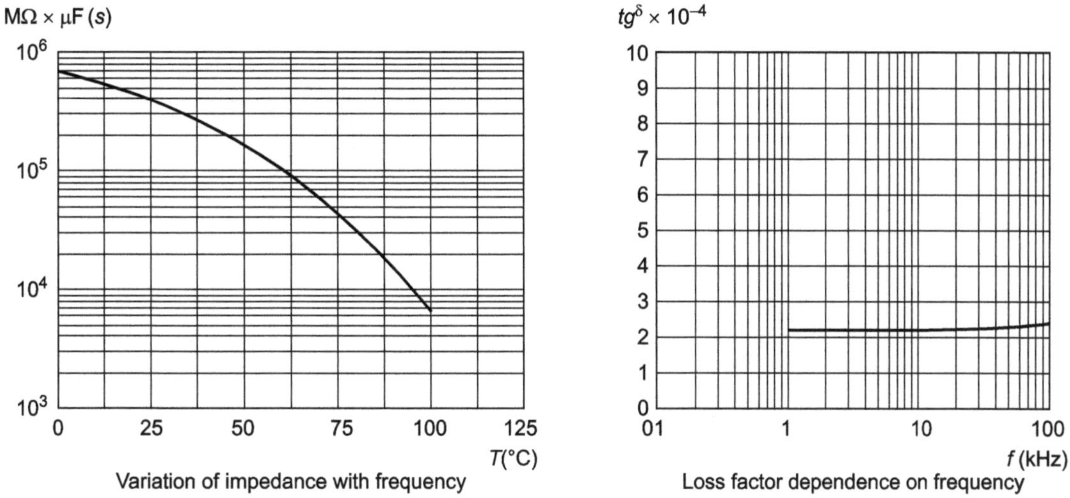

Fig. 3.3 *Dependence of impedance and loss factor of PP capacitors on frequency.*

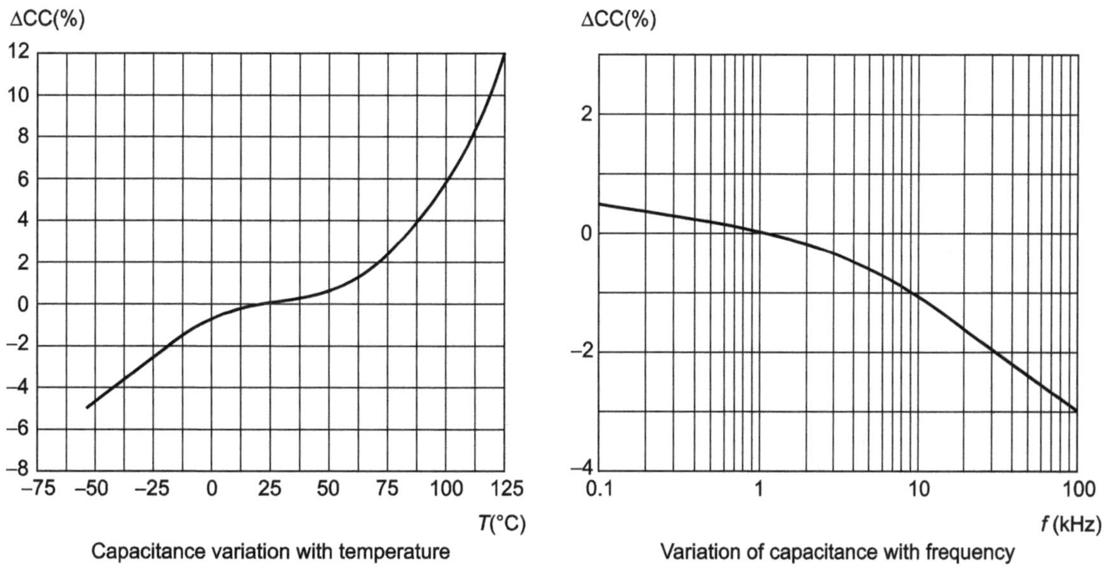

Fig. 3.4 *Dependence of PET capacitor value on temperature and frequency.*

for AC applications, the +125°C internal hot spot temperature limit is extremely misleading. It is very difficult to use the temperature range between + 85°C and + 125°C for AC applications because the allowable AC voltage drops so fast with increasing temperature.

The DF vs. temperature behaviour changes drastically with frequency. The capacitor designer has to conduct the temperature rise tests under the worst electrical and thermal conditions to verify that a proposed polyester capacitor is indeed suitable for that specific application. Since a very small ambient temperature change can result in a large increase in DF, the polyester capacitor behaviour can be very sensitive to the thermal environment. However, there are many AC applications where polyester capacitors represent the best solution from size, cost and electrical performance options.

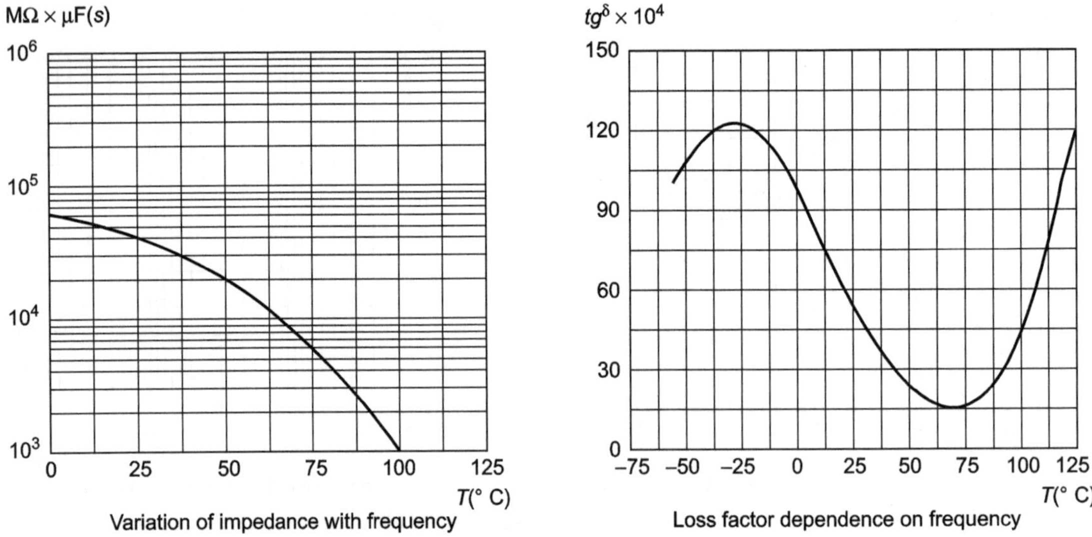

Fig. 3.5 *Typical characteristics of polyester capacitors.*

3.5 CORONA DISCHARGE AND PARTIAL DISCHARGE

A corona discharge is usually a relatively steady glow or brush discharge in air. The electric field gradient (volts/µm) in air, perpendicular to a dielectric, will be greater than the electric field gradient in the dielectric, by a multiple of the relative dielectric constant unless there is a surface charge on the dielectric. A corona can be present in air voids in the dielectric at normal working stresses, due to the difference in relative permittivity of air and the dielectric. Operation of the dielectric system in a corona can rapidly cause permanent damage and eventually lead to failure. However, proper design parameters utilizing an adequate safety factor will allow long operational life of the dielectric system. In a capacitor, discharge may also occur over the film surface at the extreme ends of the winding element over non-metallized surface of dielectric to the end spray or the other side (between two electrodes), if the distance over the surface is insufficient.

The corona inception voltage (CIV) is the lowest AC RMS voltage at which corona discharge occurs as the voltage is increased from zero. CIV is traditionally expressed in the RMS value of a sine wave. Continuous discharge occurs when sinusoidal waves or recurrent transients are applied which are above CIV regardless of the amount of DC voltage component of the applied voltage.

While a corona involves a relatively steady glow in air, partial discharge within a solid insulation system is not visible. An important part of research and development in capacitor reliability has been on the partial discharge phenomenon and its role in capacitor design. When the voltage across the plates of a capacitor dielectric system is slowly raised, a level is reached where a multitude of partial discharges begin to occur at a consistent voltage level. This is referred to as the partial discharge inception voltage (DIV) of the dielectric system. Partial discharges are very short-duration, minute current pulses that have been observed to occur in dielectric systems under high electrical stresses.

Partial discharge is a localized insulation breakdown of a small portion of solid or liquid insulation under high voltage stress. Voltage stress beyond a limit causes partial breakdown of the dielectric by sparking across air voids in the dielectric, and does not result immediately in complete insulation breakdown. However, it causes rapid deterioration of the film and/or metallization due to the hot spot temperature resulting from the heat concentrations during the discharge. Presence of impurities lowers the DIV and causes early failure of capacitors.

One must avoid any voids or spaces in capacitor winding which can trap air or moisture. However, even air pockets inside the plastic film created during its manufacturing process will give rise to partial discharges. The result is a continuous electric discharge through these air spaces, causing degradation, partial evaporation of metallization at these points and eventual burning of film. The capacitor will gradually have increased loss factor and will fail.

Corona and partial discharge have the following effects on capacitors:

- Deterioration of dielectric surface at the ends, leading to ultimate failure,
- Bombardment of the dielectric,
- Charring of the dielectric surface by the heat produced by the arc,
- Chemical reaction caused by new compounds such as ozone produced by arc-like conditions.

In case of continuous occurrence, carbon tracks in the dielectric are conductive-cumulative damage may lead to insulation breakdown of film in a relatively short time, and can eventually cause the capacitor to short out and fail. For film/foil capacitors this will result in a short circuit. For metallized capacitors this will result in removal of the film metallization edges, and thus progressive capacitance loss.

The corona inception voltage is affected by humidity and dielectric thickness. All PP films made for metallization are treated specially to withstand the corona effect.

3.6 COMPARISON OF PP AND POLYESTER FILM CAPACITORS

[1] Polyester

[A] Advantages

- High dielectric constant, and available as the thinnest film, resulting in the smallest size for a given capacitance value and rated voltage.
- Lowest cost
- High dielectric strength results in wide operating temperature up to 125°C

[B] Limitations

- Dissipation factor and capacitance variation with temperature change;
  - At higher temperature, the dissipation factor hovers at 1%.
  - The capacitance varies about ±5% over the range of – 55 to +85°C

Capacitance change from 0 to +50°C is only ±1%, but beyond this range the variation is very large and rather unpredictable.

Polyester capacitors are not suitable when close tolerance (≤ ±5%) is needed.

- Power dissipation precludes the use of polyester film capacitors for high current and high frequency AC applications.

[C] Applications

- Ideally suited for electronic circuits for coupling, decoupling and bypass applications.
- Rugged physical and excellent dielectric strength suited for discharge or energy storage application.

[2] Polypropylene

[A] Advantages

- Lower dissipation factor makes it a good choice for high voltage, high-frequency AC, large current applications.
- High insulation resistance, low dielectric absorption makes it ideal for precision DC applications.
- Excellent stability over frequency in KHz range and temperatures up to 105°C.

[B] Limitations

- Lower dielectric constant results in a large size; this is because ultra-thin films are not available in polypropylene.
- Maximum operating temperature: 105°C.

[C] Applications

- Lower loss makes it a natural choice for AC applications.
- Well suited for high power applications (SMPS).
- Useful in precision applications and when close tolerance is needed (filtering, timing, sample-and-hold, integration).
- Ideal in audio applications.

3.7 POLYPHENYLENE SULPHIDE (PPS)

This film is for precision capacitance and wide temperature applications. PPS has replaced PC capacitors in most applications. PPS capacitors are able to operate from –55°C to 150°C and hold capacitance change to less than 1% over the extremes of the range. Figures 2.7 and 2.8 show excellent stability of PPS properties over the entire range of temperature from –55°C to 125°C and up to 100 kHz. Polyphenylene sulphide is the preferred precision-capacitor dielectric and is the dielectric film for modern chip capacitors. Short-term exposure up to 260°C does not degrade capacitors made from PPS. This film is ideal for 'surface mounting' or high temperature applications.

PPS capacitors are typically used in timers and filters, and automotive and other applications in high ambient temperatures. PPS applications also include high-frequency coupling, decoupling and general high-speed applications requiring high dv/dt, such as pulse operation in SMPS and snubber applications. Exposure to both moisture and temperature cycling causes little change. In general, since PPS and PC have the same dielectric constant, the size of a PPS and a PC capacitor will be the same for a given film thickness.

The properties of capacitors made from these dielectrics are summarized in Table 3.3 below.

Table 3.3 *Comparative Properties of PP, PET and PPS Capacitors*

Dielectric	*Change – 25 to +85°C*	*Change/ Year*	*Typical DF*	*IR*	*Size 1 μF/100 V*
MPET	+5%	0.4%	0.5%	30 GΩ	1.48 CC
MPP	– 3%	0.1%	0.1%	100 GΩ	2.13 CC
PPS metallized	±0.5%	<1%	0.2%	3 GΩ	1.47 CC
Polyester film/foil	+5%	0.4%	0.5%	100 GΩ	6.55 CC
Polypropylene film/foil	– 3%	0.2%	0.05%	200 GΩ	11.63 CC

4

METALLIZED FILMS

4.1 METALLIZED FILMS IN CAPACITORS

Use of metallized films has virtually taken over most capacitor applications in the past few decades. The advantages offered by metallized films have proved to be of immense value in major areas by way of low mean time between failures (MTBF) and long life. They are able to take on much more stress compared to conventional paper/foil or film/foil designs, and have helped reduce capacitor sizes considerably.

Metallized polypropylene (MPP) film used for AC capacitors is very thin in nature, of the order of 4 to 10 microns in thickness. Typically, an 8-micron thin film used for a 440 V AC capacitor bears a working voltage stress of 55 V per micron (i.e. 55 kV/mm). The peak voltage stress for this sinusoidal voltage is $55 \times \sqrt{2} = 77$ kV/mm. Voltage stress during a 1-minute HV test is 1.5 to 2.15 times this value. With 7-micron film, working stress will go up to 63 V/μ AC and test voltages about 125 V/μ AC.

To imagine the magnitude of this voltage, consider that a normal PVC wire used in AC mains has a typical insulation thickness of 0.5 mm for 250 V AC rating, which works out to 500 V/mm (0.5 kV/mm AC). Further, breakdown voltage of dust-free dry air (or vacuum) is 3 kV/mm (DC). These are insignificant values compared to stresses of capacitor dielectrics.

The dielectric film has to be of very high purity to work under such stresses, and also be practically free of weak spots and pinholes etc. With improvement in materials and manufacturing techniques, including development of sophisticated machinery, AC working voltage stresses have increased from 40 to 42 V per micron to 60 to 65 V per micron over two decades, and stresses of higher magnitudes are being used in modern capacitors.

The thin film, and its high working stress, demand that the materials used in capacitors retain their purity during manufacture and use. Dust particles in normal air, of the order of 4 to 10 microns, are invisible to the naked eye. Any particles entering the dielectric of comparable thickness get embedded in the layers, and create weak or conducting points. Hence the winding room has to be absolutely dust-free and other working areas must also be meticulously maintained.

The PP film is quite susceptible to changes in temperatures, and a change in temperature during storage or use can cause warping and creases. A working temperature of 25 – 27°C and a relative humidity not exceeding 55% to 65% is generally recommended for film storage and the working environment. These conditions are maintained at the time of manufacture and processing of films, and have to be observed at the time of their use in capacitor manufacturing process for best results.

4.2 METALLIZED ELECTRODE

The metallized electrode coating was first developed using vacuum deposited aluminium. Though aluminium has high stability in normal storage conditions, demand for greater capacitance stability in AC applications led to the development of zinc coating. Zinc gives much more stable capacitance value over time, but has very low storage life and is also highly susceptible to heat, air and moisture. Hence, zinc aluminium alloy metallization was developed, and is now generally used. The thickness of the metal deposit has to be accurately controlled. If it fails to evaporate and isolate a defective spot, a permanent short circuit will form and the capacitor will go out of service.

Aluminium is hard to beat for cost and reliability. Hence it is still the preferred choice for DC capacitors. Zinc has been used for its better healing properties, but it is more vulnerable to corrosion. Zn-Al alloys are now used in large, high-voltage film capacitors for more reliable self-healing. A layer of aluminium is first coated, and immediately thereafter, a zinc coating is made on the film. Aluminium partially comes to the surface during process, and improves storage capability (shelf life) of the film. These films have better storage stability compared to those with plain zinc metallization.

The coating of the metal electrode layer is extremely thin, just around 0.02–0.03 μm, and it cannot be measured by normal mechanical methods. Hence, an indirect method of measuring the resistivity of the electrode surface is used. Resistivity is measured in ohms per unit area, and is given as 'ohms per square' or 'ohms sq^{-1}' (Ω^{-1}). This is because the resistivity measured over any perfectly square area is the same, irrespective of the unit of length. This can be easily verified mathematically.

The metallized film has a metal-free edge clearance at one end – usually 2 to 2.5 mm for 440 V capacitors. In every capacitor, the two layers of film will have free margins at opposite edges, and metallizing on the films is on the opposite edges. Round coils are wound and the edges are sprayed with zinc. A conducting lead is soldered onto these surfaces. There is a microscopic layer of air between these layers. The coils are wound tight. They are further shrunk under heat treatment. This reduces the air thickness between layers very significantly.

The non-metallized free margin plays a significant role in deciding the safety factor and as insulation between two metallized electrodes. The higher the capacitor voltage, bigger is the margin. The smallest margins used today can be as low as 0.5 mm in some very low voltage DC capacitors, while 2–2.5 mm is common for 250–440 V AC power applications. Power capacitors may see margins going up to 5 mm, while electronic capacitors may use 1–1.5 mm margins.

Efforts to improve the quality and performance of films led to the development of **tri-layer metallization.** In this process, an extremely thin silver deposition is made on film before metallizing with zinc and aluminium. Silver has better adherence to base film, and this gives a distinct advantage over the normal zinc alloy films by way of better stability. Using silver in addition to aluminium and zinc, the three-layer metallization overcomes the shortcomings of conventional types of metallization and improves capacitance stability and shelf life of zinc alloy metallized film. The function of silver is to improve adhesion of zinc to the base film. Three-layer zinc-alloy metallization with reinforced edge facilitates superior end spray adhesion, thereby improving the surge withstanding capability of sprayed elements. Al and Zn metallized film, multilayer metallized film with coating of Al, Ag and Zn-Al solves the problem of rapid oxidization of Zn while maintaining all the merits of Zn.

4.3 SELF-HEALING

The process by which the electrical properties of a metallized capacitor are restored rapidly after a local breakdown (including partial discharge), to the values before breakdown, is termed self-healing. Self-healing is absolutely critical for long-term reliability for high-energy film capacitors.

Since all the defects in a single layer of MPP can be healed during the manufacturing process, a single-layer capacitor can be formed quite comfortably with higher operating dielectric voltage stress.

Fig. 4.1 *Self-healing process.*

In the event of a weak spot on the film appearing during manufacture or subsequently in service, the resultant energy discharge through the weak spot vaporizes the surrounding coating. The weak spot gets isolated, as shown in Fig. 4.1, and the capacitor continues to remain in service, practically intact. These capacitors are hence called 'self-healing capacitors'. (The area around the weak spot so cleared is de-metallized, and the active area of the capacitor gets reduced with every self-healing operation, causing a small drop in capacitance value.)

If the fault is cleared without sufficient energy, or a series of clearings occurring successively, the area around the cleared spots has lower insulation resistance due to incomplete vaporization of metal, and becomes a leakage current path. Every such clearing adds to the loss factor, and in the long run, may develop thermal runaway, damaging the unit completely, or even causing a fire.

4.4 PLAIN AND HEAVY EDGE METALLIZATION

Normally for all DC applications and some AC applications, metallization is uniform (plain) over the film surface, and the metal used is aluminium, being cheapest and very stable, with excellent storage life. However, with the advent of zinc for metallization, a search for better self-healing properties led to the development of heavy edge film, where the edges are made thick to enable good contact with sprayed metal and also to have better current density at current collecting areas, and metallization thickness on active area is reduced considerably. The evaporation of surface metal is much more efficient because of thinner coating, while heavier zinc or Zn-Al

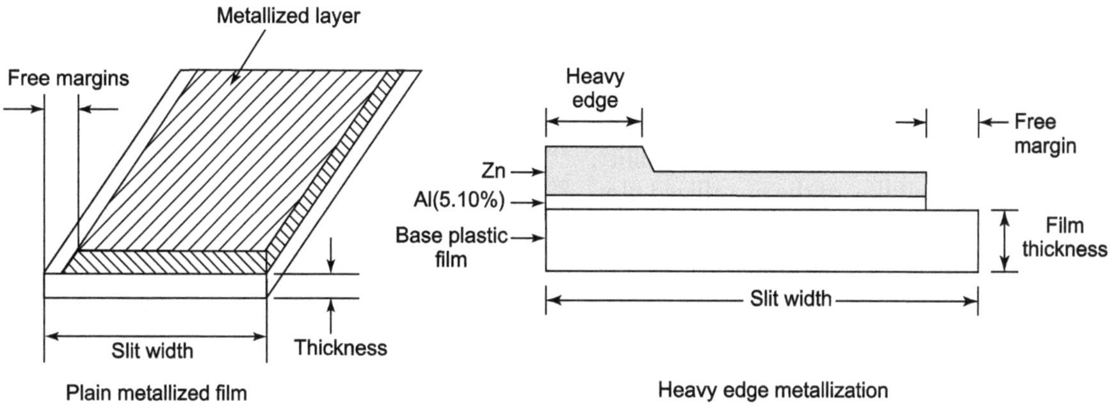

Fig 4.2 *Plain and heavy metallized films.*

coating at the edge makes for normal and reliable end contact with sprayed zinc metal. Only aluminium metallization is normally used for BOPP or polyester film in DC applications for easy handling of capacitors manufactured with oil impregnation.

4.5 SURFACE RESISTIVITY OF METALLIZED FILMS

For plain metallization 2 to 4 Ω^{-1} is usual, whereas for heavy edge films, 2 to 4 Ω^{-1} on edges and 6 to 10 Ω^{-1} on main (active) area, are found to be generally acceptable. These are found to give optimum performance for AC capacitor applications. The average resistivity measured over the entire width of film is between 5 and 9 Ω^{-1}. These parameters can change on the lower side (thicker coating) 1 to 2 Ω^{-1} and 2 to 4 Ω^{-1} respectively for special application capacitors subjected to heavy current surges (commutation and auto ignition, for example). Heavy edge resistivity is usually half to one-third of active area resistivity. Usual metal free margins lie between 1 and 2.5 mm, and heavy edge widths of 2 to 3 mm are common. The resistivity, apart from giving a measure of proper coating thickness, also gives an idea of the purity of metallization, since any oxidation of metal surface by atmospheric effect or moisture causes an increase in resistivity.

The resistivity of active area, heavy edge area and the average resistivity need to be measured separately at manufacturer's end, and instruments are specially designed for this purpose. For certain critical applications like auto ignition or commutation capacitors, the resistivity needs to be much lower, almost half of the above figures (i.e. heavy metal coating) to take care of heavy current densities or discharge currents.

The metallized layer is very sensitive to handling and if rubbed, the metal can be simply erased out of the film. Hence any physical contact with the metallized end of film roll or the wound element must be avoided. The film should be held from its winding core (plastic or metal cylinder) while handling or loading on machines. Further, it should not be exposed to any dust or moisture in storage/handling. The film also develops creasing or warping at higher temperatures, and it should be preferably acclimatized at winding temperature for a few hours before use to get best results.

The stability of metallization can be judged by following the change in resistivity over time, and in zinc alloy metallization, is a function of aluminium content in the alloy. As shown in Fig. 4.3, an increase in aluminium content in metallization increases the stability, and hence the shelf life, of the metallized film. Film with pure Al metallization is the most stable, while pure zinc shows a very sharp deterioration in storage conditions.

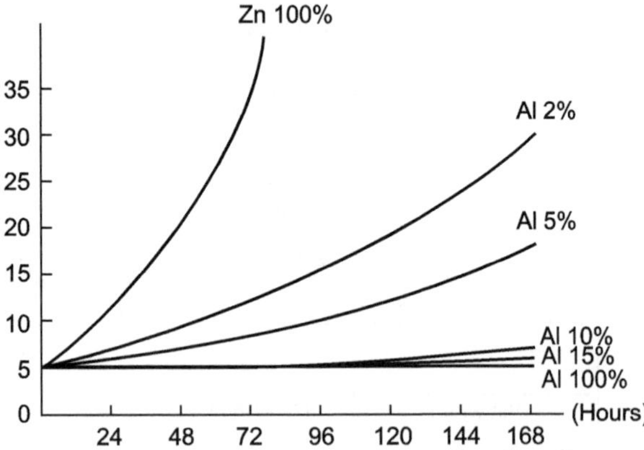

Fig. 4.3 *Films resistivity increase over time as a proportion of aluminium at 40°C (*Source : Tibcon*)*

The capacitance stability over the lifetime of a capacitor is far superior with zinc metallization, with practically no change over 1000 hours, and as the percentage of aluminium in metallization increases, the drop in capacitance goes up, as seen from Fig. 4.4. An Al percentage of 2% to 10% keeps the drop within 1%, while beyond this, the drop increases sharply.

Similarly, the loss factor of a capacitor undergoes no change for a pure zinc metallization, and as the percentage of aluminium in metallization goes up, the loss factor stability gets worse, as seen in Fig. 4.5. Here also, the increase in tan delta remains within fair limits for aluminium content up to 10%. It is thus seen that from the point of view of both capacitance stability and tan delta deterioration, Al percentage must be below 10%. Actual results may vary depending upon manufacturer and process refinement.

Zinc alloy with varying composition gives different results, and further there is some variation in different manufacturers' products depending upon the process. Today films are available with sloped metallization, where the metallization thickness decreases with the distance from heavy metallized edge. The current density on the film decreases away from the heavy edge, and the

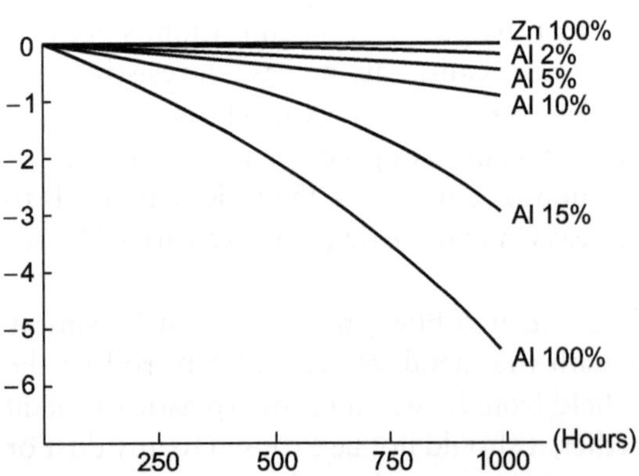

Fig. 4.4 *Capacitance stability for different proportion of aluminium C.*

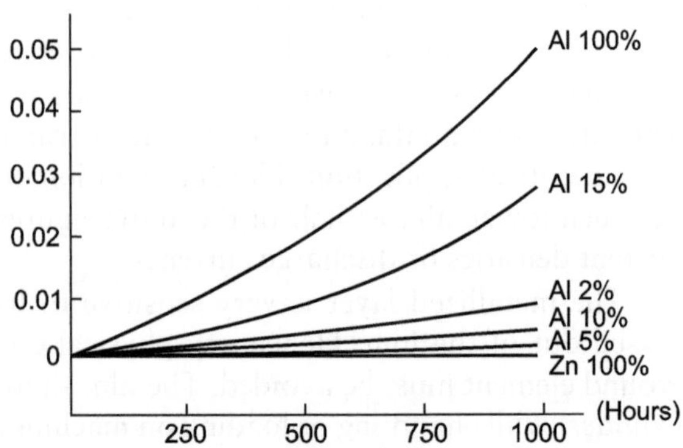

Fig. 4.5 *Tan delta stability over time as aluminium Contents C.*

self-healing can take place at lower current densities at weak spots. This improves the failure recovery mechanism of the capacitor.

Comparison of Metallization Compositions

A comparative behaviour of different types of metallization is given in Table 4.1 below.

Table 4.1 *Comparison of Zn, Zn-Al Alloy and Al Metallized Films*

	Shelf life	*Self-healing property*	*Capacitance stability*	*Atmospheric stability*
Zinc	Poorest	Best	Best	Poorest
Zinc/Al	Moderate	Better	Better	Poor.
Aluminium	Best	Poorer	Lower	Best

Film resistivity standards as followed for different metallization types are shown in Table 4.2. This is standard practice and developed after experimentation and study of field results.

Table 4.2 *Resistivity Standards for Zn/Zn Alloy and Al Metallization*

	Standard resistivity	
Film type	Active area	Heavy edge
Zinc /alloy	$7.5 \pm 1.5\ \Omega^{-1}$	$3.0 \pm 1.0\ \Omega^{-1}$
Aluminium	$1-2\ \Omega^{-1}$	$2-4\ \Omega^{-1}$

Resistivity values much lower than above indicate heavier thickness, resulting in difficulty in proper short clearing. Higher values are the result of insufficient metallization, and the capacitor will not be able to carry the current safely. However, in sloped metallization, metallization thickness is heaviest near the heavy edge contact side, and goes on reducing towards the non-metallized free margin. This is because the current from the entire width of film is collected at the heavy edge, and current densities are consequently lesser as we move away from the current collecting heavy edge. The logic in sloped metallization is to make short clearing easier at any spot by relating it to the current density at that spot.

4.6 SEGMENTED FILMS

One method to improve self-healing is to 'segment' the metallization. Instead of metallizing all of the film, the metallization is laid down in many rectangular or diamond-shaped patches which are connected by narrow metallized 'bridges'. **Patterned film, or segmented film** has been developed, where the metallization is not continuous, but a series of rectangular or crossed sections of metallized sections, which are joined to each other by small restricted paths. These paths act as fuses, and in the event of a weak spot appearing in a section, the self-healing takes place in these pathways and the entire section is isolated. This eliminates the possibility of low resistivity paths being developed, and the capacitor retains good dielectric properties. The segmented design also ensures that in the event of failure at a point on the dielectric, the corresponding area is isolated and the capacitor continues working normally, reducing the chances of low resistivity paths and consequent thermal runaway.

Several designs of segments are available for various capacitor applications, depending on film size and reliability desired. Segmented film capacitors have a conventional winding structure with unique segmented metallized electrodes. These electrodes form a parallel connection of many small value capacitors. Should dielectric breakdown occur due to over-voltage, only the faulty small segment is disconnected resulting in an open circuit, thereby preventing the entire capacitor from burning.

The segmentation pattern in the metallized electrode allows localized fusing and disconnects during high voltage surges. However, self-healing by this mechanism can result in a progressive loss of capacitance, a potential problem in some applications. Hence a potential downside is limited useful lifespan due to capacitance drop should excessive clearings and defusing occur.

Capacitor technology in the higher volume power applications starts at the higher input voltages. The pressure to downsize the input EMI filters has caused a redesign of many capacitors from PET film to segmented PP film. Segmentation of the metallized patterns on thinner polypropylene has allowed higher values and a slight decrease in the package size of the 'X' capacitors. The trend toward higher voltage ratings per micron thickness has led to increased capacitance density.

The segmented film design has some drawbacks compared with regular metallized film:

- Loss of active area, e.g. diamond pattern consumes 10% more material
- Tan δ increases depending on the chosen resistance and pattern.
- Self-healing tests of several standards do not detect the fuses going off and this is misinterpreted as if no self-healing took place.
- The de-metallizing equipment in the winding machines may not always be compatible with segmented film.
- The winding quality of films thinner than 2 microns may be reduced compared to standard metallized films.

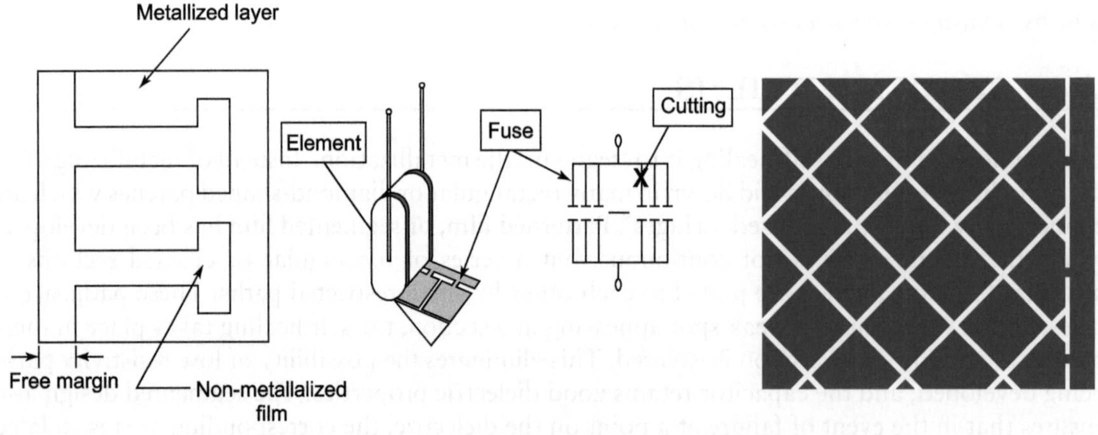

Fig. 4.6 *Segmented metallized films.*

4.7 WAVE-CUT FILMS

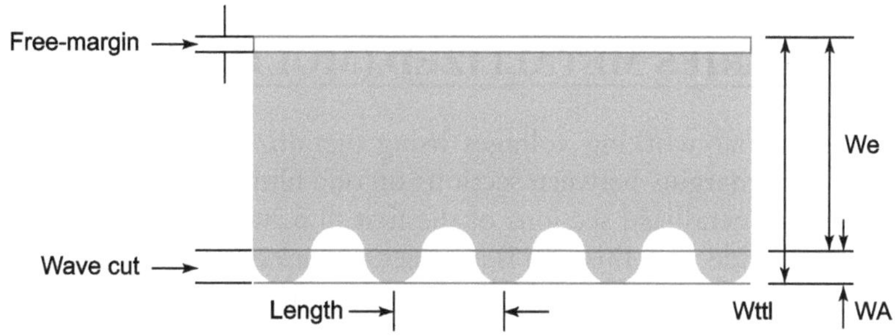

Fig. 4.7 *Wave-cut metallized film.*

When a zinc layer is sprayed on the ends of winding, it makes contact with the metallized edge of film. The edge is not in continuous contact with it, but since the spray is in granular form, the contact takes place at a series of points along the length. The collection of current is also accordingly through these points, creating points of concentrated current densities. Films have now been developed to offer better end contact with the sprayed metal layer. This is done by cutting the film edge in a wavy shape instead of straight. This creates hills and valleys on the surface, and sprayed metal enters the spaces thus created. Figure 4.7 shows the wave shape edges of these films. The spray thus holds the metal even from the sides, creating a much stronger bond. This adds to the stability of the metallized layer in service.

This reduces the current density, and associated thermal stress, thereby increasing the current capacity of end surface for the same area. Reliability of the capacitor goes up by way of reduced failures on this count. The wave cut distributes film tension layers more evenly, and causes less mechanical stress in pressing and heat treatment during capacitor manufacture. Since the current density is reduced, it may be possible to reduce film thickness for some capacitors.

However, very large winding offsets do not derive any benefit from wave cutting over straight cut films. Too small offsets reduce the contact area, and the wave may even stick out of the winding. The offset has to be equal to the wave amplitude to derive full benefit of better ESR, dv/dt and mechanical strength.

Table 4.3 gives a general description and practical wave-cut film dimensions.

Table 4.3 *Description of Wave-Cut Film* (Source : Steiner, GmbH)

Amplitude/ Peak to peak (= 2 WA)	*Amplitude*		*Wave*		*Remark*
	WA	*Tolerance*	*Length*	*Tolerance*	
0.15 mm	0.075 mm	0.15 ± 0.08 mm	1.50 mm	0.15 ± 1.00 mm	Mini wave
0.30 mm	0.15 mm	0.30 ± 0.20 mm	1.50 mm	0.15 ± 1.00 mm	Small wave length
			2.50 mm	2.50 ± 1.50 mm	Standard wave
0.40 mm	0.40 mm	0.80 ± 0.25 mm	5.00 mm	5.00 ± 2.00 mm	For wider films

Wave-cut metallized film was developed to enlarge the effective end surface area in order to improve the adhesive strength during the metal spray process. This in turn is effective for the improvement of the capacitor durability. Wave-cut metallized film is generally used for X2 type capacitors.

4.8 SERIES METALLIZED (MULTISECTION METALLIZED) FILMS

To get higher working voltages using metallized films, the metallization is done in sections, with clear margins between sections on one film. In the second film, clear margins appear at the centres of metallized sections of the first film, with metallized ends appearing on both edges. A winding made with this film is in effect a series combination of multiple capacitors to give a high corresponding voltage. This gives the benefit of ease of construction of capacitors, and since only the extreme ends are metallized, more reliable connections and reduced interconnections between capacitor sections give an added advantage. The capacitor is thus more reliable with better life expectancy.

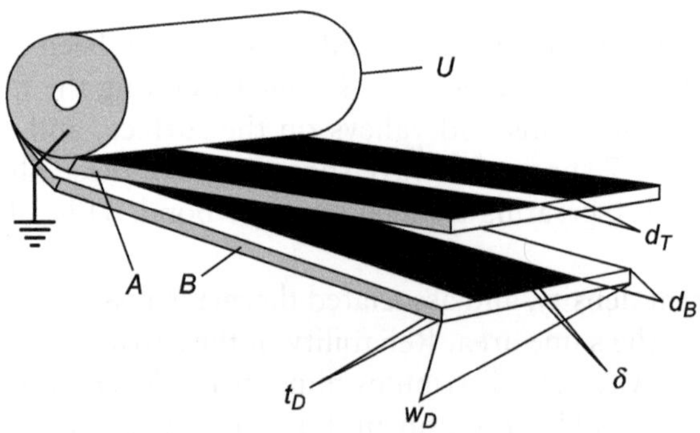

Fig. 4.8 *Series metallized film.*

Typical Specification for a series metallized film could be :
Dielectric film thickness t_D - 6 µm
Dielectric width W_D - 55 mm
Metallization thickness - 20 nm
Axial track distance d_T - 5 mm

In power capacitors, double dielectric capacitors are becoming common, and this type of film could be a preferred choice for obvious reasons. A series connected 2-section capacitor using this film appears as shown in Fig. 4.8. It is possible to have multiple series sections for higher operating voltages.

4.9 STORAGE OF METALLIZED FILMS

The nature of metallization and its susceptibility (particularly that of zinc) makes it necessary to store the metallized films in sealed condition until needed. They should also not be exposed to heat during storage. The winding process has to be carried out in an air-conditioned and dust-free atmosphere, preferably under controlled humidity. The ideal conditions would be temperature of 23 to 27°C, and a relative humidity not exceeding 65%. Further, once elements are wound, care has to be taken to avoid effects of surrounding atmosphere and moisture. It is further preferable that a pair of reels once unpacked should be used up on the same day.

4.10 MLP CAPACITORS

Multilayer film capacitors evolved from stacked film types to the true surface mount chip styles are available today. Polymer dielectrics having no voltage or current dependency are the most stable system at higher voltages (including 24 and 48 V). Ultra-thin films in the 1.4-micron range for 100 V applications have led to highly stable film chips in large capacitance values.

High frequency ESR and ESL is comparable to the ceramic chips in X7R dielectric while the electrical stability is superior. Further reduction in size may happen by developing even thinner films and through the use of barrier coatings on these ultra-thin plastic films. Reduction in unit cost is also getting further attention.

4.11 LIMITATIONS OF MPP FILMS

Moisture getting in between layers oxidizes the thin metal deposits in its thickness fully. An oxidized boundary isolates a healthy section of a deposit, resulting in rapid or continuous fall of capacitor value. This happens mainly in loosely wound capacitors or poorly made elements. The winding ends are normally sealed with resin or the coils are impregnated with an insulating liquid to prevent this.

1. Aluminium at the sprayed end forms a non-perfect joint with zinc spray – it corrodes more in the presence of moisture. This cuts off the entire healthy metallization below from the conducting edge, resulting in rapid fall of capacitance. Heavy edge metallization helps prevents this. Zinc deposit over aluminium (Zinc alloy metallization) on the entire film surface also prevents this to a great extent. Nevertheless, care has to be taken to avoid or minimize exposure of metallized elements to moisture or oxidizing atmosphere.
2. Consider a large air void between layers and an irregularity in the form of a sharp point. As the voltage increases across the dielectric, electron streams originate from this sharp point and cut through the air path. This is the beginning of a partial discharge. It creates hot spots and eventually causes failure of the element.
3. The most critical portion of the metallized capacitor is the edge gap. The full applied voltage appears across this gap between metallization at the end of margin of one film to the sprayed end of the other. It is spread on a very thin metallization base 0.2 to 0.3 microns thick as against 5 to 6 microns thick foil in extended foil traditional capacitors. The voltage stress is very high – leading to instant or even sustained partial discharges, should the voltages cross the air gap strength. Normally a 2.5 mm gap across a 0.3 micron base can sustain AC voltages up to 440 V + 10%. This makes these capacitors unsuitable where there are steady high voltages or sudden and continuous voltage fluctuations.

4.12 METALLIZED VS. NON-METALLIZED CAPACITORS

[A] Advantages of Metallized Film Capacitors

- The metallized coating on films is extremely thin, of the order of 0.02 to 0.03 microns. Hence a large reduction in capacitor volume is achieved compared with a unit with 6-micron thick aluminium foil electrodes.
- In the event of a fault during manufacture or subsequently in service, the resultant energy discharge through the weak spot evaporates the surrounding coating. The fault gets isolated and the capacitor continues to remain in service, practically intact.
- Because of this self-healing property, minor film breakdowns during the operation of a capacitor will not lead to failure of the capacitor as it would in the case of a film/foil capacitor.
- Self-healing also permits the use of a single-layer dielectric, as against a minimum of two layers used in film/foil capacitors. Further, with improvement in materials, increasing voltage stresses are being used. The total number of layers is only two against a minimum of six required in film/foil capacitors.
- Both these factors viz., absence of separate Al foil electrodes and single-layer thin film construction, reduce the capacitor size to almost half. The cost is also correspondingly much lower.
- In case of failure, high fault current results in overheating and shrinking of film away from sprayed end connections, causing the capacitor to fail in open-circuit mode. Foil type capacitors invariably fail in short-circuit mode.
- Sprayed end connections in metallized capacitors result in these capacitors being non-inductive.
- Use of only two layers allows winding machines to be simpler and it has become possible to design high-speed automatic machines.

[B] Advantages of Film/Foil Capacitors

- These sustain higher voltage fluctuations with time than metallized types. Because of inherent lower resistance of electrodes, for a given voltage they stand more pulses per second.
- For the same reason, they can handle higher surge currents.
- Capacitance value does not undergo any degradation with time.
- Extended foil non-inductive capacitors are used at higher frequencies in preference to metallized films.
- Close tolerance small value capacitors are generally made with film/foil construction. Thicker foil electrode allows more accurate winding. Metallized elements may in most cases be too small to have adequate area for metal spraying on ends.
- They also have a better insulation resistance and dissipation factor than metallized capacitors.

Different applications of capacitors, their end use, working ambient temperatures and other environmental conditions, intended lifespan and reliability expectations must be considered in designing a capacitor. Capacitor design is essentially a balancing act between all these parameters, along with costs.

[C] Typical Applications of MPP Capacitors

- Distribution lines where voltage and load variations over a 24-hour period are moderate. Typical examples are a city with a large spread of various loads served by substations with automatic on load tap changers.
- Automatically controlled capacitor banks with built-in over-voltage, under-voltage, over-current and power failure controls and with current limiting chokes on each step.
- Heavily overloaded rural and other distribution and supply system operating at perennially low voltages.
- Where cost is prime consideration, but with attention to the first and third factors above.

[D] Circumstances where MPP Capacitors are not to be Recommended

- On load with widely fluctuating currents such as rolling mills, arc furnaces, workshops with heavy presses and similar impulse type energy drawing machines, welding machines, etc.
- Locations where higher incidence of harmonics is expected.
- Hazardous areas (oil installation, new power generators or generator bus ducts) where explosions are not acceptable. Generally MPP Capacitors are more explosion-prone than other types of capacitors.
- Areas with high short circuit level for distribution networks. (This is likely to affect self-healing.)
- Supply systems with wide daily voltage fluctuations – where the night-time voltages shoot beyond the guaranteed limits, or heavy transients are expected.

4.13 DRY VS. IMPREGNATED OIL FILLED MPP CAPACITORS

MPP capacitor elements are made with metallized BOPP polypropylene film. The film has a smooth non-porous surface. Winding is done in such a way that no air pockets are left between layers of adjacent film, and the layers are tightly wound. The elements are then cured in the oven, which shrinks it into a hard solid mass.

A stagger is provided between two films of elements to allow the metallized end of one film to extend over the free non-metallized margin of the other film by about 0.5–1.5 mm. The zinc metal sprayed on the end makes contact with the surface formed by the edge of this metallized film. Hence an air gap is unavoidably left between the sprayed metal and the non-metallized margin of the second film.

Subsequent processes for encapsulation or impregnation are carried out basically to protect the element from the effect of air and moisture and to hermetically seal it from the surrounding.

During resin encapsulation, while the element is sealed and outside air is excluded, the air between layers of film and sprayed metal remains trapped in the capacitor permanently. This air being very small in quantity, and being sealed inside the element, has a miniscule effect on capacitor film metallization for a short duration and thereafter it does not deteriorate the film or spray metal any further, so long as corona voltage is not exceeded. Whatever oxygen and moisture is trapped is used up.

In vacuum-impregnated oil-filled capacitors, the air from the capacitor is completely evacuated. In the process the air below the sprayed metal is also evacuated, since the sprayed metal is not a solid metal, but a porous mass. The oil then enters the capacitor and fills up this space as it slowly penetrates under the sprayed metal surface. The oil serves an important function: it cools the dielectric and contact area immediately in the event of self-healing discharges or sparks at metallized ends, does not allow the film to melt and also controls the damage due to such discharges.

The choice of suitable high-grade oil (or other impregnating medium) is important as it remains permanently in contact with film and sprayed metal, and these materials should not interact with each other chemically or otherwise. It is also important for the same reason that the oil should be of high purity, fully dry and free of any suspended or dissolved impurities or gases. It is thus seen that ideally a properly oil impregnated capacitor will have an edge in long-term performance life.

From the above discussion it is also clear that the time between successive processes in capacitor manufacture must be very minimal. This will completely exclude the effect of air and moisture on the capacitor life. Whatever effect the surrounding air has on the element before sealing, will certainly affect the long-term performance. Table 4.4 below summarizes the comparative properties of film/foil and metallized film capacitors.

Table 4.4 *Comparison of Film/Foil and Metallized Capacitors*

Parameter	*Film/foil*	*Metallized capacitor*
Current capacity	High	Lower
ESR	Low	Higher
Volume	Higher (50–150%)	Lower
Inductance	Present	Low
Insulation resistance	High	Lower
Heat dissipation	Low	Higher
Tan delta	Low	Higher
Self-healing	No	Yes
Failure mode	Short	Open (usual)
Dielectric layers	Minimum 2	Minimum 1
Capacitance stability	High	Lower

In the analysis of performance and reliability, film capacitors come out better than most other types of capacitors for a large area of applications.

5

TYPES OF CAPACITORS

Capacitors can be mainly divided into various types and categories depending upon

- Type of dielectric and electrodes used and
- Their function and end use.

5.1 CATEGORIES BASED ON DIELECTRICS

Based on the dielectric, basic categories are:

- *Electrostatic capacitors:* Use insulating material between electrodes to act as dielectric. These are non-polar in nature.
- *Electrolytic capacitors*: Use solid or liquid electrolytes and have higher capacitance values. The dielectric layer is an oxide film formed on a metal plate surface. They are inherently polar due to their construction.
- *Electrochemical (or EC) capacitors:* The dielectric layer forms naturally with applied voltage. This class has been developed in the past two decades, and its applications are growing by the day. This is dealt with in a separate chapter.

5.1.1 Electrostatic Capacitors

These capacitors use an insulating material between electrodes to act as dielectric. Different dielectric materials are used depending upon the application and size of capacitors, generally as shown in Fig. 5.1

Electrostatic capacitors are characterized by extremely low ESR and ESL and very low impedance at higher resonance frequencies, going into the megahertz range. Their extremely low time constant RC makes them the logical choice for most high-speed circuits. Being inherently stable at high frequency compared to electrolytics, replacing electrolytic with electrostatic capacitors would improve the performance and reliability of the output filter sections. Electrostatic capacitors are further divided into (a) Fixed and (b) variable capacitors.

[A] Fixed capacitors

These may be further classified based on dielectric used and the electrode material.

(a) Paper / foil capacitors. These are impregnated with oil / wax / jelly.

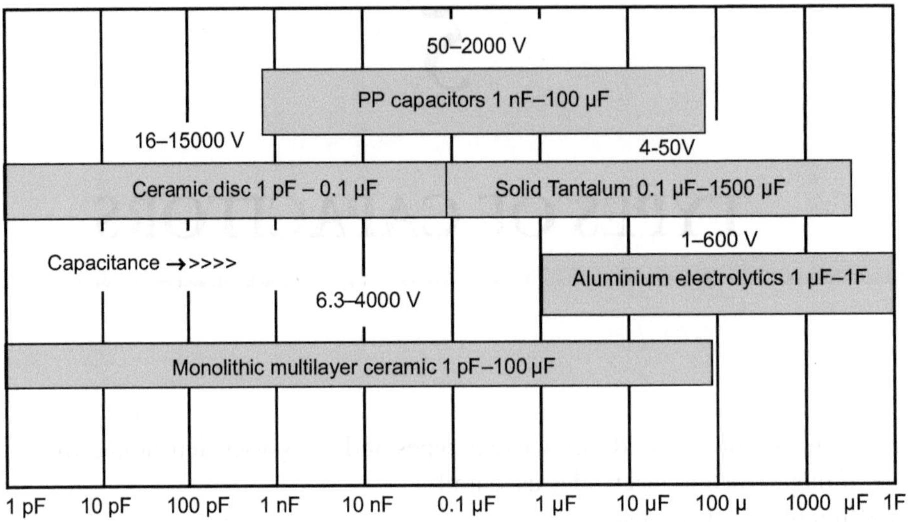

Fig. 5.1 *Capacitance and voltage range of various dielectrics.*

(b) Polypropylene film /foil capacitors. These are impregnated with oil/chemicals.

(c) Mixed dielectric capacitors, where a PP film and paper combination is used with oil impregnation.

Fig. 5.2 *Oil-filled capacitors.*

The above types are generally used for high voltage AC shunt capacitors and high reliability LT shunt capacitors. Their use in fan and motor capacitors has been almost fully taken over by metallized film capacitors. These are also preferred in high current surge applications and environments where heat is encountered.

(d) PP and polyester (PET) film/foil capacitors: PP/foil and PET/foil capacitors (non-inductive) are widely used where circuits may be exposed to moisture. Their resistance to moisture penetration is superior. Film capacitors are applied in electronic circuits requiring blocking, buffering, bypassing, coupling, tuning and timing.

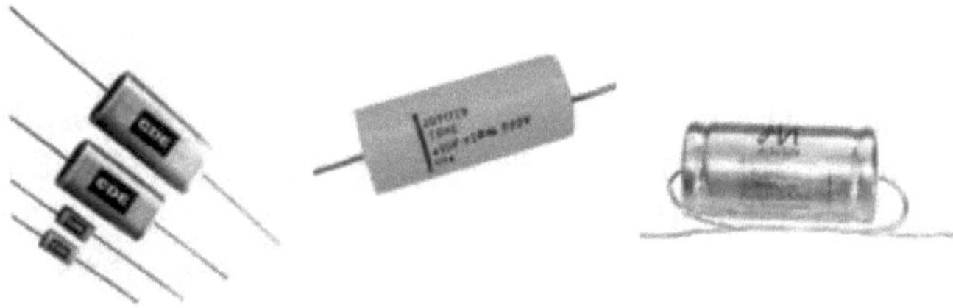

Fig. 5.3 *Film/foil capacitors.*

(e) Metallized paper/metallized PP film (MPP)/metallized polyester (MPET) capacitors: These may be dry construction or impregnated with liquid. These are ideal where great amounts of heat may be present in a circuit.

Fig. 5.4 *Range of metallized capacitors.*

These capacitors possess a unique self-healing property, whereby they eliminate momentary short circuits in their dielectrics caused by surrounding circuit elements. The ability to self-heal permits these capacitors to have higher voltage ratings for a given film thickness.

(f) MKV capacitors, where paper metallized on both sides acts as electrode in an impregnated self-healing capacitor. The term MKV comes from Metalisert Kunstoff Velustrom, a process patented in Germany. The dielectric could be paper, or more commonly, plastic film (MPP). Impregnation is done with non-PCB oil. Electrodes are of paper metallized on both sides and PP foil serves as a dielectric. The system is impregnated with mineral or synthetic oil. MKV capacitors are suitable for higher power loading and higher ambient temperature.

(g) MKP self-healing capacitors: The metal layer evaporates in case of voltage breakdown. The insulating surface area thus formed is very small and does not affect the functionality of the capacitor. Capacitor windings are inserted into an aluminium container equipped with overpressure disconnector. MKP capacitors are made of one-side metallized PP film. Contacts with winding are made by zinc spraying. This is dry configuration without impregnant.

(h) Polycarbonate/polystyrene film/foil capacitors: These were used in DC circuits for their stability and temperature performance, but PEN and PPS are replacing them due to superior properties and availability.

(i) PPS/PEN capacitors: These use new dielectrics in place of PC and PS.

(j) Mica capacitors: These find use in applications such as high-frequency filtering, bypassing, blocking, buffering, coupling and fixed tuning.

(k) Ceramic disc capacitors
(l) Multilayer ceramic capacitors
(m) Vacuum capacitors: These capacitors have the lowest possible dielectric constant and are limited to capacitances of 10^3 pf (10^{-3} μF), can range up to 50 kV (50×10^3 volts)

These can carry huge currents up to 100 amperes. Vacuum capacitors are extremely useful because their lifetime, barring any particle contamination in the vacuum chamber, is indefinite.

Fig. 5.5 *Vacuum capacitor.*

[B] Variable capacitors

Variable capacitors may be

- Moving parallel plate capacitors (gang condensers)
- Trimmer capacitors
- Diaphragm type, used as sensors

Variable capacitors are invaluable in the design of electronic equipment. These are generally employed to provide a range of capacitance and are commonly used in applications where exact capacitance values cannot be obtained using normal design procedures, or need to be frequently adjusted as in radio set tuning. These capacitors are usually constructed such that varying the capacitance is accomplished by adjusting the metal plates in the capacitor.

Screws provided on trimmer capacitors increase or decrease effective plate area thereby causing an increase or decrease in capacitance. The most widely used trimmers are ceramic, glass, air, plastic and mica.

5.1.2 Electrolytic Capacitors

These use solid or liquid electrolyte and have higher capacitance values. They are inherently polar in nature, i.e. a voltage can be applied at their terminals only in one direction, positive of supply to the positive terminal of capacitor. These can be any of the following types.

- Aluminium electrolytic capacitors
- Tantalum capacitors
- Niobium oxide capacitors. This is a recent development, and these capacitors are used in place of tantalum due to certain advantages.

Applications of electrolytic capacitors

Representative applications of electrolytic capacitors include

- Blocking and DC bypass
- DC filters and rectifiers
- Energy discharge application
- Photoflash, strobe, military (laser, radar)
- Audio systems
- AC motor start: These are basically electrolytic capacitors connected back-to-back, remaining in circuit for a few seconds in the start coil of a motor.

A comparison of various common types of capacitors with different dielectrics is made in the following table:

Table 5.1 *Comparison of Dielectrics in Capacitors*

Type	*Capacitance range*	*Max. voltage*	*Max. op. temp. °C*	*Tolerance %*	*Ins. res. MΩ*	*Remarks*
Electrolytic aluminium	1 μF–1 F	3–600 V	85	+100 to – 20	<1	Popular, large values, high leakage, large tolerance
Tantalum	0.001–1000 μF	6–100 V	125	±5 to 20	>1	
Ceramic	10 pF–1 μF	50–1000 V	125	±5 to 100	1000	Popular, small, cheap, poor tolerance
Mica	1 pF–0.1 μF	100–600 V	150	±0.25 to ±5	100,000	Excellent performance, used in high frequency circuits
PET	500 pF–10 μF	50–600 V	125	±10	10,000	Popular, cheap, good performance
Paper	500 pF–50 μF	100,000 V	125	±5 to ±20	100	-
PP	10 pF–10 μF	100–600 V	85	±0.5	10,000	High quality, very accurate, used in signal filters
PC	100 pF–10 μF	50–400 V	140	±1	10,000	High quality, very accurate
Glass	10–10,000 pF	100–600 V	125	±1 to ±20	100,000	Long term stability
Paper/foil	0.1–20 μF	200 V–10 kV	85	±1 to ±20	Good	Large, high voltage filters, long life

5.1.3 Glass Capacitors

Glass capacitors are used where the ultimate performance is required for RF circuits. These offer a very high level of performance. Typically a glass capacitor will have a relatively low capacitance value. Their use is mainly in radio frequency circuit design.

The Cassini-Huygens spacecraft launched by NASA on October 15, 1997, reached the sixth planet from the sun on July 1, 2004 and is currently studying Saturn and the surrounding moons. Glass capacitors played a vital role in the wake up circuitry and deployment of the Huygens space probe on reaching its destination. They facilitated separation of the probe from the spacecraft on December 25, 2004, and allowed the probe to begin a 21-day trip to the surface of Titan, Saturn's largest moon. On Huygens' descent to the surface of Titan, a CYR10 glass capacitor enabled the central computer to wake up from its 7-yr hibernation to take readings and measurements of Titan's atmosphere and images of the surface floor.

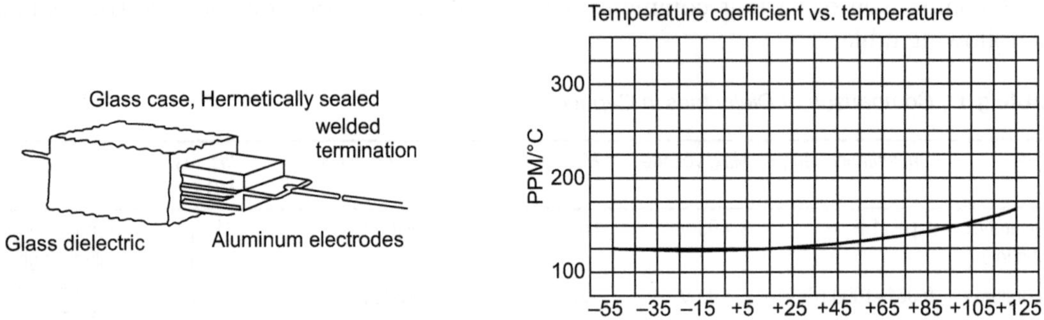

Fig. 5.6 *Typical glass capacitor construction, and its temperature stability characteristics.*

These capacitors also find use in jet engine sensors, medical monitoring systems, audio equipment, and high-temperature oil exploration equipment, as also some military applications where a very high reliability and stability is desired.

Glass capacitors offer several advantages over other types. In particular glass capacitors are applicable for very high performance RF applications:

- **Low temperature coefficient:** Figures of just over 100 ppm C^{-1} are often obtained for these capacitors.
- **No hysteresis:** Glass capacitors retain the same capacitance through extreme temperature swings.
- **Zero ageing:** Their original values are stable over long periods of time.
- **No piezo-electric noise:** This can solve the problem of piezo-electric effect in microphones or oscillators.
- **Extremely low loss/high Q:** Virtually no dielectric loss, so very high Q circuits can be built using them provided the other components (e.g. inductors) are not lossy.
- **Large RF current capability:** Glass capacitors which are suitable for use in RF high power amplifiers, etc. as they can withstand large values of currents.
- **High operating temperature capability:** Suitable to operate at very high temperatures, even up to about 200°C without damage or performance shortfall.

5.1.4 Electrochemical Capacitors (EC Capacitors)

These contain one porous electrode like activated carbon, and a separator from the container, while an electrolyte fills up the carbon electrode and also makes contact with the other electrode, viz. container. The dielectric layer is formed by way of an oxide layer naturally at the contact area between carbon pores and electrolyte with applied voltage. The electrode being porous, and the dielectric constant of oxide higher, capacitance values of very high order, in farads, is achievable. The voltage rating is low, of the order of 0.8 – 1.4 V, but the energy stored in these capacitors is very high compared to conventional capacitors.

A variant of these using two porous electrodes, and electrolyte filling both carbon electrode volumes, with a separator in between, is in use in several applications. A layer of oxide is formed at each electrode interface with electrolyte, forming two layers of oxides. Hence they are known as electrochemical double layer capacitors (EDLC). The voltage can be 2.3 – 2.8 V, and higher cell voltages have been developed. These capacitors, known by various names like ultracapacitors, supercapacitors etc. are increasingly finding use as energy storage devices in several applications.

5.2 CAPACITOR CLASSIFICATION BASED ON THEIR USAGE

Capacitors are used extensively in most electrical and electronic circuits, and serve different functions at different applications. The general classification of capacitors based on their usage goes on the following lines:

5.2.1 Fan and Motor Run Capacitors

The capacitors, connected in auxiliary winding, create a rotating magnetic field by creating phase shift between main and auxiliary winding fields. Capacitor-run motors are more efficient and have a better power factor compared to other types.

5.2.2 Motor Start Capacitors

High capacitance values in auxiliary winding while starting a motor give good starting torque. Once the motor picks up speed, the capacitor is removed from the circuit by means of a centrifugal switch. Then the motor can run without a capacitor or with a run capacitor depending on design. These have all along been AC electrolytic capacitors, but lately metallized film capacitors are coming into use due to superior properties and longer life. The motor start capacitors have duty cycle of 3 s at a time, and a maximum of 20 start operations per hour.

5.2.3 Lighting Capacitors

Most lighting circuits use gas discharge lamps (sodium/mercury vapour lamps), fluorescent lamps or CFL, and are highly lagging loads. Capacitors are used in these circuits to improve the power factor.

5.2.4 Power Factor Correction Capacitors (KVAR)

These are used for improvement of power factor in a supply system by providing leading reactive current (and leading KVAR) to compensate for lagging currents of loads. This improves the efficiency of the system and releases generator capacity to take up additional active load, while at the same time giving better stability to the system. These are connected at the motor or load terminals, near the mains with power factor control panels, and also on utility systems at various points for optimum utilization of the system. These could be connected in parallel with supply (shunt capacitors), or in series (series capacitors).

5.2.5 Commutation/Snubber/IGBT Capacitors

These are used in commutation motors and traction systems, or similar circuits in order to bypass heavy surge currents and eliminate sparking. The RMS currents as well as peak currents are decided by dv/dt in each cycle at current cut-off, and hence currents in capacitors are quite high.

In most Silicon controlled rectifier (SCR) inverter designs, it is required to turn off the SCR during the non-conducting mode to enable the SCR to fire more rapidly. Normally the SCR is turned off by discharging the capacitor into the SCR with a reverse polarity and hence normally commutation capacitors are high current, high dv/dt and high value capacitors.

5.2.6 Capacitors for Constant Voltage Transformers (CVT)

These capacitors are usually located in enclosed spaces and near a transformer, and are subjected to radiated heat from the transformer. They have to be specially designed keeping this in view, apart from the fact that they generally operate very close to their rated voltages. They also face harmonic currents over long periods. Capacitors in aluminium cans are usually preferred so that the heat from the transformer on one side of the capacitor gets dissipated over the entire surface area, thus keeping temperatures more uniform inside the capacitor, avoiding differential thermal stresses.

5.2.7 Furnace Capacitors

Induction furnaces work at high frequencies, mainly 400 Hz, and heating is achieved by inductive currents in materials. Loads are highly inductive, and need balancing capacitors to work at these frequencies. The heat associated is quite high at these frequencies, and capacitors often need cooling by water or forced air. Sometimes water pipes go through the body of the capacitor for effective cooling. Figure 5.7 shows a water-cooled furnace capacitor.

The terminations need to be kept short since currents are high and quick response time is important. The external connections are often brought directly from the winding element.

Fig.5.7 *Furnace capacitor.*

5.2.8 Auto Condensers

Most petrol driven vehicles use capacitors in ignition circuits, which when discharged through the ignition coil primary, give high voltage for spark for igniting the mixture of air and petrol. The repetitive charge discharge rate is very high, and service temperatures are also treacherous, considering they are located near engines. The design needs careful consideration of these factors.

Fig. 5.8 *Auto condensers.*

5.2.9 Fan Regulator Capacitors

Conventional resistance type ceiling fan regulators are on the way out, due to their being inefficient and energy consuming. Up to 30% to 35% energy is taken up by the regulators. Capacitor type regulators score over electronic ones due to their reliability and longer life. The capacitors need special attention as they discharge into one another to cause current surges during speed change. The capacitors are made in flat configuration and encapsulated.

5.2.10 Energy Storage Electrolytic Capacitors

These are generally DC electrolytic capacitors of 150, 300 or 450 V rating and capacitance values between 30 and 600 μF. They are charged through an electronic circuit using a battery, and discharged into the flashbulb to give out instant flash. Some large electrolytic capacitors of a few thousand μF and voltages over 450 V are used in banks on ships for energy storage applications.

Other types of capacitors found in industry are:

- **Energy storage high voltage capacitors**
- **Filter capacitors (power)**
- **RF interference suppression capacitors – X and Y type.**
- **Tuning capacitors** – These are variable capacitors used in radio, TV and instruments for fine-tuning a circuit to the desired frequency.
- **Coupling capacitors** – Used for coupling of AC and DC circuits, or two branches of circuits by separating DC bias levels.
- **Bypass capacitors** – Used for bypassing/filtering certain frequencies.
- **Potential dividers** – These are in use for high voltage measurements.
- **High voltage capacitors** – Used in high voltage applications. These could be film/foil, mixed dielectric, ceramic or vacuum capacitors depending upon the application.
- **SMD capacitors** – Circuits using Surface Mount Devices (SMD) use SMD capacitors suitable for direct soldering on PCBs without wire leads.

Fig. 5.9 *High voltage capacitors*

Each of these capacitors demands a different range and type of capacitor. Even within a particular type the appropriateness for a given application will decide the specific suitable dielectric. Ceramic capacitors require a no-surge current screen. Film and foil types are especially suited for high-current applications, and metallized types have a self-healing feature that improves reliability.

Construction details and designs may differ according to the end use requirements and the nature of duties or work environments of capacitors.

5.3 APPLICATIONS IN THE ELECTRICAL INDUSTRY

The capacitor appears as an integral part in most electrical and electronic circuits. It is versatile in nature, performing a variety of functions in different applications. In AC electrical circuits, capacitor current leads the applied voltage in electrical phase, as opposed to an inductor. Most electrical loads use motors, transformers etc., which are inductive in nature. A capacitor generates leading power factor load and may be used to compensate the resultant lagging power factor. This can be achieved in many ways in electrical supply circuits.

5.3.1 Fan and Motor Capacitors

In single-phase electric induction motors generally two separate sets of windings are used, viz. main and auxiliary winding. A capacitor is used in series with auxiliary winding to create a phase difference between currents of the two windings. This in turn creates a rotating magnetic field around the rotor, making it rotate.

This method of creating a phase difference between the windings for running the motor is much more efficient compared to other available methods and hence capacitors find extensive use in the electrical industry for capacitor run motors for electric fans, motors, pumps, air conditioners and a host of other applications.

Some motors use high value capacitors in auxiliary windings only during the starting, which are disconnected through a centrifugal switch once the motor attains the required speed. These are mostly of electrolytic type and are called motor start capacitors. Capacitor-start capacitor-run motors are also commonly employed.

There are generally three different reasons why a capacitor is used with a ceiling fan or motor:

(a) The capacitor is used to give a phase shift and create a rotating magnetic field, so the motor will start turning more easily, against the inertia of the blades or load.
(b) It is sometimes used in the speed control circuitry.
(c) The capacitor may be used for power factor correction.

Power factor correction capacitors will be dealt with in a separate chapter. The discussion here will pertain to capacitors used to start and/or run single-phase induction motors, and those used for speed control. These motors are not self-starting, but have an auxiliary starter winding driven by an out of phase current of near 90°. Once started, the auxiliary winding is optional.

Three types of AC induction motors are available:

(a) The auxiliary winding of a permanent-split capacitor motor (PSC) has a capacitor in series with it during starting and running.
(b) A capacitor-start induction motor has a non-polarized electrolytic capacitor in series with the auxiliary winding during starting, which is disconnected once the motor starts.
(c) A capacitor-start capacitor-run motor typically has a large non-polarized electrolytic capacitor in series with the auxiliary winding for starting, then a smaller non-electrolytic capacitor during running.

[A] Permanent split capacitor motor

Run capacitors are designed for continuous duty, and they are energized the entire time the motor is running. Run capacitors are rated in a range of 2–70 microfarads (μF, or mfd); sometime even higher, with voltage classifications of 400 V or 440 V AC. A single-phase electric motor needs a capacitor to energize a second-phase winding. PSC motors are frequently used in blowers, fans (ceiling, table or exhaust fans) and other cases where a variable speed is desired.

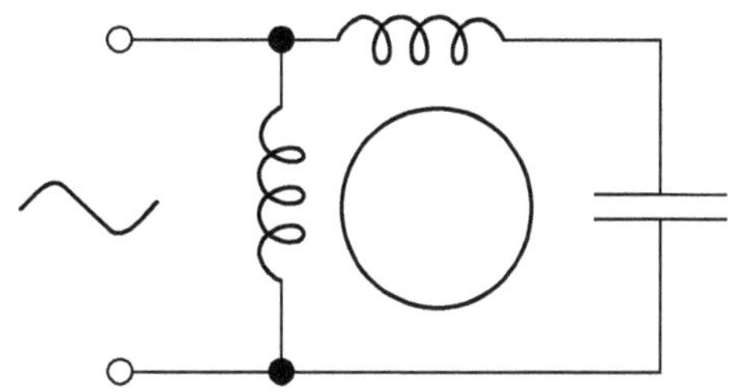

Fig. 5.10 *Permanent split capacitor motor circuit.*

A capacitor is connected in series as in Fig. 5.10 with the start windings and remains in the circuit during the run cycle. The start windings and run windings are nearly identical in this motor, and reverse motion can be achieved by reversing the wiring of the two windings, with the capacitor connected to the other windings as start windings. By changing taps on the running winding but keeping the load constant, the motor can be made to run at different speeds. Also, in case all six winding connections are available separately, a three-phase motor can be converted to a capacitor start and run motor by making two of the windings common, and connecting the third via a capacitor to act as a start winding.

A permanent split capacitor (PSC) motor has a run-type capacitor permanently connected in series with the start winding. This makes the start winding an auxiliary winding once the motor reaches running speed. Typical starting torque of PSC motors is low, from 130% to 150% of rated load. However, unlike split-phase induction-run motors, PSC motors have low starting currents, usually less than 20% of rated load current, making them excellent for applications with high cycle rates. This motor configuration works well up to 1/4 horsepower (200 W), generally for smaller motors. The direction of motor is easily reversed by switching the capacitor in series with the other winding.

Voltage across capacitor terminals is much higher than the supply voltage because of the series inductance of auxiliary winding, and relative lower values of capacitors. If a wrong run capacitor is installed, the motor will not have an even magnetic field. This can make the motor noisy, increase energy consumption, and cause performance to drop and the motor to overheat.

PSC motors have several advantages. They need no starting mechanism and so can be reversed easily. Designs can be easily altered for use with speed controllers. They can also be designed for optimum efficiency and high power factor at rated load, and are considered to be the most reliable of the single-phase motors, mostly because no starting switch is needed.

[B] Capacitor-start induction-run motors

A larger capacitor may be used to start a single-phase induction motor via the auxiliary winding if it is switched out by a centrifugal switch, as in Fig. 5.11, once the motor is up to speed. The auxiliary winding may be made of more turns of heavier wire than used in a resistance split-phase motor to mitigate excessive temperature rise. The result is more starting torque for heavy loads like air conditioning compressors.

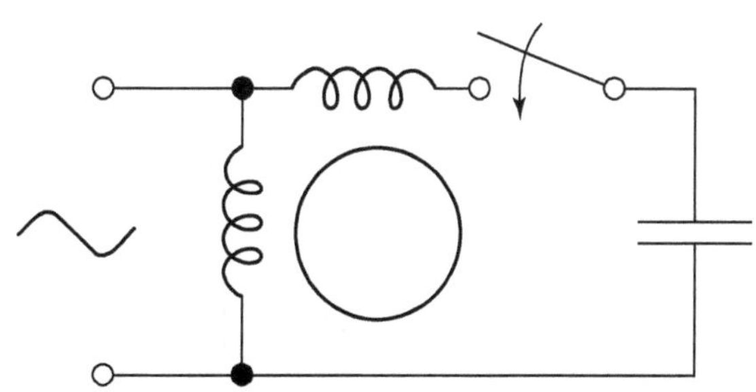

Fig. 5.11 *Capacitor-start induction-run motor circuit.*

These capacitors briefly increase motor starting torque and allow a motor to be cycled on and off rapidly. Start capacitors generally have values between 40–60 μF and 200–250 μF, for voltage rating of 230 V. A start capacitor stays energized long enough to rapidly bring the motor to three-fourths of full speed and is then taken out of the circuit; these large values of capacitors are usually designed at near supply voltage, or slightly higher, and provide much higher phase shift to give large starting torque.

During starting, the capacitor is connected in series with the starter winding. The current through this winding leads the voltage V, while the current in the main winding lags the applied voltage V across the circuit. The larger the phase difference between the two windings, better the resultant rotating magnetic field.

When the motor reaches about 75% of the full load speed, the centrifugal switch opens, disconnecting the starter winding and the capacitor from the main winding. The torque developed by a split-phase induction motor is directly proportional to the sine of the angle between winding currents. The angle is 30° in case of split-phase motors, but in case of capacitor-start induction-run motors, it is about 80° and the increase in the starting torque is nearly twice value developed by a standard split-phase induction motor.

Since the capacitor is in series with a heavy start winding, it creates more starting torque, typically 200% to 400% of rated load. This allows higher cycle rates and reliable thermal protection. The capacitor-start induction-run motors are more expensive than comparable split-phase designs because of the additional cost of the start capacitor and centrifugal switch disconnector. But the application range is much wider because of higher starting torque and lower starting current. These can be used for a wide range of belt-driven applications like small conveyors, large blowers and pumps, as well as many direct-drive or geared applications. These are the workhorses of general-purpose single-phase industrial motors.

[C] Capacitor-start capacitor-run motor

Capacitor start and run normally refers to motors with a capacitor permanently connected to the secondary winding, with a second larger capacitor connected in parallel, via the usual centrifugal switch, to give a good starting boost. These motors are normally designed to improve the torque curve in comparison to standard single-phase motors so they are much harder to stall and better able to run up to speed under load. The additional complexity of the capacitor-run motor is justified for larger size motors.

This type combines the best of the capacitor-start induction-run motor and the permanent split capacitor motor. It has a start-type capacitor in series with the auxiliary winding like the

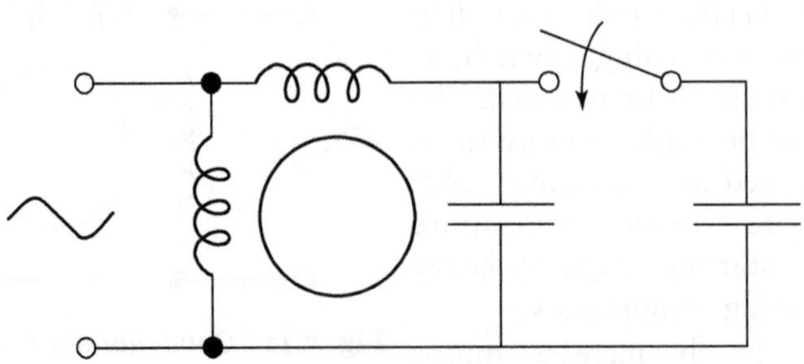

Fig. 5.12 *Capacitor-start capacitor-run induction motor.*

capacitor-start motor for high starting torque, and like a PSC motor, it also has a run-type capacitor that is in series with the auxiliary winding after the start capacitor is switched out of the circuit. This allows high breakdown or overload torque.

Another advantage of the capacitor-start capacitor-run type motor: It can be designed for lower full-load currents and higher efficiency. Among other things, this means it operates at lower temperature than other single-phase motor types of comparable horsepower. The only disadvantage to a capacitor-start capacitor-run motor is its higher price, mostly the result of more capacitors, plus a starting switch. But it is able to handle applications too demanding for any other kind of single-phase motor. These include woodworking machinery, air compressors, high-pressure water pumps, vacuum pumps and other high torque applications requiring 1–10 hp.

A motor starting capacitor may be a double-anode non-polar electrolytic capacitor which could be two + to + (or – to –) series connected polarized electrolytic capacitors. Such AC rated electrolytic capacitors have losses so high that they can only be used for intermittent duty (1 s on, 60 s off or 3 s on, 180 s off). A capacitor for motor running must not be of electrolytic construction, but a lower loss polymer type. Examples of motor capacitors are: A 36 µF 440 V run capacitor, or an 80–100 µF 250 V start capacitor.

The capacitors for the above motor applications are discussed below in detail.

(1) Motor run capacitor

These are usually MPP capacitors, of dry or impregnated construction. They are designed to serve continuously in the run circuit of capacitor-start capacitor-run motors. They withstand higher voltages, in the range of 400–440 V AC.

Fan capacitors, governed by IS 1709 in India, are rated between 1.5 and 10 µF, and are rated 400/440 V. Dry MPP construction in plastic cans is nowadays more common (Fig. 5.13), but many applications demand use of aluminium or steel cans (Fig. 5.14). The temperature rating is normally 70°C/85°C. Motor capacitors have similar voltage and temperature ratings, but may go to 72 µF or even more. These are governed by IS 2993. Both are internationally covered by IEC 252. Tolerances on these capacitors are close, normally ±5%, or sometimes even closer, or unsymmetrical (e.g. – 3% to +1%), depending upon motor design parameters.

Fig. 5.13 *Fan and motor run metallized capacitors in plastic cans.*

Motor capacitors may be impregnated with liquid, gel or gas for longer life, or may be dry resin filled. Their self-healing property means they last longer and failure rates are low. They are often provided with a fail-safe mechanism by way of burst-proof construction-pressure sensitive interrupters, thermal fuses, or double groove mechanisms are common. This protects the capacitor and surroundings in the event of end-of-life failure by avoiding explosion and fire risks. The capacitors are generally in round cans, but rectangular cans for larger ratings are also common. Metallic cans are unavoidable for most oil-filled

Fig. 5.14 *Motor capacitors in metallic cans.*

capacitors. However, plastic can construction for oil filled capacitors has been lately developed and patented by an American company.

(2) **Dual run capacitors**

A dual run capacitor supports two electric motors, such as in a large air conditioner unit, with both a fan motor and a compressor motor in the outdoor heat pump. The dual capacitor has three terminals, labelled 'C' or 'FAN' or 'HERM' for the common, fan and hermetic (pressure/compressor) electric lines.

Fig. 5.15 *Dual capacitors.*

Dual-run capacitors are commonly used for air conditioning, to help in the starting of the compressor and the condenser fan motor. Dual capacitors come in a variety of sizes, depending on microfarads (μF), such as 40 plus 5 μF, and the voltage. These also find common use in washing machines. A 440 V capacitor can be used in place of a 370 V one, but not vice versa. The microfarads must stay the same within 5% of its original value.

(3) **Motor start capacitor**

The electrolytic start capacitor helps motors achieve the most beneficial phase angles between start and main windings for maximum locked-rotor torque. It is disconnected from the start circuit when the motor reaches about 75% of full-load speed. The start capacitor is designed for short-time duty. Extended application of voltage to the capacitor will cause premature failure, if not immediate destruction.

Fig. 5.16 *Motor start capacitors.*

Electrolytic capacitors, by their nature, have wide tolerances. Common ratings in India are 40–60 μF, 60–80 μF, 60–100 μF, 100–120 μF, 120–150 μF, 150–200 μF and 200–250 μF. Sometimes 20–30 μF and 250–300 μF are also used. These capacitors have high tan delta, anywhere between 1.8% and 5%, depending upon manufacturer and technology used. The dielectric heating is therefore very high, and the capacitor duty cycle is short, about 1.7%, with 3 s on and 3 min off, and with a maximum limit of 20 starts per hour.

The centrifugal switch has to disconnect the capacitor within this 'on' time limit. In case the switch does not open, the capacitor fails, and along with it, the start winding also burns out. An electrolytic capacitor is provided with burst-proof construction, usually in the form of a rubber vent which punctures in the event of pressure build up, and allows gases to escape. The capacitor fails in open circuit mode. The vent mechanism is usually tested by observing the time taken by the vent to burst, when a rated supply is directly connected across capacitor terminals. The burst time generally varies between 3 and 5 min for a properly constructed capacitor.

Nowadays, metallized self-healing capacitors are also being used as starting capacitors. These have the advantage of very low dissipation factor from 0.03% to 0.4% depending upon the film used. Voltage ratings are 250 V in place of 230 V, and they can take surge voltages up to 375 V. The manufacturing tolerances on capacitance are close, from 5% to 10%, ensuring more consistent torque. The burst vent may not be provided in view of capacitor properties and self-healing mechanism, and the switch operating time. Withstand time at rated voltage can go beyond 30–45 min.

Metallized capacitors open up another possibility for motor manufacturers by further optimizing their design of starting winding, allowing more current, and saving copper in the process. The overall reliability of the motor goes up by way of much lower failure rates.

(4) Fan regulator capacitors for speed control

Conventional resistance type ceiling fan regulators are on the way out, and being replaced by capacitor type regulators. Capacitors have lately replaced resistors in earlier generation fan regulators. The capacitors being added as series impedance in steps are free of watt loss. The regulator is small in size, and safe for concealed wiring. In switching between positions, capacitors discharge into one another while switching. Capacitors have to be designed keeping these surge current requirements in view.

These regulators have an advantage over electronic ones due to their reliability and longer life. The capacitors need special attention as they discharge into one another during speed change. These use MPP or MPET films. Manufacturing processes, end spraying and connections need special attention. The capacitors are made in flat configuration and encapsulated. Loss factor should be low, and the ambient temperature can go higher in case of concealed wiring. With all capacitors, the regulator design has to provide for series and parallel resistors so as to keep surge and discharge currents within limits, and also to discharge a capacitor every time it goes out of circuit during regulator operation.

A typical basic circuit for a 5-step regulator is shown in Fig. 5.17. In two positions, capacitors discharge into one another while switching. Capacitors are made using MPP and MPET films keeping these surge current requirements in view. The circuit also has to make allowance for surge current limiting by way of series resistors with capacitors, as also to discharge them when not in circuit.

(Series resistors are not shown.)

1 P – C1 – Fan
2 P – C2 – Fan
3 P – C2 + C3 – Fan
4 P – C1 + C2 – Fan
5 P – Fan

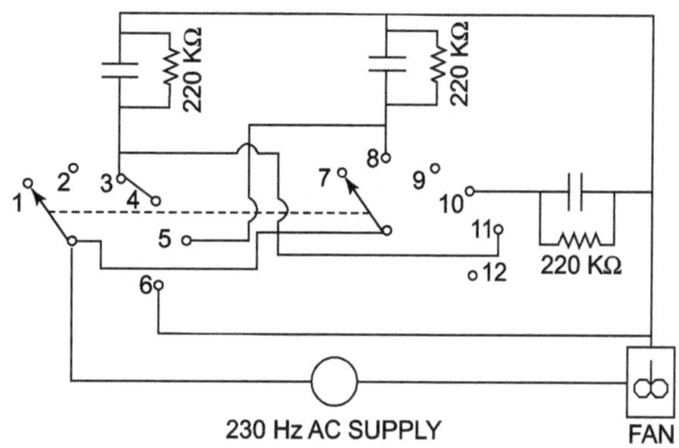

Fig. 5.17 *5-speed fan regulator circuit.*

Capacitors for fan regulators are available in different configurations and combinations. Figure 5.18 shows two such constructions. Capacitor type regulators offer a number of advantages over old resistance type as well as electronic type regulators:

- No heat loss in regulators. (Resistance type old designs lose 30% to 35%of fan wattage as heat loss.) No heat sinks needed.

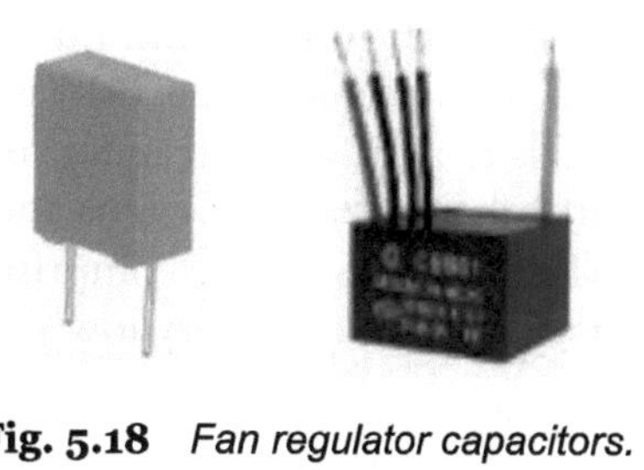

Fig. 5.18 *Fan regulator capacitors.*

- ➢ No heating of regulator casings. Plastic casings of resistance type regulators often used to melt in time due to heat.
- ➢ Given the above, these are ideal for concealed wiring.

5.3.2 Constant Voltage Transformers

Capacitors also find use in constant voltage transformers and voltage stabilizer circuits feedback systems to maintain steady voltage. These generally face higher temperatures, and need to carry more transients in service.

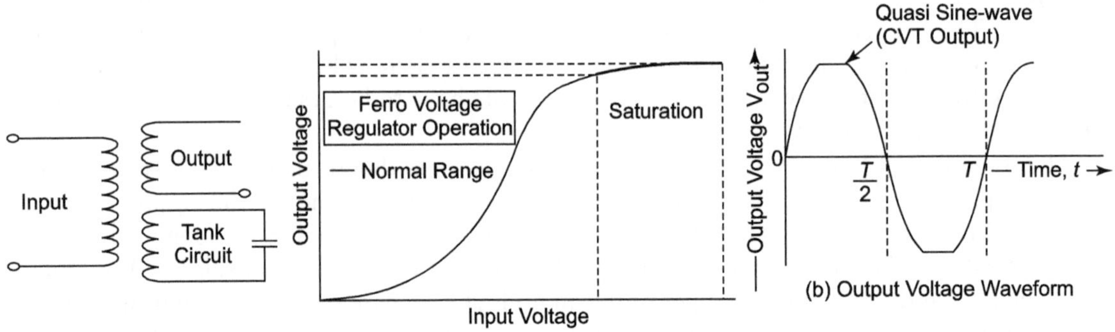

Fig. 5.19 *(a) CVT schematic. b) CVT operation (Source: UST, Inc.). (c) CVT waveform.*

The CVT uses the unique principle of ferroresonance (operation of a transformer in the region of magnetic saturation). A schematic of a ferroresonant transformer is shown in Fig. 5.19(a). Most ferroresonant transformers incorporate an LC tank circuit, shown in the figure, tuned to the AC frequency to effectively filter out many of the distortions produced in such operation near saturation (shown between dotted lines) in Fig. 5.19(b). Winding current and magnetic flux are proportional to the input and output voltage, respectively. Hence relatively large changes in input voltage have little effect on output voltage, this being the fundamental purpose of an automatic voltage regulator. In the saturation region of the curve, a large change in input voltage results in a small change in output voltage. The resultant output voltage is near sinusoidal as shown in Fig. 5.19(c).

5.3.3 Capacitors in the Lighting Industry

Discharge lamps are the most convenient light source, and have a very high efficiency level. At the same level of absorbed power, a discharge lamp provides a light emission much higher than that of a traditional incandescent lamp. Life of a discharge lamp is also much longer. Size and ratings depend upon lamp type and size, location and circuit used. They could be rated for 250 V or 440 V AC in India and Europe, while in countries with 110 V AC mains, 110 V and 220 V capacitors are common. Most lighting capacitors have a resistance connected across their terminals to reduce the terminal voltage to below 50 V within one minute of disconnection, so as to allow their safe handling, as and when necessary (in case of lamp failures, maintenance etc.).

Most lighting loads, comprising fluorescent lamps and sodium and mercury vapour gas discharge lamps, offer very low power factor inductive loads to the supply. Capacitors are used in many lighting fittings and at proper locations to compensate for low power factor and improve it to an acceptable limit.

Fluorescent lamps

The discharge occurs in inert gas (neon, krypton, xenon, etc.). Ultraviolet radiation is emitted and converted into visible light. It is possible to obtain different light tones including the solar one. Due to the high quality of the light and the low level of glare, these lamps are particularly suitable for domestic and industrial use.

Metal vapour lamps (mercury vapour, sodium, high and low pressure)

The modest nearly monochromatic quality of the emitted light is counterbalanced by a very high lighting performance. Typical applications are street lighting and flood lighting.

(a) Working principle

An electric arc is induced between two electrodes suspended in the above gases or metal vapours. The arc is supplied by AC supply; therefore in case of a standard 50 Hz main it starts and extinguishes 100 times per second.

The electric arc is unstable, increasing the current, when its resistance diminishes. Therefore it is necessary to stabilize the arc by connecting series impedance, in the form of ballast or reactor. Discharge lamps (fluorescent lamps, halogen-metal vapour lamps, mercury vapour lamps and sodium vapour lamps) require suitable chokes or leakage reactance transformers for ignition and also for current limitation during discharge. Due to their inductance, the power factor is, depending on the type, between 0.5 and 0.7.

To improve the power factor to above 0.9, capacitors which are designed for the widest range of operating, climatic and thermal conditions should be used. Capacitors can be for single lamps or group correction (parallel types), as well as for series use (where the choke and capacitor are in series).

Parallel capacitors are rated for 250 V~ / – 40°C to 85°C, with a capacitance tolerance ±10%, whereas series capacitors are rated 440 V/450 V~/ – 25°C to 85°C, capacitance tolerance ±5% (or sometimes closer). Capacitors in light fittings can retain their charge for a considerable time after they have been disconnected. EN 61048/IS 1569 stipulate the use of special discharge resistors, to ensure that 1 min after disconnecting, the voltage across the capacitor is reduced to less than 50 V.

(a) Single-lamp and group correction by means of parallel capacitors

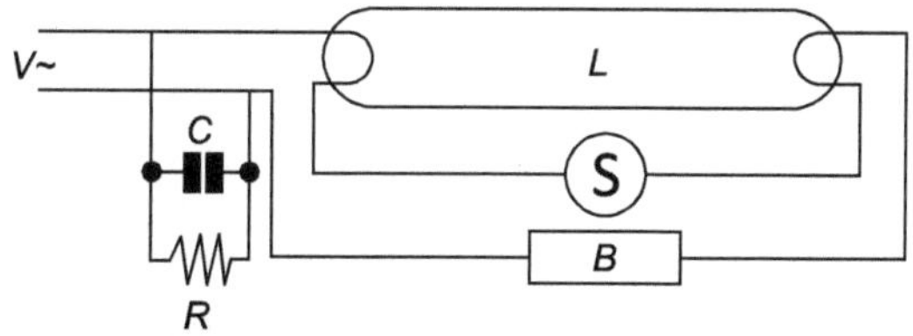

Fig. 5.20 *Single tube light wiring with parallel capacitor.*

L = Lamp, S = Starter, B = Ballast,
C = Capacitor, R = Discharge resistor

Figure 5.20 shows the connection for a single tube light. The capacitor's function is of power factor correction; it does not intervene in the functioning of the lamp/group of lamps. The capacitor is directly connected to the mains, in parallel to the ballast-lamp group. Its voltage is the same as the network voltage.

Each separate lamp (single-lamp correction) or a group of lamps controlled by a common switch (group correction) can be power factor corrected by shunting a capacitor across the line. If group correction is employed, then the correction capacitor must have a value equal to the sum of the capacitance values that would be required to correct individual lamps.

(b) Power factor correction using series capacitors

With this type of power factor correction one choke is overcorrected so that the reactive power handled by the capacitor is sufficient to compensate for two chokes. Connections are made as shown in Fig. 5.21. Due to the series-connection of choke and power factor correction capacitor; the voltage developed across the capacitor while the lamp is energized exceeds the line voltage. The series capacitor must be rated for an appropriately higher voltage, typically 400/440 V for 230 V mains supply.

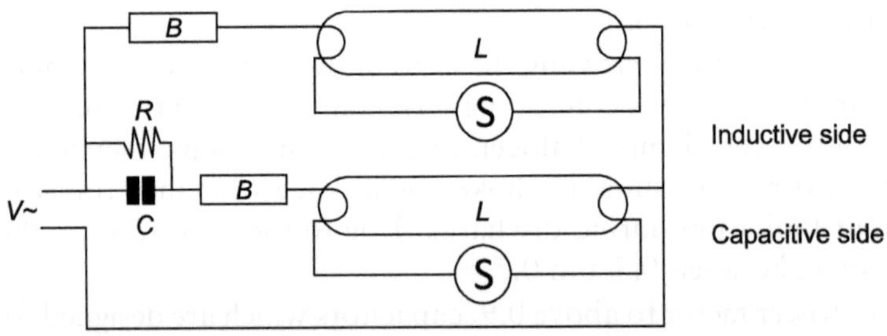

Fig. 5.21 *Twin tube wiring diagram with series capacitor.*

The power factor correction of the group and the elimination of the stroboscopic effect are simultaneously obtained, since the ignition and the extinguishing of the arc in the two sides of the circuit (inductive and capacitive) are in opposite phase. The combined light produced by the two lamps is almost completely free of flicker, since the current through one lamp is always a maximum whilst that through the other is zero. The capacitor is directly carrying the current of one lamp. The presence of the inductance B causes a voltage increase at the capacitor terminals much higher than the mains voltage (i.e. mains 250 V AC and capacitor 440 V AC).

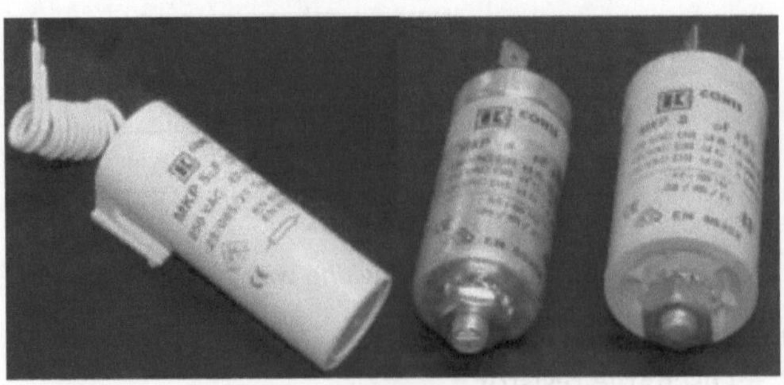

Fig. 5.22 *Lighting capacitors.*

Also, this type of power factor correction is more economic than parallel correction and is the only permissible way if the supply carries an audio-frequency signal used for central control purposes. For the correction of series power factor the capacitance is critical and is suggested by the manufacturer of the ballast. In this application the capacitance tolerance must be close (±5%).

(c) Typical working conditions

Capacitors are generally located inside the fittings, where the temperature is always high. The duty is quite heavy, since the lamp may remain continuously switched on. At the moment of the switching, some remarkable transient over-voltages can also take place. The discharge resistor R is indispensable; if it is not incorporated in the capacitor or in its assembling fitting, it must be included in the power supply.

Typical ratings of lighting capacitors will be as under:

Rated voltage	250 V/440 V 50/60 Hz
Capacitance tolerance	±10% for 250 V, ±5% for 440 V
Temperature range (T_{max})	–40°C ... +85°C
Thermal class	40/85/10 (IEC 68)
Limit values of relative humidity	65% annual means
	85% maximum value on 30 days
	75% on all other days
Reference standards	IEC 566, BS 14017
	IEC 1048/9, IS 1569
Test voltage: Between terminals	2.15 U_N ac
Test voltage: Terminals to case	3 kV ac

Common capacitor ratings for lighting capacitors – fluorescent, mercury vapour and sodium vapour lamps are given in Appendix C.

5.3.4 Power Factor Correction Capacitors

One major use of capacitors is in electrical utilities and industry, for compensating heavy inductive load (kilo–volt–ampere reactive, or KVAR) by supplying leading KVAR. The inductive load draws a heavy current from the supply line and power systems. It is of no use for active power consumption, but only increases the line current unavoidably, causing a heavy drag on the system. Compensation by capacitors takes care of this by supplying leading KVAR, thus reducing the line current, as shown in Fig. 5.23, helping better and more efficient use of generation and distribution capacity. Hence capacitors, known as power factor correction capacitors (or KVAR capacitors), are used in all generation, transmission and distribution systems.

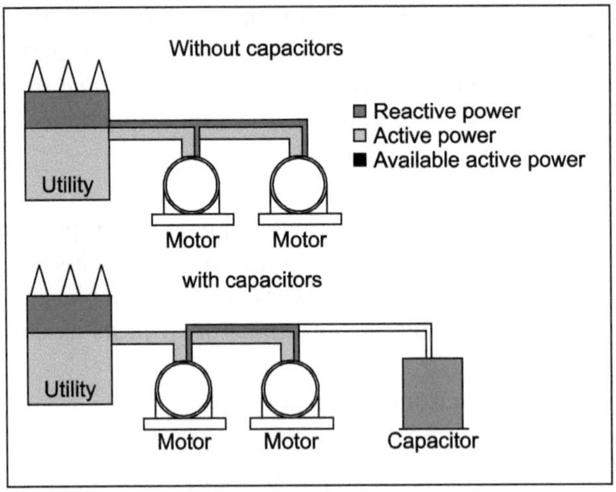

Fig. 5.23 *Lagging reactive power compensation.*

Most industrial loads, being of inductive nature, cause a low power factor (lagging). Capacitors are used in industries at supply mains and near inductive loads like motors, transformers, inductive furnaces etc. to maintain a healthy power factor.

It is mandatory in almost all countries to maintain a load power factor of 0.85 or 0.90 and above (unity power factor means a pure resistive load). Capacitors are used for this application all over the world.

Power capacitors are dealt with separately in a separate chapter in detail, as these constitute a very important segment of capacitor usage.

5.3.5 Other Applications

(a) Auto ignition condensers: Petrol driven vehicles make use of capacitors in the ignition circuit for giving high voltage spike to strike a spark on each cycle to ignite the petrol, and also in other circuits such as flashers, horns etc. These will be discussed in a separate chapter.

(b) Filter capacitors are used in all rectification, tuning, timing and filter circuits, and also in most electronic equipment to serve various functions. Filters could be half wave, full wave, RC or LC filters. Some filter combinations are shown below.

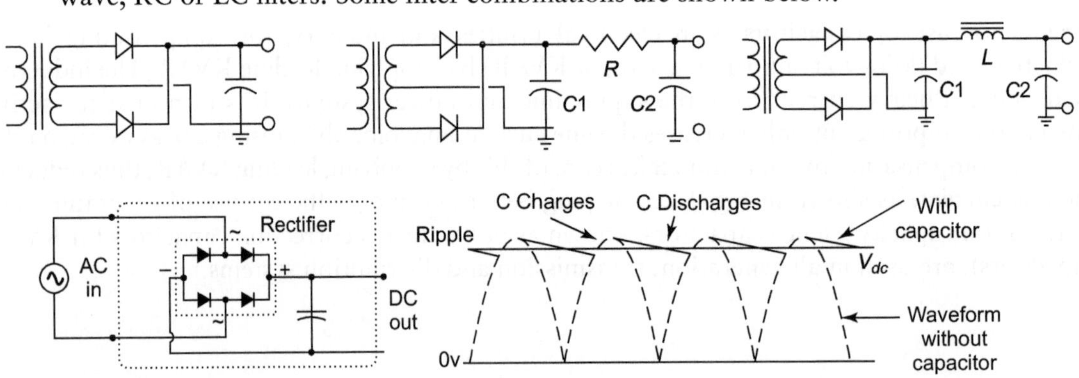

Fig. 5.24 *Rectifier–filter circuits with output waveform showing effect of capacitor*

(c) XY capacitors are used as radio frequency interference protection as also for effective grounding of electrical and electronic circuits, generators and instruments. These form an important class and are discussed in a separate chapter.

(d) Snubber (damping) and commutation capacitors protect high power semiconductors by limiting rate of rise caused during the switching of voltage in thyristor and IGBT circuits. RMS currents are low, but repetitive peak currents are very high, limited by the dv/dt of electronic circuits. Switched in parallel to a thyristor, these capacitors are designed to quench the conductive state of the thyristor. Repetitive peak current may substantially exceed the RMS value for the same reason.

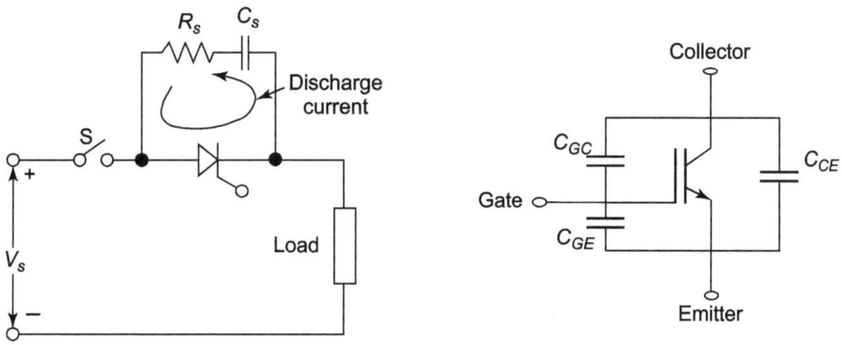

Fig. 5.25 *Snubber and IGBT capacitor circuits.*

Both these types of capacitors have low inductance and very high current ratings at rated voltages (220/440 V) due to high dv/dt. Non-metallized capacitors are preferred, but in many applications, specially designed metallized capacitors are in use. Current densities of elements are very high, and measures like providing more end surface area and extra thick end spray, in case of metallized electrodes, or thicker and extended metal electrodes are adopted. Figure 5.26 shows some designs of snubber and IGBT capacitors. The terminations and internal leads of these capacitors need to be heavy and shorter in length to carry the heavy currents. Inductance of winding and leads has to be as low as practically possible, necessary for high dv/dt ratings.

Fig. 5.26 *Snubber and commutation capacitors.*

(e) The conventional aluminium foil electrode in AC power frequency capacitor is being replaced by a thin layer of metallized coating on PP. For DC and for various electronic applications, other dielectric materials like polyester, polycarbonate and PTFE are used with metallized coating. These are most widely used today because of their excellent self-healing properties. These capacitors are dealt with in the chapter on power electronic capacitors.

(f) DC link capacitors: DC link capacitor of pulse width modulation (PWM) converter acts as energy storage and filter for DC link voltage. DC link capacitor is loaded with the rectified mains voltage. The rectifier allows energy flow only from the mains to the DC link. In case of braking (reverse-energy flow from motor to DC link) it is necessary to protect the capacitor from over-voltage. A chopper loads the DC link with the braking resistor to convert the excess energy into heat. Recent developments with an inverter instead of a simple rectifier allow energy flow in both directions.

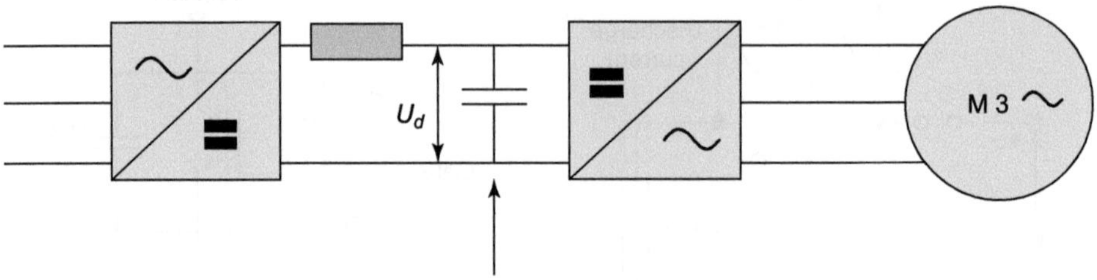

Fig. 5.27 *DC link capacitor schematic.*

(g) Furnace capacitors and water-cooled capacitors

Fig. 5.28 *Water-cooled furnace capacitors.*

These operate at higher frequencies of 400 Hz or more, and are used for induction heating furnaces. The heat generation is quite high because of frequencies involved, and they are provided with fins or water circulation tubes to dissipate the heat.

6

POWER FACTOR CORRECTION CAPACITORS

6.1 POWER FACTOR IN SUPPLY SYSTEMS

Most system loads in electric utilities, right from generation equipment, up to the final consumer systems, are inductive in nature. They load the system with a highly lagging current component. The result is a low power factor, with large overall current forced on the equipment, cable, transformers switchgear etc. This inductive component of current constitutes a lagging load, (called apparent or reactive load) and is represented as inductive KVAR. This component increases the supply line current considerably, causing underutilization of available generation and distribution capacity at all levels.

Modern electric power utilities face many challenges due to ever increasing complexity in operation and structure, one problem receiving wide attention being voltage instability. Lack of new generation and transmission facilities, overexploitation of existing facilities, coupled with increasing load demands aggravate these problems in modern power systems. The main cause of voltage instability is the inability of the power system to meet the demand for reactive power. Voltage instability is the cause of system voltage collapse, when the system voltage decays to a level from which it is unable to recover. Voltage collapse may lead to partial or full power interruption in the system. It becomes imperative to provide adequate reactive power support at the appropriate location to solve voltage instability problems.

Static Voltage Stability

Static voltage instability is mainly associated with reactive power imbalance. The loading capacity of a bus in a system depends on the reactive power support that it can receive from the system. When the system approaches the maximum loading point (or voltage collapse point), both real and reactive load components increase rapidly. Hence the reactive power supports have to be locally adequate. The only way to save the system is to reduce the reactive power load (or add additional reactive power) prior to reaching the point of voltage collapse.

Capacitors are connected across the system at several places to compensate the lagging power factor by offering leading or capacitive KVAR in sufficient magnitude, to reduce line currents effectively and to achieve optimum load conditions so as to make maximum use of the equipment and the system. The lower the power factor, the higher are the currents for the same active load (MW or kW). For better utilization of the system, it must be maintained at a lagging power factor of 0.95 and above.

Most utilities nowadays impose penalties for low power factors, and also offer incentives for higher power factor above 0.98. It makes sense to install capacitors on all systems to reduce line currents, lowering line losses, and also lowering switchgear requirements in many cases on this account. Voltage fluctuations arising on systems are also taken care of with capacitors installed in position. System and generation capacities are released for addition of extra load.

In electric power distribution, capacitors are used for power factor correction. Such capacitors often come as three capacitors connected as a three-phase load, although single/two-phase power factor capacitors are not uncommon. The capacitors are normally connected in delta, while it is general practice to have star connected capacitors for high voltage systems. Usually, the values of these capacitors are given not in microfarads, but in a more useful form as a reactive power output in kilo–volt–ampere reactive (KVAR). They serve to counteract inductive loading from devices like electric motors and transmission lines to make the load appear mostly resistive to the supply side. Individual motor or lamp loads may have capacitors connected with them for power factor correction, and larger sets of capacitors (usually with automatic switching devices) are installed at a load centre within a building or in a large utility substation.

6.2 BENEFITS OF POWER FACTOR CORRECTION

1. Power factor correction reduces the reactive power in a system and overall power factor of system is improved.
2. Effective installation use: An improved power factor means an electrical installation works more economically (higher effective power for the same apparent power), and costs come down.
3. Improved voltage quality: Capacitors generally improve the voltage regulation by decreasing voltage drop and absorbing some impulses.
4. Fewer voltage drops, since lagging currents and hence overall current in the system are reduced. Capacitors cause a rise in voltage at the load centre due to this effect.
5. Optimum cable dimensioning: Cable cross-section can be reduced with improvement of power factor (less current). In existing installations for instance, extra or higher power can be transmitted.
6. Smaller transmission losses: The transmission and switching devices carry less current, corresponding to only the effective power, thus reducing the ohmic losses in the leads.
7. Lower power losses up to capacitor locations, improving system efficiency.

While shunt capacitors offer several advantages at all voltage levels, they should be put in and out of circuit at appropriate times to get the most from the system. When load levels are high, a shunt capacitor system is beneficial. When the load is off, however, the capacitor must be disconnected, as it can do more harm otherwise. An excess of capacitance in service can lead to higher voltages, excessively leading power factors and resonance problems.

6.3 POWER FACTOR CORRECTION

[A] Individual power factor correction of motors

Capacitors connected directly across motor terminals should just compensate for the no-load magnetizing currents of motors. If a loaded motor power factor is sought to be compensated, a capacitor of higher rating beyond a limit may pose a danger to the motor.

When the motor is slowing down, it gets over or self-excited. Over-voltage occurs when a motor and its capacitors are disconnected from line and the motor continues to rotate for some time. The motor acts as a generator, with the capacitor supplying the magnetizing currents. Voltages across its terminals can go to over 140% to 160% with improper capacitors, causing damage to both the motor and the capacitor. Capacitors across motor terminals should be generally as per guidelines mentioned in the tables given in Appendix D.

Reactive power is required by an asynchronous motor for the magnetic field. The amount of reactive power consumption of a motor depends on various parameters such as power rating, loading, rated speed, and design. The capacitor output should not exceed 90% of the apparent power of an asynchronous motor under no-load conditions. This is important to avoid dangerous self-excitation of the motor. A measurement of motor current under no-load conditions can be easily performed or may be obtained from the manufacturer. Appendices D1 through D3 give the power factor of partially loaded motors, guidelines and standards for capacitor selection for power factor correction of motors.

[B] Individual power factor correction of welding machines

Resistance-welding machines are usually characterized by a large intermittent kilo–volt–ampere demand at low power factor, from 0.3 to 0.45, which is sometimes difficult to handle on the available power circuits due to regulation problems. Welding loads present quite a challenge to the designers of power factor control systems because of the rapidly fluctuating current taken during the welding process. Connection of capacitors for power factor improvement is also not an easy task.

The output of capacitors for single and two-phase welding transformers and resistance welding machines needs to be up to 50% of the nominal transformer capacity. For welding rectifiers, a capacitor output of about 10% of the nominal capacity of the transformer/rectifier is sufficient. Three-phase welding machines can be provided capacitors with around 33% of their KVA rating. Appendix E-2 gives recommended capacitor ratings for welding transformers.

It is standard practice to switch the capacitors on and off with the welding transformer to avoid loading the supply with capacitive current when not necessary. A specially designed current sensing relay and a fast discharging arrangement for the capacitor is deployed to adjust to the welding process at a fast rate. The capacitor used for power factor is specially provided with an arrangement for quick discharge, so that it can be switched on again in a short time. This allows the controller to follow the welding cycle properly. Fuzzy logic current sensing relays are currently in vogue for accurate welding controls. Single-phase, two-phase as well as three-phase welders can be controlled in this way. The load current is measured by the relay, and a contactor is switched on or off depending upon the fuzzy logic programmed into the controller.

The contactor is used to switch the capacitor for improving the power factor. A fixed capacitor is used to compensate the inductance of a transformer.

[C] Individual power factor correction of power transformers

The voltage regulation of a transformer is the percentage change in the output voltage from no-load to full-load. As power factor is a determining factor in the secondary voltage, it influences voltage regulation. This means the voltage regulation of a transformer is a dynamic, load-dependent number. The number of primary windings would not change; the number of secondary windings would not change, but the voltage regulation varies with the power factor.

Ideally, there should be no change in the transformer's output voltage from no-load to full-load. In such a case the voltage regulation is said to be 0%. To get the best performance out of a transformer, voltage regulation should be the lowest possible. This may be calculated from load conditions and the result saved as a troubleshooting and predictive maintenance benchmark. If the percentage change is too high, a check is needed for power factor correction for the loads on that transformer.

The direct connection of the capacitor to a power transformer, which is jointly switched in and out, is feasible and permissible both at the high voltage side and the low voltage side. In cases where harmonics exist in the line, the line should be checked to determine whether the capacitors and the power transformer create a resonance. Care should be taken not to over-compensate the power transformer during low load operation in order to avoid an unacceptable rise in voltage.

For power factor correction of transformers only the no-load reactive power has to be covered. The required capacitor output for three-phase transformers depends on the short-circuit voltage and is between 3% and 12% of the rated transformer output. In case harmonics are present on the high voltage side, the capacitor can form a series resonance circuit with the inductance of the transformer. Therefore the capacitor output has to be selected very carefully in collaboration with power utilities and the transformer manufacturer. A general guideline for capacitor ratings for different sizes of transformers is given in Appendix E-1.

[D] Group power factor correction

With individual PFC, the capacitor is individually assigned or switched to a load. With group PFC, the power factor of a load group is determined with varying power configuration. Multiple capacitors are automatically switched in or out by a VAR controller. The task of both application types is to improve the power factor and thus achieve a reduction of the reactive power. With group power factor correction, the physical arrangement of the contactors and capacitors is mostly in the proximity of the low-voltage transformer, e.g. in the low-voltage sub-distributor. At this point, it is important to observe that the operating voltage and the short-circuit rating are higher during a fault. Calculation of capacitor ratings for improvement from a known existing power factor to a targeted improvement has been simplified by means of a table, or a nomogram, as shown in Appendices F-1 and F-2.

The power factor correction capacitor is connected to the secondary distribution system which feeds a number of individual motors, operating either continuously or at intervals. The motors and the capacitors are each switched in and out separately and are monitored by separate protective devices. The capacitors can be switched in or out individually or in groups. A single bank of power capacitor equipment can be used for a large group of small power consumer centres, e.g. motors. The reactive power consumption of such a group is widely variable, and a bank should be divided into several stages. The discussion here will exclude rotating machinery like synchronous compensation motors.

[E] Advantages of static shunt capacitors in group correction

- Low dielectric loss.
- Simple transport and construction. Large banks are built from small units.
- Any bank can be enlarged by simple addition of new units.
- The bank can be located near the consumer.
- Supervision and maintenance is simple. A damaged unit can be replaced easily.

Percentage apparent power (KVAR) reduction (or network unloading) related to power factor improvements are as per Table 6.1 below.

Table 6.1 *Apparent Power Reduction with Power Factor Improvement*

Actual power factor	*Improved power factor*	*Apparent power reduction %*	*Improved power factor*	*Apparent power reduction %*
0.60	0.90	33	0.95	37
0.65	0.90	28	0.95	32
0.70	0.90	22	0.95	26
0.75	0.90	17	0.95	21
0.80	0.90	11	0.95	16
0.85	0.90	6	0.95	11
0.90	0.90	-	0.95	5

Network transmission losses in the form of heating are remarkable. Losses depend on the square of line current. These also get reduced with power factor improvement due to reduction in overall current.

6.4 SHUNT CAPACITORS

These capacitors are connected in parallel to the load, and directly across the supply mains. Hence the name, 'shunt capacitors'. These are by far the most commonly used for reactive compensation. Capacitors are available from 1 KVAR to 100 KVAR units, rated at 415/440 V for normal supply mains. Units for medium and high voltages are used on higher voltages. These are most often meant for indoor use; although pole-mounted KVAR capacitors are also available. High voltage capacitors for outdoor installations are rated up to 11 kV. Most capacitors for 440 V supply mains are now being made with MPP film. For highly fluctuating loads like sugar mills, rolling mills etc., film/foil capacitors or mixed dielectric capacitors are preferred on account of their surge current capacity and capacitance stability. MPP capacitors are normally used in most applications on account of their lower costs and self-healing ability.

Inductor coils are usually provided in series with capacitor elements, of a few micro henries to limit surge currents within manageable limits. Often means are provided like louvres to allow air-cooling. The series reactor coils and discharge resistors are sources of heat generation, and the heat has to be dissipated. Smaller capacitors may have cable/wires coming out for connections, while most units are provided with bolt and nut arrangements for connections. Brackets may be provided for wall mounting or panel mounting, and cable cover is provided over the terminals. Many capacitors are provided with internal or external fuses for protection. These are particularly necessary for higher voltages.

Units of shunt capacitors are connected in parallel to form banks of higher KVAR ratings. In these cases, a bus is provided per phase for making parallel connections, and often series reactors are connected between capacitor terminals and bus bar. In case of capacitors connected in supply systems or automatic power factor correction panels (APFC panels), it is customary to add a series reactor, about 6% of capacitor reactance, to limit the surge and harmonic currents and protect the capacitor. However, the reactor voltage being in opposite phase by nature, actual voltage across capacitor terminals becomes higher than supply voltage. Capacitor KVAR output goes up, and it is necessary to keep this increased voltage within capacitor ratings. Actual capacitor KVAR so connected will be higher than the system requirement by a factor equal to the series inductor KVAR.

It is possible to have single or two-phase capacitors rated 250 V or 440 V, to be connected across welding transformers or lighting load. For calculations, 1 KVAR at 440 V gives a capacitance value of 16.5 μF, while at 415 V, it comes to 18.5 μF. At 240 V (single phase), the capacitance value comes to about 56 μF per KVAR. Considering the tolerance on capacitance values, which is –5% to +10%, the average target value is 3% above this level. This throws up a value of 5.65 μF per phase per KVAR of 440 V capacitor, and 6.3 μF per phase per KVAR for 415 V capacitors for three-phase units. Values for any given KVAR rating can be calculated from this relation. When measuring capacitance across two terminals of a finished three-phase delta connected capacitor, the measured value on the meter shows one capacitor across terminals, plus two other capacitors in series coming in parallel (corresponding to other two phases). Hence the measured capacitance is 1.5 times the value per KVAR, viz. about 8.22 μF between any two terminals per phase for a 440 V rated delta connected capacitor.

[A] Important considerations in use of shunt capacitors

Shunt capacitors are subjected to some of the most rigorous working conditions anywhere in industry. They are always subjected to full KVAR capacity, limited only by the applied voltage. Unlike motors or transformers, their load cannot be reduced or increased, and hence they are always fully loaded so long as they are connected to system voltage. They have a tendency to offer feedback effect in the system in a way that the line voltages tend to rise on light loads. Predominantly capacitive loads are also harmful to the system and the system KVAR has to be maintained at a lagging power factor of 0.9 and higher.

For this purpose, KVAR capacitors have to be disconnected from the supply often depending upon load conditions. A capacitor, at the time of switching on, offers a temporary short-circuit like condition to the supply terminals, causing a heavy current inrush. Further, sudden voltage fluctuation and transient conditions also result in heavy current surges in the capacitor. In metallized shunt capacitors, these current surges cause large current densities to occur at the metallized ends and also on the metallized electrode surface of capacitor elements.

When a capacitor is switched on and connected to supply, while other capacitors are already connected, this new capacitor offers a momentary short-circuit condition for all these capacitors (and the supply), and they get momentarily discharged through the new unit. In metallized shunt capacitors, these current surges cause large current densities to occur at the metallized ends and also on metallized electrode surface of capacitor elements. A metallized capacitor has a tendency to self-heal under condition of transients, and hence the capacitance value and correspondingly KVAR output goes down with time. This factor becomes predominant for fluctuating loads like sugar factories, rolling mills, drawing units, cable industry etc.

On account of these factors, non-metallized KVAR capacitors have come to stay as they are extremely stable and reliable for power factor correction, whenever frequent switching of capacitors like those in capacitor banks is involved. A capacitor bank consists of a number of capacitors connected in parallel to each other. These capacitor banks may be made partially ON and OFF depending upon load conditions, in such a way as to maintain required power factor. The capacitors may be connected in regular three-phase systems, or even on single phase or two-phase loads.

Fig. 6.1 *KVAR capacitor units and capacitor bank.*

[B] Capacitors used for utility and load power factor correction

These can be of the following types:

- MPP dry capacitor for power systems (shunt/KVAR capacitors)
- MPP oil impregnated capacitor for power systems.

In these self-healing capacitors, with every self-healing event, the capacitance value goes down in an infinitesimally small measure. Hence, MPP capacitor has a tendency to lose capacitance value over the years. The manufacturing processes have to be meticulously planned and carried to keep the good capacitance stability in service.

- Mixed dielectric shunt capacitor

These use a paper plus PP dielectric combination impregnated with oil, along with aluminium electrodes.

- PP/foil (All PP) shunt capacitors:

Only PP film and aluminium foil construction impregnated with oil are used in these capacitors.

The list is not exhaustive and covers some of the prime areas of use.

The foil-type non-self-healing capacitors are very stable and do not undergo capacitance change in their service life. PP/foil capacitors are the most expensive but best in performance, while dry MPP capacitors are the most cost effective. The cost of different types of capacitors increases from (i) to (vi) above.

6.5 SERIES REACTORS

Reactors are used in series with capacitor banks for two reasons:

- To dampen the effect of transients during capacitor switching, and
- Control the natural frequency of capacitor banks and system impedance to avoid resonance and to sink harmonics.

In many applications, current limiting reactors are all that is needed for safe operation of capacitor banks and the network.

Modern systems of speed control and voltage control frequently use thyristor drives, which use chopping circuits, causing highly distorted wave shapes. The system uses complicated electronic systems to correct the wave shape to near sine wave, while keeping the power factor near unity.

6.6 SERIES CAPACITORS

Series capacitors are connected between transmission point and load, and become a part of series impedance of the transmission line, as in Fig. 6.2. Any change in load automatically affects the capacitive voltage drop and partly compensates for the change.

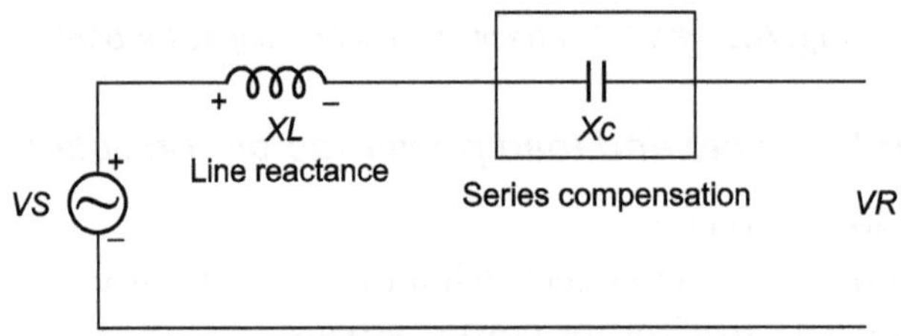

Fig. 6.2 *Series capacitor in a line.*

By adding series capacitance, it can be seen that the receiving line end voltage will be closer to the sending line end voltage. This decrease in voltage drop across the line allows more power to be transferred over the line for any given sending line end voltage. Practical installation of a series reactor can be seen from Fig. 6.3.

The benefits of series capacitors are that they:

- Increase power transmission capability
- Improve power system transient and steady state stability
- Reduce system losses
- Improve voltage profile of the lines
- Optimize power flow between parallel lines

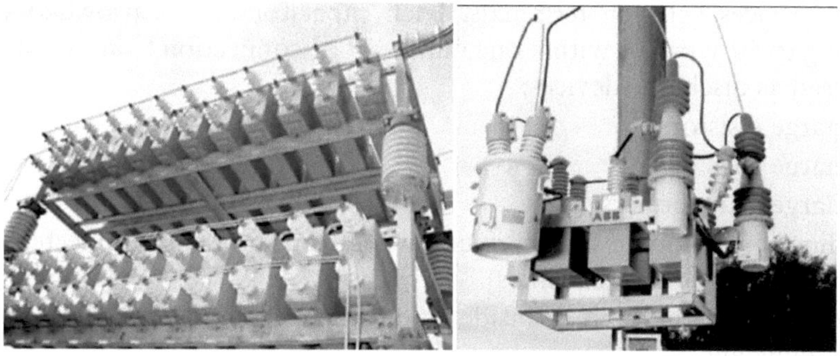

Fig. 6.3 *Series capacitor banks.*

6.7 CONSTRUCTION OF KVAR CAPACITORS

Basically, shunt capacitors are three sets of capacitors connected internally or externally in a star or delta three-phase system, with common terminations or bushings brought out for cable connections. The three sets of capacitors may be winding elements in a common container, or independent capacitors may be mounted in a common box and interconnected. Sometimes the three separate sets of terminals are brought out, making it a six-terminal capacitor. The interconnections are made by the user on site.

The size and rating of capacitors may be decided by the KVAR required and voltage rating. The capacitors may be of any of the representative types below:

- Paper/foil/oil – This is now almost obsolete.
- PP/Paper/foil/oil – Mixed dielectric non-metallized.
- Hazy PP film/foil/oil – All PP capacitor.
- MPP (dry) – MPP in metal or plastic can, resin encapsulated.
- MPP (impregnated) – MPP impregnated in oil.
- MKV (double metallized paper/PP or paper/oil) – Electrode formed by metallizing on both sides of paper without margin, and oil impregnated.
- MKP (double metallized PP/PP/oil) – Electrode formed on PP film both sides metallized, used with plain/hazy film dielectric, oil impregnated.
- Hazy PP film/foil + both sides metallized film, oil impregnated.
- Segmented film, resin encapsulated.

Essential parts of shunt capacitors are:

- Capacitor elements/capacitors and their interconnections.
- Internal insulations.
- Impregnation fluid (in oil-filled capacitors).
- Bushings/terminals/lead wires.
- Cover, where provision for cable connection is made.
- Mounting provisions.

- Discharge devices between terminals. Each capacitor unit is provided with means for discharging to 50 V or less within one minute of disconnection from supply. The following may be used as discharge devices:
  - Discharge resistors
  - Discharge coils
  - Discharge transformers
  - Windings of motors or transformers (where capacitors are connected directly against motor or transformer terminals)
- Series inductors – these may be inside or outside capacitor units. Their purpose is surge current dampening.
- Earthing provisions with marking.

Specifications and Terminology: The following standard systems are used for capacitor specifications and calculations:

Rated output of a capacitor: Reactive power derived from the rated values of capacitance, voltage and frequency. This may be calculated from three single-phase capacitance measurements, ammeter voltmeter method or KVAR meter.

Tolerance: Capacitance and output shall have a tolerance of (a) –5 to +10% for units up to 100KVAR, (b) – 0 to + 10% for units and banks above 100 KVAR, with the condition that in three-phase units, the ratio of maximum to minimum values measured between any two terminals shall not exceed 1.05.

The basic formula for individual capacitor per phase is

$VAR = V \times I = V \times \omega CV = 2\pi fCV^2 = 314\ CV^2$ for 50 Hz, where C is in farads. In a balanced system, actual VAR is 3 times this figure, and KVAR = VAR/1000.

For a three-phase delta connected system, $VA = \sqrt{3} \times VI$, where I is the line current in each phase of the balanced system. The line current is approximately 1.31 amp per phase per KVAR value of capacitor in a balanced system for 440 V supply. Table 6.2 below gives the method for calculation for different connections and configurations of capacitors, both by current measurement method and capacitance reading method.

Table 6.2 *Reactive Outputs for Different Connections*

Type of connection	Capacitance measured between	Formula from current reading	Formula from capacitance reading	KVAR factor
Single phase	Line terminals	$VI_L/1000$	$2\pi fCV^2/10^9$	1
Two-phase 4 wire	A1-A2 and B1-B2	$2VI_L/1000$	$4\pi fCV^2/10^9$	2
Two-phase 3 wire	A1-N and B1-N	$2VI_L/1000$	$4\pi fCV^2/10^9$	2
Three-phase delta connected	3 pairs of line terminals	$\sqrt{3}\ VI_L/1000$	$4\pi fCV^2/10^9$	$\sqrt{3}$
Star connected	Three pairs of line terminals	$\sqrt{3}\ VI_L/1000$	$4\pi fCV^2/10^9$	$\sqrt{3}$
Star connected	3 pairs of line terminals and neutral if brought out separately	-	$2\pi fCV^2/10^9$	

Climatic category for KVAR capacitors: Climatic category of a power capacitor is denoted by a symbol representing the climatic temperature range it can withstand in service. Since the temperature of a place keeps varying over the year, and also with the time of day, these are taken into account when deciding the maximum category temperature. These are specified in Indian standards by symbols as in Table 6.3 below.

Table 6.3 *Climatic Category of Power Capacitors Defined in Indian Standards*

Symbol	*Maximum*	*Ambient air temperature °C highest mean over any period of*	
		24 hours	1 year
A	40	30	20
B	45	35	25
C	50	40	30
D	55	45	35

Permissible overloads on power capacitors: The capacitors in service are subjected to frequent fluctuations of voltage and transients. The load is always changing, and as a result, voltage and current are ever fluctuating, and harmonics are also present. These factors have to be within the following limits: Capacitors have to bear variations as per Table 6.4 below as specified in the Indian standards during their operation in supply systems.

Table 6.4 *Permissible Overloads for Power Capacitors*

Operating frequency	*Voltage factor (Vf)*	*Maximum duration*	*Observation*
Power frequency	1.00	Continuous	Highest average value during any period for energization periods less than 24 h, with exceptions given below.
Power frequency	1.10	12 h in every 24 h	System voltage regulation and fluctuation
Power frequency	1.13	30 min. in every 24 h	System voltage regulation and fluctuation
Power frequency	1.20	5 min.	System voltage regulation and fluctuation
Power frequency	1.30	1 min.	
Power frequency plus harmonics	Such that the current does not exceed 1.43 nominal current.		

6.8 ECONOMIC CONSIDERATIONS OF POWER FACTOR CORRECTION

Electrical power administrators of all countries with relatively developed industries promote installations of power factor improvement systems, by penal rate charging for low power factor, and many times giving incentives for power factor closer to unity. Sometimes even the reactive power is charged. Savings due to reduced power system losses and financial results based on higher system efficiency, stability as well as reduced need for new generation capacity are of paramount importance for the national economy. A power price policy, i.e. selling price level for active and reactive power, or by way of penalty for low power factor, is of crucial importance for a successful economic approach. Appendix I gives the financial benefits and payback period for power capacitors, along with savings thereby. In most countries, the investments on a low voltage power capacitor bank are recovered in 2–3 yrs. After this period, the banks earn net profit.

7

SWITCHING OF CAPACITORS

7.1 CAPACITOR SWITCHING

Capacitor banks and systems are designed with switching mechanisms to connect and disconnect them from the system as needed, often several times a day. Switching a capacitor bank is different from switching a normal load. In a capacitive load, the current waveform leads the voltage by 90°.

[1] A capacitor, at the time of switching on, offers a temporary short-circuit like condition to the supply terminals, causing a heavy current inrush. Further, sudden voltage fluctuation and transient conditions also give rise to heavy current surges in the capacitor. If the switch contacts close at the moment of zero voltage, a sudden current surge will follow, overshooting the maximum peak value. The peak current inrush magnitude is a function of the rated capacitor current and the strength of the system to which the capacitor is connected, limited by the available short-circuit current of the system. The voltage and current waveforms oscillate at a frequency much higher than the power system frequency. After a short period of time, the waveforms settle down to their steady-state values. Figure 7.1 shows the resultant waveforms under switching conditions.

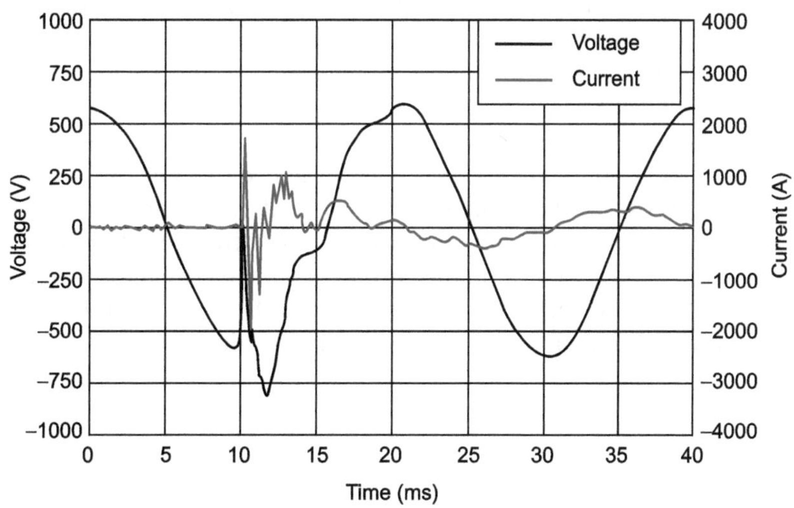

Fig. 7.1 *Line current and voltage waveform during switching in a capacitor.*

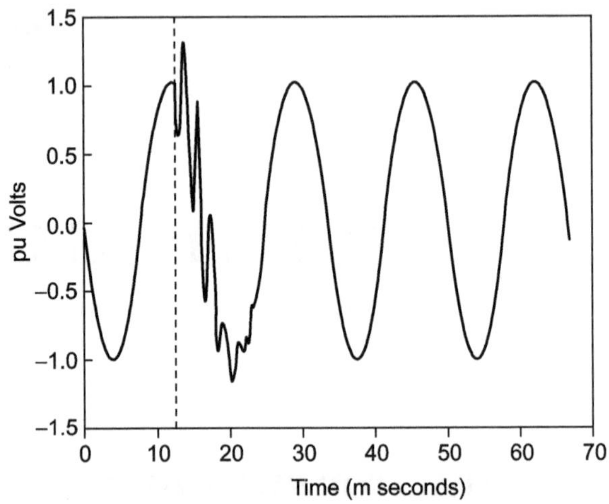

Fig. 7.2 *Secondary bus voltage waveforms during utility capacitor switching.*

[2] Switching off a capacitive load poses even more challenges, if the current is interrupted very close to its zero crossing when the voltage is at its peak value. As the switch contacts open, the charge (corresponding to the voltage at separation) trapped in the capacitor cannot be neutralized instantly. As the contacts continue to separate – a process that lasts about three electrical cycles – the dielectric strength of the gap between them increases in a near linear fashion, but the voltage difference across the contacts, (i.e. between the sinusoidal system voltage and the constant capacitor voltage) increases faster. The voltage across the parting contacts may exceed the dielectric strength of the gap between the contacts, causing an insulation breakdown, and resulting in an arc that re-establishes current flow. This re-strike of current flow may occur after a quarter of a cycle of initial interruption. When the current is re-established, it becomes a high-frequency, exponentially decaying sinusoidal wave. The high-frequency oscillations give rise to high-frequency voltage fluctuations, similar to that of the capacitor switching on. Resistance present in the system quickly damps these oscillations.

At the next current zero, the arc will be interrupted again, but this time the contacts will be farther apart than during the first interruption, thereby providing a greater dielectric strength between the parting contacts. At the second interruption, the dielectric strength between the parting contacts, which are still at less than half of their ultimate separation distance, may slightly exceed the voltage difference across the opening contacts. This will allow a successful current interruption. A typical voltage waveform during capacitor disconnection is given in Fig. 7.2. (Current waveform is current flowing into capacitor bank being energized.) In some cases, a second re-strike may occur at this point, and successful interruption may take place at the third current zero.

Capacitor banks applied within distribution substations typically consist of one to four banks of switched capacitors. The switched banks are designed to come on and off automatically based on power factor, VARs, and/or voltage. Due to load variations, a number of switching operations will occur daily. Each switching event is followed by a low-frequency decaying ring wave transient that can result in power quality problems for nearby industrial and commercial loads. A typical power factor correction system is shown in Fig. 7.3, with three capacitor banks of 1.5 MVAR each, along with their series reactors and individual switches.

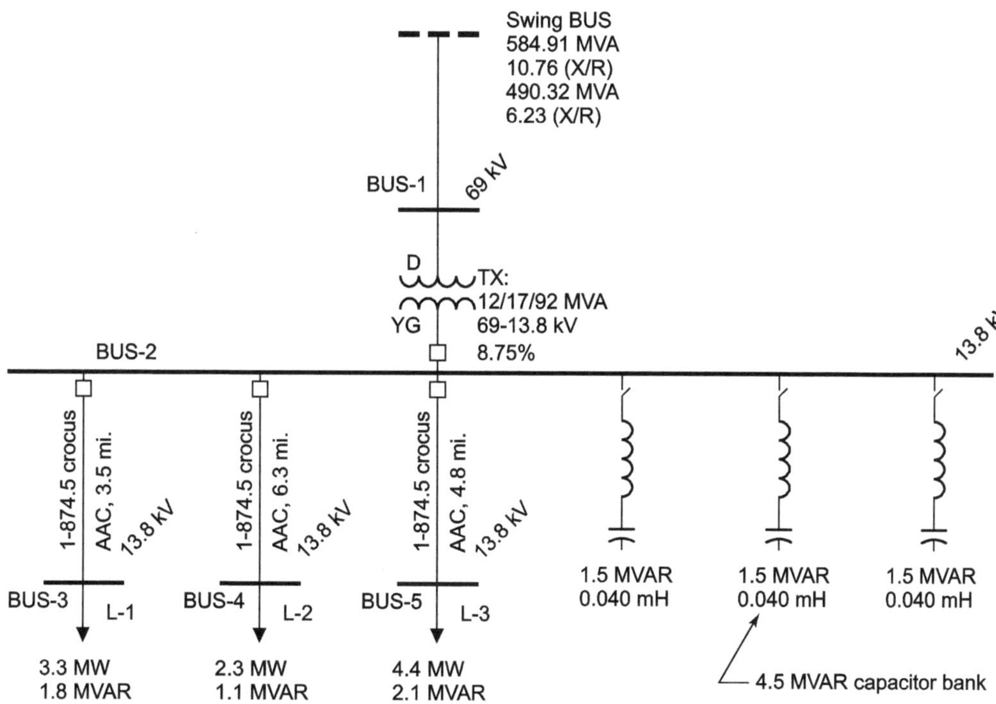

Fig. 7.3 *Typical utility substation showing capacitor banks and adjacent distribution loads.*

7.2 SINGLE BANK SWITCHING TRANSIENT

Figure 7.4 (a) shows the voltage transient that will occur for the closing of the first 1500 KVAR capacitor step of Fig. 7.3, while no other steps are energized. Due to switch variations, and possible pre-strike conditions, phase A and phase B vacuum switches are assumed to close prior to the phase C switch. For an ungrounded bank, the first phase switch to close will result in no current flow or voltage transient. The neutral voltage will then follow the phase voltage, and phase-to-phase voltage will be impressed across the remaining two switches. Upon closure of a second contact, a transient such as the one shown in Fig. 7.4 (b) will

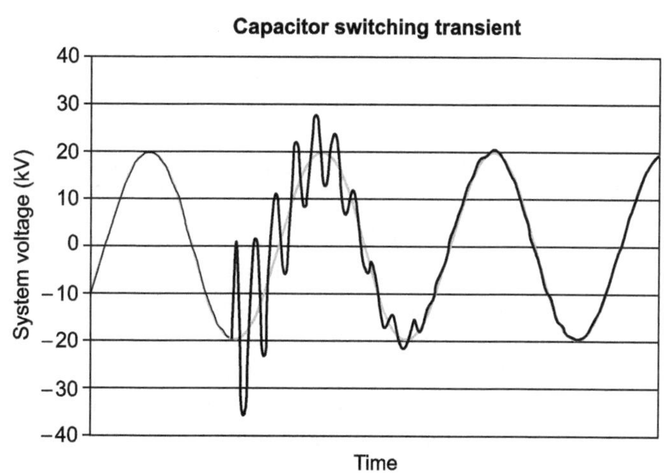

Fig. 7.4 *(a) Capacitor switching transient.*

occur. The worst-case transient will occur when the second switch closes near the peak of the phase-to-phase voltage waveform. Switches will begin to conduct near peak voltage due to pre-strike.

The transient is actually composed of a decaying ring-wave transient superimposed on the voltage waveform as shown in Fig. 7.4 (b). The duration of the decaying ring-wave transient is dependent on the system X/R ratio at the capacitor bank. High X/R ratios will result in long durations, while low X/R ratios will result in short duration transients.

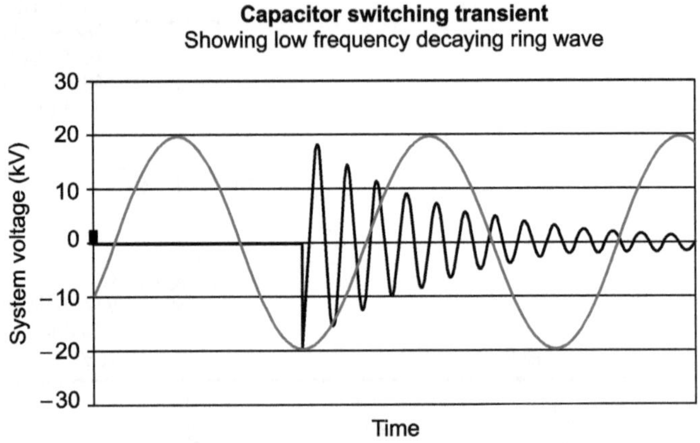

Fig. 7.4 *(b) Decaying ring wave.*

In examining the transient waveforms the following statements can be made in regard to single capacitor bank switching.

- On closing of the second contact (for an ungrounded bank), the line-to-line system voltage will be pulled to near zero volts. This is called voltage depression.
- Immediately following the voltage depression, the system voltage will attempt to recover, but will over-shoot the normal system voltage by an amount that is nearly equal to the voltage dip. Theoretically, two per-unit over-voltages can occur due to capacitor switching.
- The frequency of the capacitor transient is equal to the system's natural frequency. Therefore, large capacitor banks will result in lower frequency decaying ring-wave transients, while small banks will result in higher frequency ring-wave transients.
- The duration of the ring-wave transient is dependent upon the system X/R ratio at the capacitor bank. Systems with higher X/R ratios result in longer duration transients. Transients associated with substation capacitor banks can last as long as 30 to 40 cycles.

7.3 MULTIPLE CAPACITOR BANK SWITCHING

Multiple capacitor bank switching transients occur when a capacitor bank is energized in close proximity to a capacitor bank that is already energized. Such a switching operation is common in multi-step automatic capacitor banks, as shown in Fig. 7.5. Upon energization of the uncharged bank, the adjacent charged bank dumps a high-frequency high-magnitude current into the uncharged bank. This high-frequency high-magnitude current is limited by the impedance between the capacitor stages (resistance and reactance of bus work, fuses, vacuum switches, etc.). Most banks have to be supplemented with transient inrush reactors to reduce the

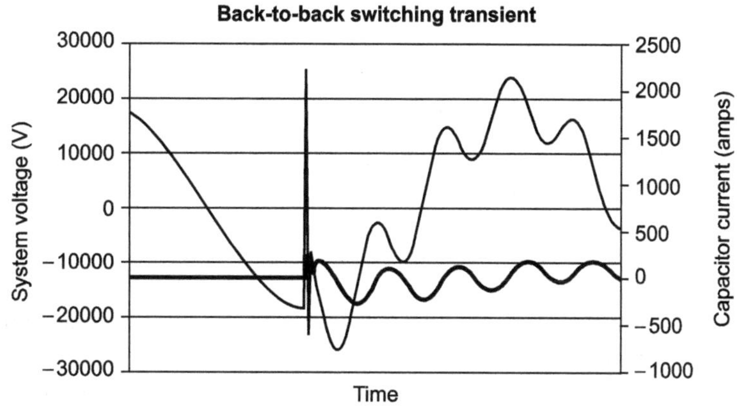

Fig. 7.5 *Voltage and current waveform associated with back-to-back capacitor bank switching.*

magnitude of the transients to within the vacuum switch and fuse ratings. The high magnitude current is not seen by the power system as it occurs between the parallel banks.

The following should be noted in regards to back-to-back capacitor bank switching:

- The system voltage still experiences a low frequency decaying ring-wave transient.
- The voltage depression is not to zero volts, as was the case for single capacitor bank switching transients.
- The system voltage over-shoot is reduced to an amount equal to the voltage depression.
- Multiple zero-crossings are still possible.

In purchasing and specifying capacitor banks and harmonic filter banks, the cost associated with nearby electrical equipment malfunction or damage should be evaluated against the cost of additional equipment to eliminate switching transients. Capacitor banks and harmonic filter banks in the 2.4 kV through 34.5 kV voltage range can be equipped with zero voltage closing controls to nearly eliminate switching transients. These controls operate their associated vacuum switches so that contact closure occurs at the zero-voltage crossing point.

7.4 POWER QUALITY CONCERNS

Capacitor bank energizing transients are becoming increasingly more important with the growing number of capacitor bank installations in power systems. This is because capacitor bank switching is one of the most frequent utility operations, potentially occurring multiple times per day and hundreds of time per year throughout the system, depending on the need for system voltage/VAR support from the banks. There are a number of important concerns when capacitor banks are applied at the transmission system voltage level, like transient-related currents and voltage transients at the capacitor bank substation and neighbouring substations. These may include:

- Phase-to-ground over-voltages, phase-to-phase over-voltages and over-voltages due to voltage magnification (i.e. exciting system resonances or dynamic over-voltages).
- Power quality impact on sensitive customer loads due to variations in voltage when energizing capacitor banks.

- This may also involve voltage magnification phenomena if there are capacitor banks on the lower voltage system in the vicinity of the switched banks.
- Capacitor bank energization inrush currents.
- Capacitor bank outrush currents due to faults in the vicinity of capacitor banks.

These transient voltages and currents can have an impact on equipment design and protection. In general, some of the equipment-specific design and protection issues include:

- Phase-to-ground and phase-to-phase insulation switching withstand to voltage stresses.
- Controlled closing for circuit breakers (pre-insertion resistors/reactors or synchronous switching).
- Capacitor bank and substation circuit breakers ANSI/IEEE C37 requirements.
- Current limiting reactor requirements.
- Surge arrester energy requirements (including capacitor bank circuit breaker re-strike events).
- Secondary metering and control issues (usually associated with high-frequency inrush currents).
- Voltages appearing on a power system associated with utility capacitor bank installations.

There are three power quality concerns associated with single capacitor bank switching transients. These concerns are most easily seen in Fig. 7.6, and are as follows:

- The initial voltage depression results in a loss of voltage of magnitude 'D' and duration 'T1'.
- The recovering system voltage will result in an initial transient over-voltage of magnitude 'S' and duration 'T2'.
- Multiple zero-crossings. For the transient in Fig. 7.6, a total of three zero crossings occur before the natural system voltage zero crossing.

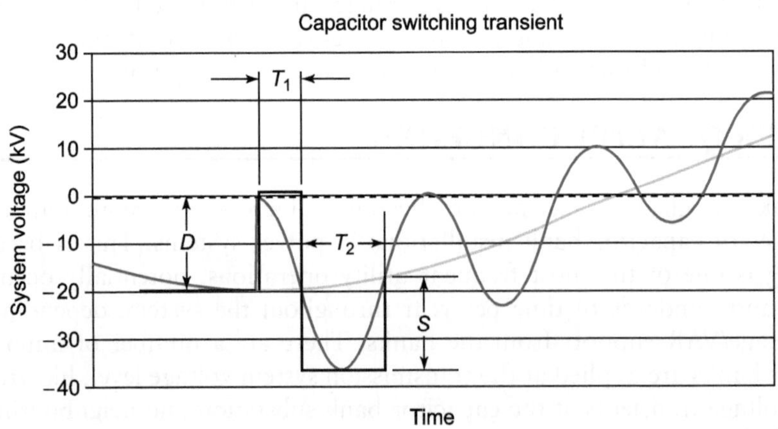

Fig. 7.6 *Height, depth, and time of voltage dip and spike shown on transient waveform.*

7.5 UTILITY CAPACITOR BANK SWITCHING

Utility capacitor bank switching can have negative impacts on power quality – especially for customer power systems. AC and DC drives, along with other electronic equipment, can be very sensitive to transient voltages. Utility capacitor bank switching transients can be magnified at low voltage capacitor locations on customer power systems, causing drives to trip, and production and other processes to stop. At a time when customers are being allowed to choose their power provider, utilities cannot afford to be seen as the cause of customer power problems.

Studying, analyzing, and preventing utility capacitor bank switching events often requires the use of sophisticated computer simulation tools. Simulations provide a convenient way to characterize transient events, determine resulting power quality problems, and evaluate possible solutions. Frequently, simulations are performed in conjunction with power system monitoring for verification of models and identification of important power quality concerns. Application considerations include capacitor bank configurations, insulation withstand levels, switchgear capabilities, grounding, over-current protection, over-voltage protection, energy duties of protective devices, and unbalance detection.

There are a number of important transient-related concerns when transmission voltage level capacitor banks are applied. Transmission system concerns include insulation withstand level, switchgear capabilities, energy duties of protective devices, and system harmonic considerations. These considerations must also include distribution systems and customer facilities.

Analytical methods provide the framework for evaluating a variety of power quality phenomena, including the impact of utility capacitor bank switching on the customer system. Typically, a model of the customer system and relevant parts of the utility system is developed to conduct transient switching surge analysis. This model can be used for simulations to predict power quality problems and evaluate possible solutions.

Analysis results include recommendations for the possible mitigation of switching problems. Analysis of various methods for controlling transient over-voltages is based on economic, control and technical considerations. Primary concerns generally evaluated for a capacitor bank application study include, apart from factors mentioned above:

- Transient over-current and over-voltage magnitudes for normal capacitor bank energizing operations.
- Various transient control methods (e.g. synchronous closing, pre-insertion inductors/resistors).
- Arrester duties for voltage magnification conditions, and during capacitor bank re-strike events.
- Phase-to-phase transients at transformer terminations.
- System frequency response characteristics (resonance).
- Ferroresonance possibilities.
- Impact of capacitor bank switching transients on customer systems.

Recommendations can then be made on the following aspects:

- Requirements for capacitor bank switching devices, including the effectiveness of pre-insertion transient limiting devices and synchronous closing control for mitigating transient over-voltages and over-currents during switching.

- Deployment of current limiting reactors for reducing both inrush and outrush currents.
- Arresters to protect against excessive transients during capacitor bank switch re-strike events.
- Protection against excessive transients at lower voltage capacitor banks, including capacitor bank switching controls or surge arresters.
- Protection against excessive transients and nuisance tripping of adjustable-speed drives at customer locations, including capacitor bank switching controls or reactors.
- Guidelines for capacitor bank design with respect to over-voltage mitigation, reactor requirements and arrester applications.

7.6 OTHER ADVERSE FACTORS FOR SWITCHING

(1) In metallized shunt capacitors, inrush current surges during switching cause large current densities to occur at the metallized edges and also on the metallized electrode surface of capacitor elements. When a capacitor is switched on and connected to supply while other capacitors are already connected, this new capacitor offers a momentary short-circuit condition for all these capacitors (and the supply), and they get momentarily discharged through the new unit. Metallized capacitors have a tendency to self-heal under conditions of such high current transients, and hence the capacitance value and correspondingly KVAR output goes down with time. This factor is predominant for fluctuating loads like sugar factories, rolling mills, drawing units, cable industry etc.

On account of these factors, non-metallized power factor correction capacitors are preferred, being extremely stable and reliable, whenever frequent switching of capacitors like those in capacitor banks and automatic power factor correction (APFC) panels are involved. A capacitor bank consists of a number of capacitors connected in parallel to each other. These capacitor banks may be made partially ON and OFF depending upon load conditions, in such a way as to maintain the required power factor.

(2) Development of semiconductor technology had adverse impact on the alternating current network. The sinusoidal waveform gets distorted by consumption of the reactive energy with the non-sinusoidal pattern of currents. Distortion can be expressed by the content of higher harmonics. The content of harmonics results in increase of capacitor current, since its impedance decreases with increasing frequency. This may cause damage to the capacitor, unsatisfactory tripping of circuit breakers and incorrect operation of the end equipment. This situation can be resolved by installation of capacitors with reactors (detuned PF correction), which attenuate the resonance circuit and such installation has also a partial filtering effect – it reduces the distortion level in the network. It is recommended in situations where the share of equipment generating higher harmonics exceeds 20% of the total load. Filtering circuits are used for removal of harmonics with a higher percentage share from the network.

(3) A capacitor at the instant of being switched on is a dead short circuit. The inrush current is limited in its peak value by system inductances up to that point, except that the circuit may now go into a natural resonance. A power contactor, by nature of its construction

and contact material, can withstand a peak current of a given magnitude, beyond which, the contactor points will weld on to themselves, leading to capacitor failure.

(4) If a capacitor is being switched on against other steps which are already on, then the other steps will discharge into this newcomer. The intervening bus bars have very low inductances and peak currents are very high – reaching 160 times the rated capacitor current or more. The contactor should be able to handle this without welding.

The following methods are available to deal with this:

- Use a liberal and proven rating for a known contactor.
- Use surge suppression choke coils on each capacitor, to introduce extra inductance and thus limit the peak current. For panels with 4 steps or more and also for panels using MPP capacitors, this is essential.
- Use a special contactor with auxiliary contacts which introduce a starting resistance at the beginning, then short it.
- A discharge resistor on a capacitor reduces the residual voltage on it after being switched off, to a safe value of 50 V within less than a minute and readies it for re-switching, should this be required. If this resistance were to burn out, re-switching will take place against a charged unit, which could prove dangerous. It is highly essential to periodically check the condition of these externally mounted discharge resistances.
- Main switch fuse is substituted by air-breakers for large banks.
- Time delay relays with an adjustable one-minute delay should be incorporated – in both APFC and manual mode to prevent re-switching of a contactor within less than one minute of switching it off.

7.7 SWITCHGEAR REQUIREMENTS FOR CAPACITOR INSTALLATION

The switchgear for capacitor must satisfy following conditions :

- Must have adequate continuous current rating.
- Must have adequate short circuit current rating.
- Circuit breakers, switches and contactors must be liberally rated in excess of capacitor rating. A general guideline, as given by PHD Chamber of Commerce and Industry, is as follows:

• Fused and unfused switches	:	165 %
• De-ion circuit breakers or equivalent	:	150 %
• Air circuit breakers	:	135 %
• Open type contactors	:	135 %
• Enclosed type contactors	:	150 %

Appendix G gives standard recommendations for cable and fuses for KVAR capacitors. Nowadays contactors specially made for capacitor duty are freely available. These may be preferred for all applications of capacitor switching as well as in APFC panels.

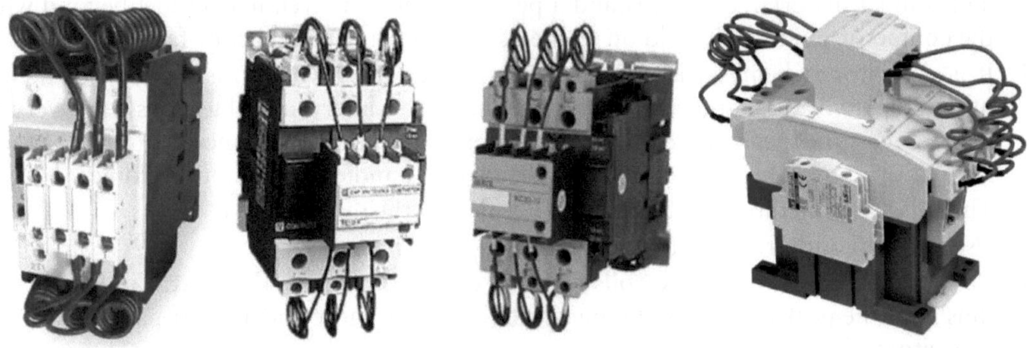

Fig. 7.7 *Capacitor duty contactors.*

7.7.1 Contactor for Capacitor Switching

During switching of a capacitor, transient current of the order of 200 times the rated current can flow, stressing the capacitor and the switching contacts immensely. This can lead to damage or welding of contacts of the contactors. Modern power factor correction systems use new generation contactors designed to switch the capacitors first through contact block of three early make auxiliary contacts in series with quick discharge damping resistor to limit inrush current to the value within contactor making capacity. Normal rated capacitor current is carried by main contacts, which, after closing after about 5 milliseconds, effectively bypass the damping resistors. The leading contacts then open up, and no current flows through them.

Benefits of capacitor duty contactors:

- Saves cost of expensive replacements.
- Minimizes the effect of inrush currents.
- Saves energy; reduced watt loss during 'ON' conditions.
- Operator safety through IP 20 shrouds on power and control terminals.
- Higher safety in operations.
- Switching of capacitor bank in parallel without de-rating.
- Less maintenance and down time.
- Higher electrical life.

These capacitor duty contactors have become common in the past few years and a number of manufacturers have them in their regular product range.

8

HARMONICS IN POWER SYSTEMS

8.1 HARMONIC DISTORTION BASICS

Harmonics are sinusoidal waves that are integral multiples of the fundamental frequency 50/60 Hz waveform, depending upon different country norms (e.g. 1st harmonic or fundamental = 50 Hz; 3rd harmonic = 150 Hz, 5th harmonic = 250 Hz). All complex waveforms can be resolved into a series of sinusoidal waves of various frequencies; therefore, any complex waveform is the sum of a number of odd or even harmonics of lesser or greater value. Harmonics are continuous (steady-state) disturbances or distortions on the electrical network and are a completely different subject or problem from line spikes, surges, sags, impulses, etc., which are categorized as transient disturbances. Any deviation from the sine-wave shape on a continuous basis is the indication of frequency other than the basic 50/60 Hz component. Distorted voltage and current waves cause voltage and current harmonics, respectively.

Current and voltage waveforms could be saw tooth, rectangular, symmetrical or asymmetrical, linear or nonlinear. Some waveforms are shown in Fig. 8.1 for clarification:

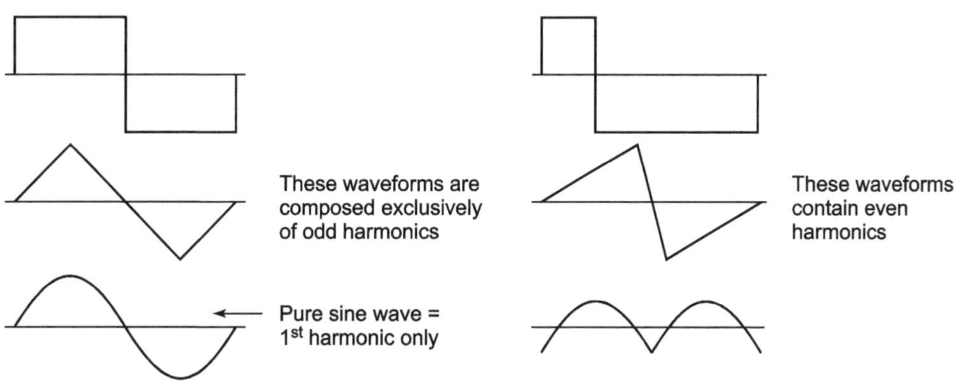

Fig. 8.1 *Some waveforms showing harmonics.*

By definition, harmonic or nonlinear loads are those that produce a non-sinusoidal current when energized by a sinusoidal voltage source. Both current waveforms are produced by turning on some type of load device. The distortion current could be produced by an electronic variable speed drive. The devices could be single-phase or three-phase. Power system harmonics are the mathematical integer multiples of the fundamental frequency within a power system. These integer multiples represent the deviation of the electrical signal from a pure sinusoid to a distorted waveform.

Even though half of the possible harmonic frequencies are eliminated by the typically symmetrical distortion of nonlinear loads, the odd harmonics can still cause problems. Some of these problems are general to all power systems, single-phase or otherwise. Transformer overheating due to eddy current losses, for example, can occur in any AC power system where there is significant harmonic content. However, there are some problems caused by harmonic currents that are specific to polyphase power systems.

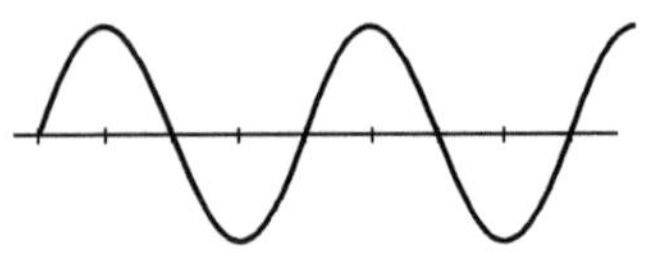

Fig. 8.2 *Non-sinusoidal currents.*

A distorted waveform can be looked upon as a combination of a number of harmonic waveforms, each one a sinusoidal one, but its frequency being multiples of the base frequency. Figure 8.3 shows a distorted wave broken up into harmonic components.

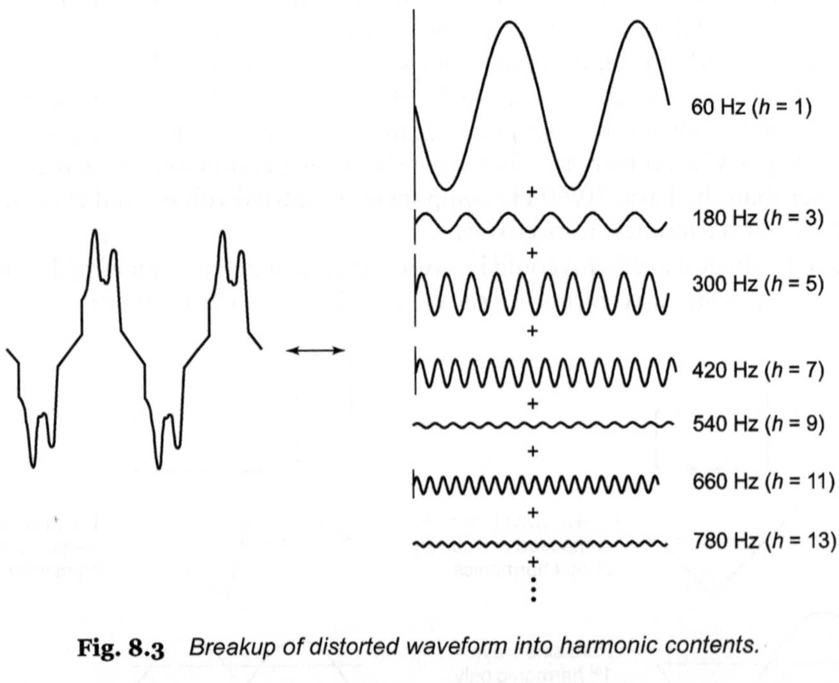

Fig. 8.3 *Breakup of distorted waveform into harmonic contents.*

One way to understand this is by imagining each harmonic frequency as a separate entity. As we add more current sources to the load, we would see further distortion of the line current waveform from the ideal sine-wave shape, and each of these harmonic currents would appear in the Fourier analysis breakup. Such a breakup can be understood with the help of Fig. 8.4.

Due to their abundance and significance in three-phase power systems, the 3rd harmonic and its multiples have their own special name: **triplen harmonics**. All triplen (triple-n) harmonics get

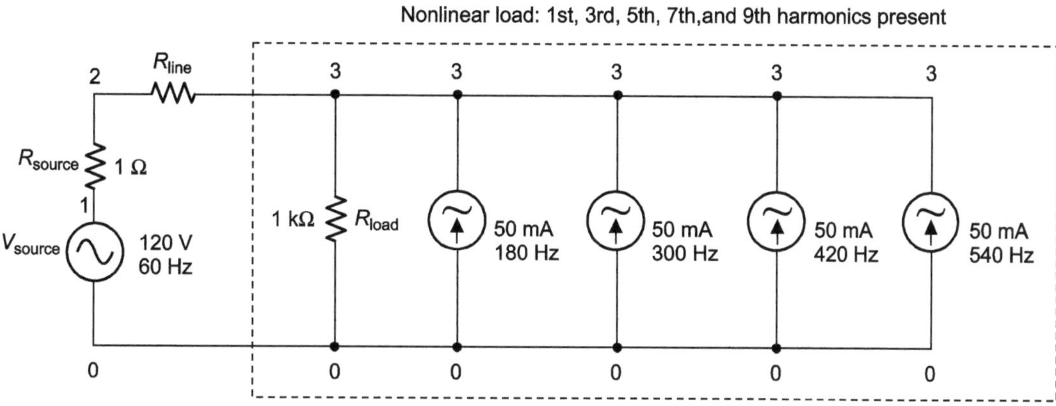

Fig. 8.4 *Total current as summing up of harmonic sources.*

added to each other in the neutral conductor of a 4-wire Y-connected load. In non-sinusoidal load systems, the triplen harmonic currents may be great enough to cause overheating of neutral conductors. This is a great nuisance, as safety concerns prohibit neutral conductors from having over-current protection, and thus there is no way for automatic interruption of these high currents.

Figure 8.5 shows the addition effect of triplen harmonic currents created at the load in the neutral conductor, ω being angular velocity of the fundamental frequency, 3ω the 3rd harmonic, 5ω the 5th harmonic, and so on.

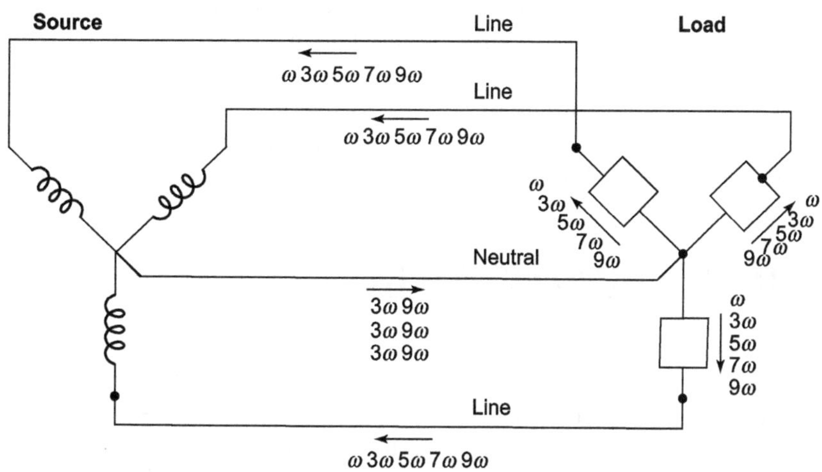

Fig. 8.5 *Triplen harmonics add in neutral conductor.*

8.2 ORIGINS OF HARMONIC DISTORTION

With AC power systems the source voltage waveform from an AC generator (alternator) is ideally supposed to be a single-frequency undistorted sine wave, without any harmonic content. It is a known fact that generator output voltage in industry or at home is invariably non-sinusoidal.

The nonlinear components draw current disproportionately to the source voltage, causing non-sinusoidal current waveforms. Distortion also occurs in transformers, whose primary winding magnetization current is usually non-sinusoidal due to the B/H saturation curve of the core, and in electric motors, when magnetic fields in the core operate near saturation levels. Even incandescent lamps generate slightly non-sinusoidal currents, as the filament resistance changes throughout the cycle due to rapid fluctuations in temperature. Harmonic disturbances are further created by loads such as:

- Arc furnaces
- Thyristor controlled drives
- Oversaturated transformers
- Uninterruptible power supplies (UPS)
- Variable frequency drives
- DC variable speed drives
- Energy saving lamps
- Welding equipment
- Nonlinear loads
- Battery chargers
- Electronic equipment with SMPS

All these loads draw highly distorted currents from the distribution system. These currents result in voltage distortions everywhere, resulting in further problems. Single-phase distorting loads like electronic ballasts and single-phase input UPS units are more troublesome since they draw triple-n harmonics: predominantly third harmonic. Triple-n harmonics add in the neutral, overload the neutral conductors and form a possible fire hazard in systems with excessive distorting loads.

These devices have become quite common in our everyday life and in industries. One way to mitigate the distortion is to install reactors for such harmonics so as not to exceed capacitor current limits. Transient over-voltage also needs to be controlled using series reactors or with internal capacitor protection.

8.3 DG SETS AS HARMONIC SOURCE

An AC voltage source is ideal if (i) its source impedance approaches zero (ii) its frequency is constant, independent of load changes (iii) its voltage is constant and independent of extraneous load conditions, and (iv) its open circuit voltage is pure sinusoidal. A Diesel Generator (DG) set is poor on all four counts. A utility supply point is much better on all the four counts. The frequency of utility supply may be considered practically constant, whereas DG set output voltage changes in response to every minor or major load change. The source impedance/reactance of a DG set for harmonic frequencies and sudden transients is usually three to four times that of a similarly rated power distribution transformer. For example a 500 kVA 11 kV/440 V distribution transformer may have 4.3% leakage impedance (500 kVA G set will have about 16% transient reactance) and for a DG set to have the same reactance as a 500 kVA transformer, it has to be 2 MVA DG set, which is impractical monetarily.

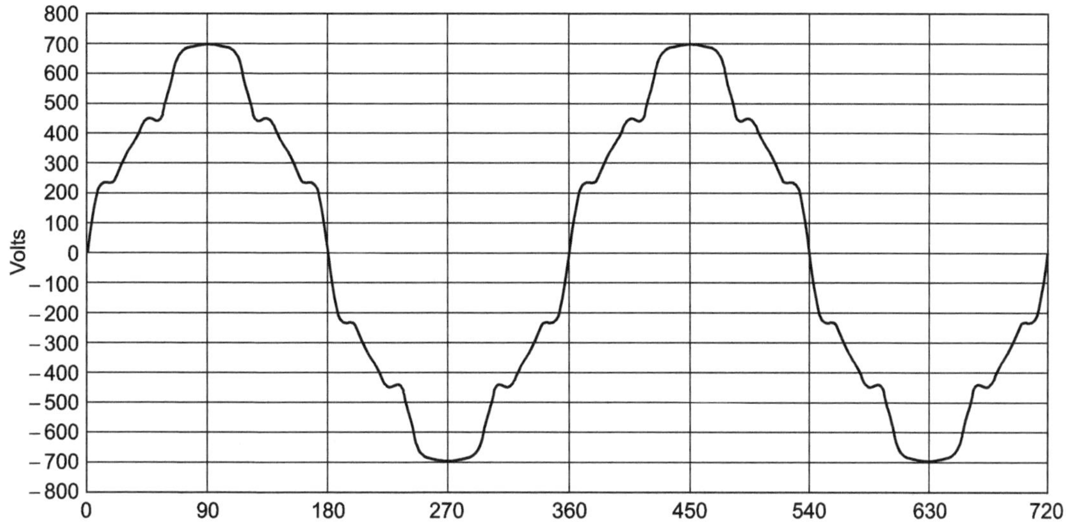

Fig. 8.6 *DG set voltage waveform.*

Large synchronous generators at utility generating stations are optimized to generate almost pure sinusoidal voltage on no load. DG sets are not. Usually DG sets are designed to eliminate the fifth harmonic; but that leaves the third harmonic in the phase voltage unattended. If the user does not connect single-phase voltage, the line voltages will not contain the third harmonic – but single-phase equipment is common and if such equipment is sensitive electronic equipment, the third harmonic content in the voltage of most DG sets is problematic.

8.4 HARMONICS DUE TO ELECTRONIC SOURCES

The ever-increasing demand of industry and commerce for stability, flexibility and accuracy of control in electrical equipment has led to the development of low-cost semiconductor control devices. With the wide use of rectifier circuits (Fig. 8.7) for UPS systems, static converters and AC and DC motor control, these modern devices have replaced older generation mercury arc rectifiers and created new and challenging conditions for system designers today.

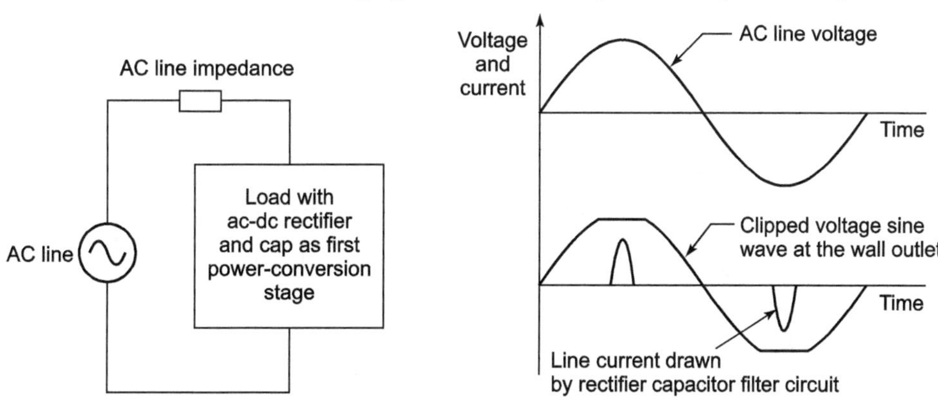

Fig. 8.7 *Harmonics from a rectifier.*

The new solid state devices have brought significant improvements in control designs and efficiency, but the harmonic currents these devices produce can cause a disturbance on the supply network, and adversely affect the operation of other electrical equipment including power factor correction capacitors.

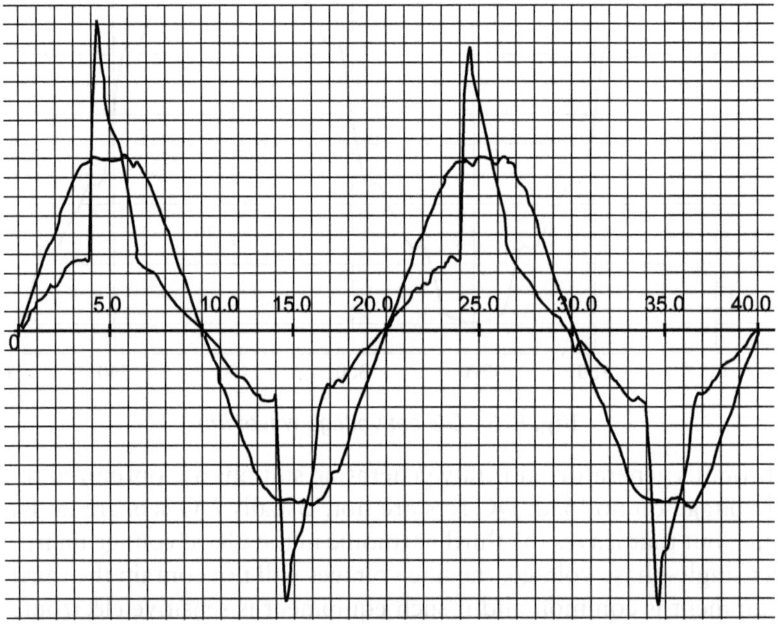

Fig. 8.8 *Most power supplies for DC devices like laptops exert a nonlinear load on the mains, their current draw often looks like this.*

In amplifiers and some other electronic equipment, the THD depends on the load or the output.

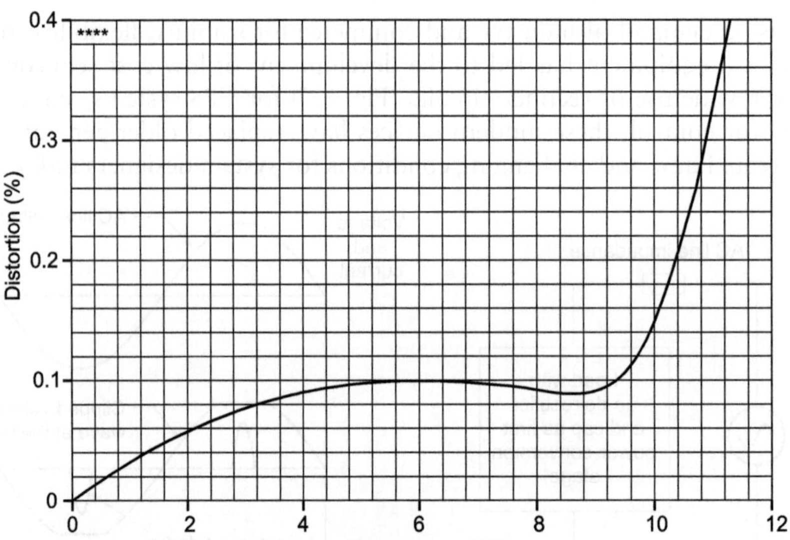

Fig. 8.9 *Total harmonic distortion plotted against output power for 10W amplifier.*

The total harmonic distortion, measured with an input signal at 1 kHz at output power of 10 Was plotted against output power in Fig. 8.9 is less than 0.2%, and nearer to 0.15%.

8.5 PROBLEMS CAUSED BY HARMONICS

- The third harmonic and multiples of the 3rd harmonic in neutral grounding systems may necessitate de-rating of neutral conductors.
- Noise that leads to erroneous operation of control system components like tripping of circuit breakers and fuses.
- Overloading of transformers, capacitors, electric motors, fluorescent lighting ballasts and other electrical distribution equipment, causing overheating and failures.
- Nuisance tripping of circuit breaker or blown fuses.
- Harmonics means additional system losses.
- Measurement errors of energy counters.
- Damage to sensitive electronic equipment.
- Malfunction of computers and other electronic equipment.
- Electronic communications interference.
- Series and parallel resonance.

8.6 HARMONIC CONTENT

Thyristors, SCR converters or rectifiers are usually referred to by the number of DC current pulses they produce in each cycle, the most common being 6 pulse and 12 pulse. Many factors can influence the harmonic content but typical harmonic currents, shown as a percentage of the fundamental current, are shown in Table 8.1. Other harmonics, though always present to some degree, need not be considered, being very small.

Table 8.1 *Typical Harmonic Content in Thyristor Output*

Order of harmonic	*Typical percentage of harmonic current*	
	6 pulse	12 pulse
1	100	100
5	20	-
7	14	-
11	9	9
13	8	8
17	6	-
19	5	-
23	4	4
25	4	4

An idea of harmonic components may be obtained from a distorted waveform component breakdown given in Fig. 8.10, showing waveform of harmonics up to 11th harmonics, and actually consisting of harmonics up to 19th harmonics.

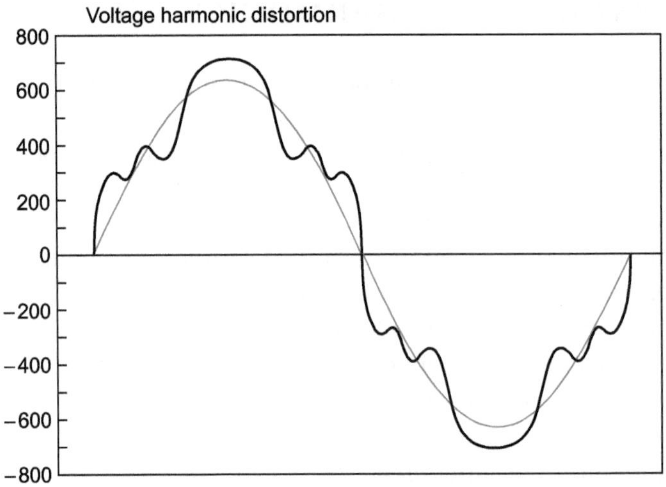

Fig. 8.10 *This very badly distorted (18% THD) voltage waveform has 15% 5th harmonic, 6% 7th harmonic, 6% 11th harmonic and 4% 13th harmonic.*

Total Harmonic Distortion (THD)

The series of harmonic components that represent a distorted waveform are often described by a single number, total harmonic distortion (THD). This is calculated as the sum of all the harmonic components (except the fundamental), divided by the magnitude of the fundamental. It will be of interest to note that an odd harmonic added to the fundamental produces two different waveforms, depending upon the phase difference at zero crossing, as shown in Fig. 8.11. The goal is to limit voltage harmonic to 3% and THD to 5%. As a guide, **THD is the square root of the sum of measured harmonic components in percentages.**

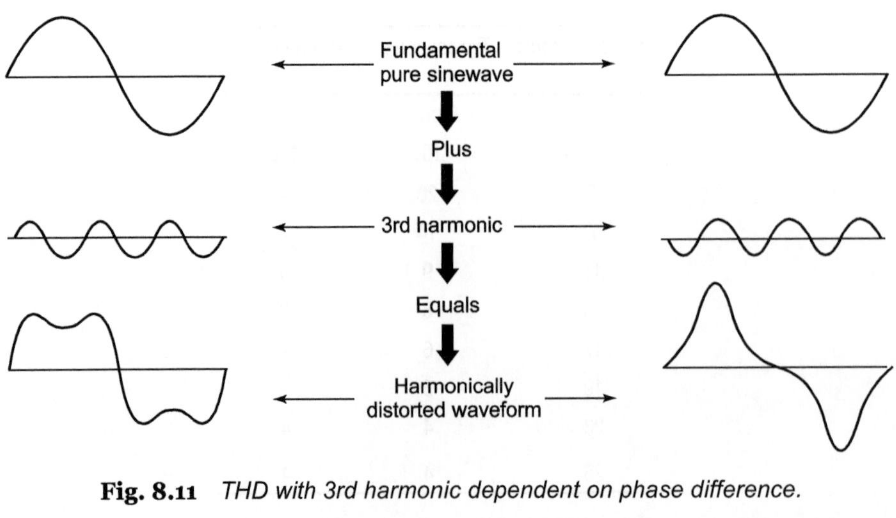

Fig. 8.11 *THD with 3rd harmonic dependent on phase difference.*

Harmonics can be a particular concern for industrial water treatment facilities, since they typically utilize variable frequency drives (VFDs) for their operations. While conventional power distribution systems have to deal with significant amounts of non-sinusoidal current, IEEE guidelines and basic software tools may be used to curtail circuits and load configurations that lead to harmonic distortion problems.

Harmonics are considered as elements of power factor because of their relationship to the power-line frequency. As Fourier components, they cumulatively represent an out-of phase current at the fundamental frequency.

8.7 HARMONIC OVERLOADING OF CAPACITORS

The impedance of a circuit dictates the current flow in a circuit. As the supply impedance is generally inductive, the network impedance increases with frequency while the impedance of a capacitor decreases. This causes a greater proportion of the currents circulating at frequencies above the fundamental supply frequency to be absorbed by the capacitor, and all equipment associated with the capacitor.

In certain circumstances, particularly near resonance, harmonic currents can exceed the value of the fundamental (50 Hz) capacitor current. These harmonic problems can also cause an increased voltage across the dielectric of the capacitor which could exceed the maximum voltage rating of the capacitor, resulting in premature capacitor failure.

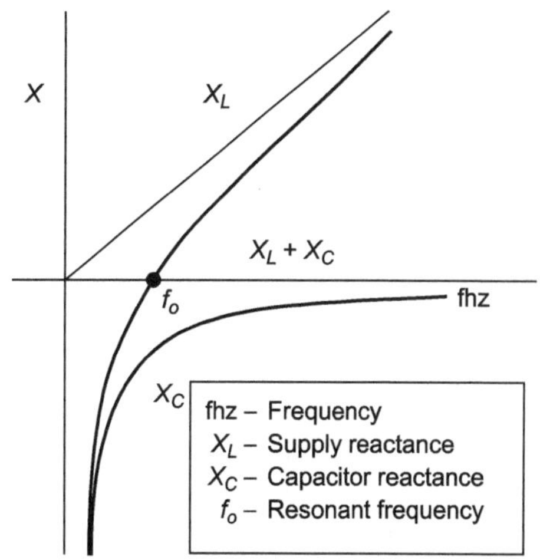

Fig. 8.12 *Harmonic resonance frequency f_o [$X_L = X_c$].*

Harmonic Resonance

The circuit selective resonant frequency is reached when the capacitor reactance and the supply reactance are equal. Whenever power factor correction capacitors are applied to a distribution network, which combines capacitance and inductance, there will always be a frequency at which the capacitors are in parallel resonance with the supply.

If this condition occurs on, or close to, one of the harmonics generated by solid-state control equipment, then large harmonic currents can circulate between the supply network and the capacitor equipment. These currents are limited only by the damping resistance in the circuit. These will add to the harmonic voltage disturbance in the network, causing an increased voltage distortion. This results in a higher voltage across the capacitor and excessive current through all capacitor components. Resonance can occur on any frequency, but in general, the resonance we are concerned with is on, or close to, the 5th, 7th, 11th and 13th harmonics for 6 pulse systems.

8.8 THE IEEE 519-1991 STANDARD

The Institute of Electrical and Electronics Engineers (IEEE) has defined acceptable limits of current and voltage harmonic distortion for various types of systems. Voltage distortion in general systems should be limited to 5% THD, according to the standard. Special, critical applications should limit the THD to 3%.

IEEE 519-1991 (*IEEE Recommended Practices and Requirements for Harmonic Control in Electrical Power Systems*) defines current distortion limits for various ratios of source short circuit current (I_{SC}) to load current (I_L) – (The short circuit current for a transformer may be obtained by dividing the full load current [A] by the output impedance [%Z].) The standard further limits total current harmonic distortion to 5% of the full load current of a transformer, unless the transformer is specifically designed for operation under loads with higher harmonic content. Many users specify IEEE voltage and current harmonic distortion limits of 5% for their entire power system. (Or, for large systems, a complete branch served by a substation transformer.) They provide details of the proposed installation, including the following:

- Source transformer KVA and %Z (or source fault current),
- Distance from the source transformer to the PCC and the conductor type (busway or cable),
- Distance from the Point of Common Connection (PCC) to the VFD and the conductor type (busway or cable),
- Number and horse power (HP) of VFDs for connection to each PCC (usually the metering or incoming service point, or the point in a plant where the equipment is connected)
- Information about the harmonics generated by other loads connected to the same PCC.

Voltage distortion limits established by IEEE-standard 519 are:

- 3% THD-v for hospitals and airports,
- 5% THD-v for general systems, and
- 10% THD-v for dedicated systems (where 100% of the load is nonlinear).

A simple one-line diagram of the VFD system can provide this information for analysis. Most VFD manufacturers have computer programs that can use this data to determine whether reactors, isolation transformers or phase shifting transformers are required for calculated compliance with IEEE 519-1991.The golden rule is **"You shall not draw so much distorted current that you upset the voltage waveform".**

- A utility must limit the **VOLTAGE** THD at PCC to under 5%
- A consumer must limit usually the **CURRENT** THD to below 5%

8.9 HARMONIC MITIGATION TECHNIQUES

[A] Harmonic analysis

The first step in solving harmonic problems is an analysis to determine the specific nature of the electrical distribution system, to establish the impedance of the supply network and the value of each harmonic current. A detailed computer analysis can then decide the capacitor, reactor and filter bank equipment needs. Three options are available to solve the problems:

1. Use of correct amount of capacitance (KVAR) in the network to avoid resonance with the source – it may be difficult, especially in automatic systems, as the capacitance is always changing. This solution usually implies connecting smaller capacitance to the system than is optimally needed for power factor correction.

2. Install reactors in series with capacitors to lower the resonance below critical order harmonics; i.e. 5th, 7th, 11th and 13th. This design tunes the resonant frequency of the system well below the critical harmonic and is called an anti-resonance bank. The capacitors then operate in a harmonic environment.
3. Filters are recommended if the harmonic distortion is above the limits recommended in IEEE 519, "Guide for Harmonic Control and Reactive Compensation of Static Power Converters". (The recommended limits for voltage distortion in IEEE 519 are presently 5% for general applications.) Tuned filters of proper size can reduce the harmonic distortion at critical frequencies and give the benefits of correcting the power factor and improving the network power quality.

Computer techniques for harmonic analysis

a) Simple radial networks can sometimes be analyzed utilizing simple calculations, though calculations become more cumbersome as the circuit complexity increases. Most common computer techniques are based on nodal admittance equations for the network.
b) Commercially available software tools facilitate harmonic analysis of complex systems and loads, and provide a graphical interface to build a variety of circuit types and multiple independent systems of different voltage levels. They also contain large amounts of data for different load and supply systems in memory, which make calculations easier.

[B] Harmonics attenuation

Harmonic current distortion is affected by the circuit impedance. Placing the same harmonic producing load at different nodes in a power system will result in different levels of distortion. This attenuation effect is used as one method of passive harmonic mitigation. Introducing a series line reactor (or inductor) at the terminals of a 100 HP pulse width modulated (PWM) adjustable speed drive (ASD) can lead to a reduction of total harmonic distortion associated with the ASD from about 81% to 38%. The THD of voltage and current can be calculated using the relations

$V_{THD} = [\sqrt{\Sigma V_h^2/V_1^2}]$ $\qquad$ $I_{THD} = [\sqrt{\Sigma I_h^2/I_1^2}]$.

The ASD operation is not adversely affected, provided the line reactor chosen for the application does not exceed approximately 5% impedance relative to the drive base.

[C] Harmonics cancellation

Cancellation of harmonics can also occur because individual harmonic components of a distorted current are affected differently when passing through normal power system transformers. The magnitude of harmonic currents, like the 50 Hz component, increases or decreases consistently with the transformer turns ratio.

The phase angle of harmonic components, however, is influenced by the type of connection of the three-phase transformer. The 5th and 7th components, for example, experience a 30° phase angle shift through a power system transformer connected delta-wye, as compared with the same current components transmitted through a wye-wye or delta-delta connected transformer.

[D] Reduction of Harmonics

Harmonics can be reduced by tuned filter or passive filter. A tuned capacitor bank consists of a series circuit of capacitor(s) and specific filter circuit reactor, similar to a detuned unit. The difference is the resonance frequency. Tuned filters are tuned very close to harmonic frequencies. The respective harmonic is absorbed by filters tuned to that particular frequency. Capacitors and reactors used for tuned filtering are specially made to withstand heavy harmonic current. Passive LC filters are used on systems with electronic power factor controllers, which are a source of harmonics due to their thyristor circuits.

When the voltage distortion rises sufficiently, variable-speed drives can turn off and disrupt operations. Also, fuses will blow causing total disruption, and capacitors occasionally fail catastrophically, and I^2R losses increase considerably. The highest squared value with harmonic currents is 2.48 times larger than that highest squared value without harmonic currents. This amounts to substantial instantaneous heating differences, resulting in blown fuses and capacitors – not intended when the capacitors were installed.

[E] Avoiding resonance

There are a number of ways to avoid resonance when installing capacitors. In larger systems it may be possible to install them in a part of the system that will not result in a parallel resonance with the supply. Varying the KVAR output rating of the capacitor bank will alter the resonant frequency. With capacitor switching there will be a different resonant frequency for each step. Changing the number of switching steps may avoid resonance at each step of switching.

Transient problems are usually solved by installing suppression or isolation devices such as surge capacitors or isolation transformers. These devices will help solve the transient problems but will not affect the mitigation of low order harmonics or solve harmonic resonance problems.

A capacitor can be protected from overloading due to harmonics using detuned reactor. A detuned capacitor bank consists of a series circuit of capacitor(s) and a specific filter circuit reactor. The resonance frequency of a detuned bank is heavily detuned, mean it is not close to any existing harmonic. The purpose of these capacitors is specifically to avoid tuning to any frequency where harmonics may be present. Hence these reactors are termed 'detuned reactors' and the bank as a whole is called a detuned bank. The basis for calculation of detuned capacitors is given in Appendix H.

[F] Overcoming resonance

If resonance cannot be avoided, an alternative solution is required. A reactor must be connected in series with each capacitor such that the capacitor/reactor combination is inductive at the critical frequencies but capacitive at the fundamental frequency. To achieve this, the capacitor and series connected reactor must have a tuning frequency below the lowest critical order of harmonic, which is usually the 5th. This means the tuning frequency is in the range of 145 Hz

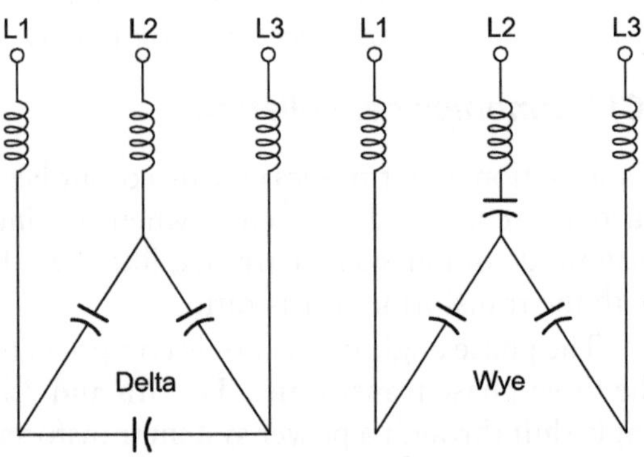

Fig. 8.13 *Reactor connections.*

to 220 Hz, although the actual frequency will depend upon the magnitude and order of the harmonic currents present.

The addition of a reactor in the capacitor circuit increases the fundamental voltage across the capacitor. Therefore, care should be taken when adding reactors to existing capacitors. It is usual to increase the delta connected capacitor rated voltage to 525 V for 440 supply voltage when used with reactors.

[G] Reduction of harmonic distortion

Harmonic currents can be significantly reduced in an electrical system by using a harmonic filter. Basically, a filter consists of a capacitor connected in series with a reactor tuned to a specific harmonic frequency. In theory, the impedance of the filter is zero at the tuning frequency; therefore, the harmonic current is absorbed by the filter. This, together with the natural resistance of the circuit, means that only a small level of harmonic current will flow in the network.

Choice of filters

The effectiveness of any filter design depends on the reactive output of the filter, tuning accuracy and the impedance of the network at the point of connection. Harmonics below the filter tuning frequency will be amplified. The filter design is important to ensure that distortion is not amplified to unacceptable levels. Where there are several harmonics present, a filter may reduce some harmonics while increasing others.

A good harmonic filter reduces the harmonics to less than 5% and overcomes all the problems related to it. Experience is extremely important in the design of such filters to ensure the most efficient and cost effective solution, while ensuring no adverse interaction between the system and the filter. Consequently, it is often necessary to use a multiple filter design where each filter is tuned to a different frequency, as shown in Fig. 8.14.

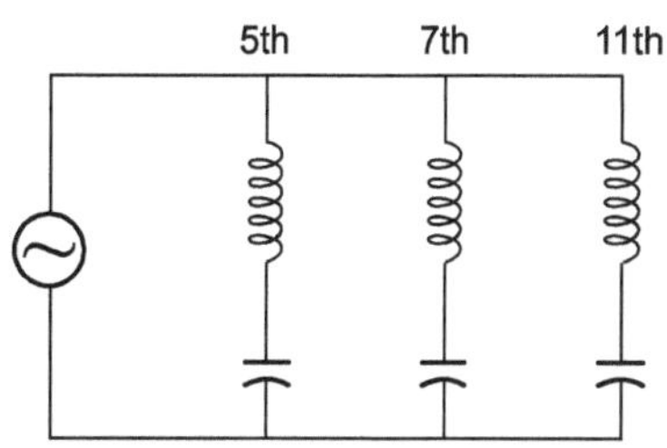

Fig. 8.14 *Filters for individual harmonics.*

[H] Load alteration

Whenever load expansion is considered, the network may change and existing filter equipment has to be evaluated with the new load condition. Two or more filters tuned to the same frequency are not advised on the same distribution system. Slight tuning differences may cause one filter to take a much larger share of the harmonic distortion. Or, it may cause amplification of the harmonic order which the equipment has been designed to reduce. When there is a need to vary the power factor correction component of a harmonic filter, careful consideration of all load parameters is necessary.

[I] A new approach – harmonic mitigating transformers (HMTs)

These are a relatively new design tool for transformers being used by some power quality professionals (including Reliant Energy, Houston, Texas) to rid commercial and industrial facilities of the power quality problems associated with harmonic currents. HMTs have the ability to attenuate harmonic currents and thus alleviate power quality problems. In addition, by cancelling certain harmonic currents, HMTs can reduce the energy losses that harmonics would otherwise cause in conventional transformers. These energy savings can offset the premium price for HMTs (which can be as much as 200%) and yield an attractive payback period. These are available in capacities from 15 to 500 kVA. All HMTs employ one or both of two approaches to combat harmonics, each of which addresses different types of harmonics. The selection of the appropriate type of HMT depends on which harmonics are present in the electrical distribution system being treated.

(i) Single-phase loads

Single-phase electronic loads generate harmonics at all odd multiples, the most prominent being the triplens. In conventional transformers, triplen harmonics are transferred to the primary (delta) winding, where they are trapped and circulate continuously. The distribution system upstream of this transformer is thus spared from having to supply triplen harmonics, but the harmonic currents cause excessive losses in the transformer. HMTs attenuate triplen harmonics by using a **"zig-zag" winding on the transformer secondary:** this is a design that places half of the turns of each phase of the secondary around two of the legs of the transformer core (in the standard design, all turns for a given phase go around just one core leg). This technique causes cancellation of the magnetic flux established by triplen harmonic currents, so little or none is transferred to the primary windings.

(ii) Three-phase loads

Harmonic problems in industrial facilities dominated by three-phase loads arise from currents flowing at the 5th, 7th, 11th, or even higher order harmonics. For these harmonics, HMTs use either **dual secondary windings or pairs of transformers** to achieve substantial attenuation of one or two of the most problematic frequencies. In either design, the two secondaries are electrically phase-shifted relative to each other. The degree of relative phase shift is selected such that the targeted harmonic currents from one secondary are close to or exactly 180° out of phase with the targeted harmonic currents from the other secondary, and thus they cancel each other.

HMTs reduce electricity costs in two ways. They reduce losses directly by minimizing harmonic currents and their related losses in the transformer primary. If the HMT under consideration has a more efficient core and/or windings than the transformer it is being compared to, then direct losses at 50/60 Hz will be reduced as well, with additional cost savings. HMTs may also contribute indirect savings if the transformers are to be located in air-conditioned space. Approximately 1 kilowatt-hour (kWh) of cooling system energy is required for every 3 kWh of heat removed. For HMTs installed in air-conditioned space, energy savings are therefore increased by about 33%.

8.10 CAPACITOR STRESS ANALYSIS

Electronic modules are available for the stress analysis of all power capacitors installed in the network, including those incorporated in filters. The analysis reports the harmonic currents and voltages of each capacitor as well as the total reactive power, RMS current, RMS voltage and peak voltage. These quantities are compared to user-defined limits and any capacitor that violates any of those limits is reported and highlighted on a one-line diagram of the network.

This includes the following:

- Bar chart plots for voltage and current distortion versus harmonic order or frequency.
- Time waveform plots.
- Impedance magnitude and phase plots versus frequency, for resonance and de-tuning analysis.
- R-X plots.
- Sensitivity analysis plots.
- Possibility to plot multiple results on the same graph.

8.10.1 VFD Applications

Another software tool is available to facilitate a common harmonic evaluation task: application of VFDs to an existing low-voltage radial power system. While this tool is not accurate for complex systems, or for systems with power factor correction capacitors or harmonic filters, it is widely applicable for evaluating VFD applications. Many designers use it for estimating the effect of variable speed drives (VSD) on a power system, as also a means of evaluating certain mitigating devices like line reactors and drive-isolation transformers, delta-wye transformer connections, and broadband filters.

Harmonics and smaller drives

The larger the VSD, the greater are the harmonics. However, a 250-hp drive is almost sure to attract much more attention from engineers than a 5-hp or smaller drive. Though this makes sense considering economics and effect on operations, there are times when a careful analysis of smaller horsepower drives becomes necessary.

On single-phase systems, drives will generate 3rd and, to a lesser degree, 5th harmonics. In commercial buildings, on four-wire systems (3-phase conductors and a neutral conductor), they contribute to the 3rd harmonic current adding up in a shared neutral. This may necessitate installing neutral conductors twice the size of phase conductors. As small drives add their contribution to the total harmonic load, any line measurements (particularly current measurements) require a true RMS meter for accuracy. In the presence of harmonics, average responding current meters can be inaccurate by as much as 40%.

On 3-phase systems, the 5th will be the predominant drive-generated harmonic. The 5th harmonic is a negative sequence harmonic: it creates reverse torque that will tend to make motors turn backward. The 5th harmonic affects motors with across-the-line mechanical starters. The across-the-line motor, driven by the much larger fundamental current, will still turn forward, but the 5th will cause additional heating and, over time, can be extremely damaging to the stator insulation. If a drive shares a bus with an across-the-line motor, such as in a motor control centre, it could damage the motor.

Note that this 5th harmonic will probably have virtually no effect on the upstream distribution system (i.e. cause minimal voltage distortion upstream), because its harmonic current is such a negligible portion of the total. But at the local level, where source impedance is at its highest, a bank of low-horsepower drives can cause enough voltage distortion at the local point of common coupling (PCC) to affect the motor loads that share that PCC.

A 3-phase VSD draws nonlinear current, which injects harmonics into the power system. The 5th harmonic will be the predominant harmonic generated by the drive.

The first line of defence against harmonics should be in the drive itself. A reactor coil, sometimes called a link inductor, is integrated into the DC link of many drives and tends to reduce current distortion on the line side of the drive. It also protects the drive from transient over-voltages (notably capacitor switching transients) that can travel to the DC link and cause DC over-voltage trips.

In some lower-cost drives, manufacturers cut costs by eliminating the reactor coil, making the drive, in effect, a 'harmonics generator'. This is especially critical when you install the drive on a bus with an across-the-line motor. In this instance, you can correct the situation by installing input line reactors or isolation transformers.

8.10.2 Motor Compatibility and Smaller Drives

VSDs can also create motor compatibility problems, especially when retrofitting drives to older motors. The high-speed switching of the insulated gate bipolar transistor (IGBT) in conjunction with long cable runs can cause over-voltage reflections (also known as standing-wave voltages or peak-to-peak or corona voltages) with peak voltages two to three times the DC link voltage.

Many drive manufacturers will specify cable not to exceed 30 m, but sometimes even this can be too long. These over-voltages tend to puncture insulation on the first few windings of the motor, causing premature failure of stator insulation. This is a problem common to high- and low-horsepower drives with PWM outputs, but low-cost, low-horsepower motors are especially vulnerable. Their stator windings are often random-wound, a less costly manufacturing process, but one that could create a high potential between adjacent wires, making them that much more vulnerable to overvoltage reflections.

At one time, low-pass filters were commonly placed at the drive output to reduce the over-voltages. In recent years, however, manufacturers have designed inverter duty motors rated at 1500 V specifically to withstand over-voltage reflections. Many drive manufacturers now specify use of inverter duty motors with their drives.

If VSDs are generating harmonics, the first thing to do is take measurements at key points in the distribution system to determine the present level of harmonics. Measurements of harmonic distortion of waveforms as well as of individual harmonics may be taken with handheld power quality analyzers, or we may view over-voltage waveforms with handheld oscilloscopes.

8.10.3 The Advantages of VSDs

Engineers have been able to cram more horsepower into smaller, more compact drives, to the point that low-horsepower drives now have more or less the same footprint as mechanical starters.

Hitherto, heat dissipation needs made it difficult to have smaller drives. A drive first converts the AC sine wave into DC and stores the power in DC link capacitor banks. The inverter then switches the DC to create a PWM signal of desired frequency. This process was a prime generator of heat.

In newer drives an IGBT has taken over the switching. Recent IGBT designs have greater current carrying capacity. IGBT switching speeds have become pretty fast, of the order of 100 to 200

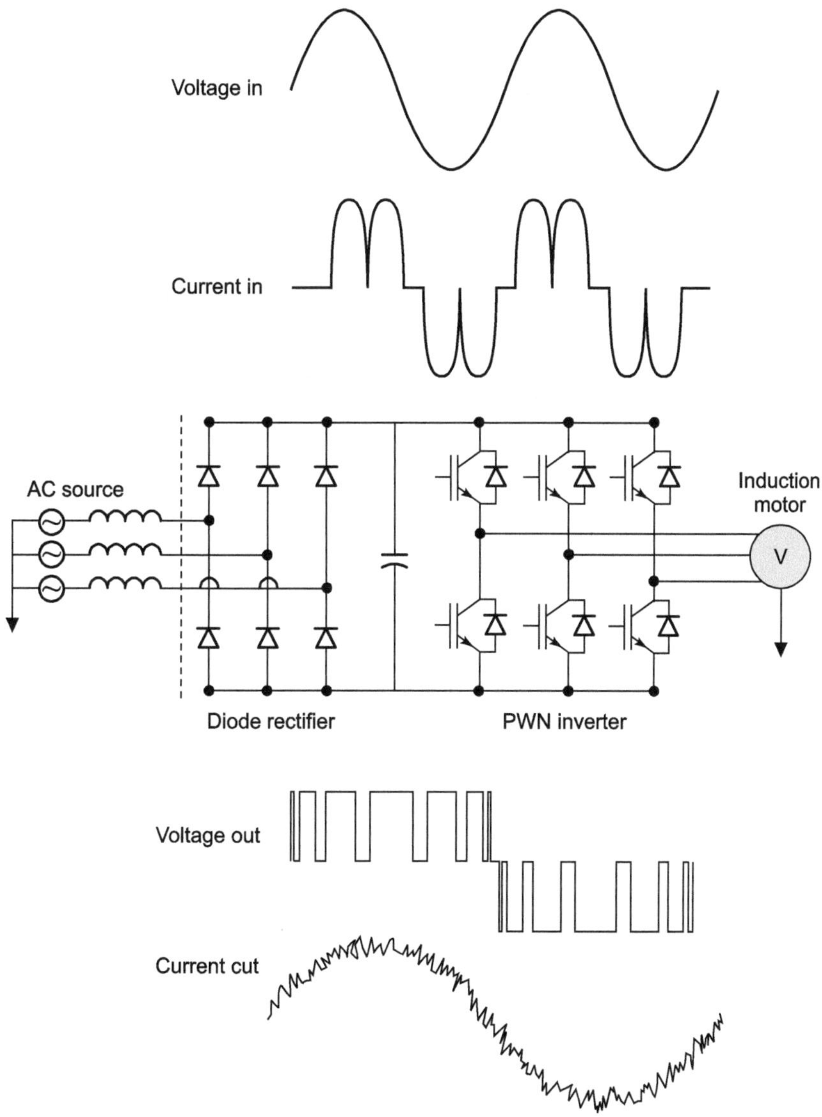

Fig. 8.15 *Current smoothing in PWM drive.*

nanoseconds. Except for a very small voltage drop across the semiconductor, these switches do not consume energy – the only wastage is by generating heat in the transition off to on and vice versa.

Higher switching speeds result in less heat loss and increased efficiency. The result is smaller heat sinks and fans and more compact drives. In addition to energy savings, drives offer many other benefits that contribute to the stability and robustness of the electrical distribution system:

- No inrush currents – they are typically limited to 110% of rated current. The inrush currents associated with starting motors can cause voltage dips or unwanted tripping of the motor. Soft start drives generally ramp a motor and load up to speed in about 20 to 30 s.
- High power factor, eliminating the need for power factor correction capacitors.
- Switching transients (spikes) are isolated when a motor is turned off (ramp down a motor over 20 to 30 s). At the final turn-off, the motor is at low-speed and low-current, the small spike is easily absorbed within the DC link section of the drive itself.
- Programmable motor control, protection, and even communication functions that are far beyond mechanical starter capacity. The drives may be programmed to reverse motor rotation, eliminating the need for an additional contactor.
- It is possible to replace single-phase motors with more rugged three-phase motors. This is because the VSD can accept a single-phase line-side voltage and output a three-phase signal on the load (motor) side. In other words, drives 'transform' single-phase to three-phase voltage.

9

POWER QUALITY MANAGEMENT

9.1 POWER FACTOR – A NEW DEFINITION

Power factor correction using KVAR capacitors is an established principle of power quality management. A number of systems are available for this purpose, and lately the use of capacitor switching contactors, timing devices and other improvements have made these systems more reliable than ever. However, advancement in technology and increasing use of electronic equipment has created new challenges and complexities for utilities and industries in ensuring reliable good quality power availability.

Modern systems of speed control and voltage control frequently use thyristor drives, which use chopping circuits, causing highly distorted wave shapes (Fig. 9.1). The system uses complicated electronic systems to correct the wave shape to near sine wave, and at the same time, keep the power factor near unity. Any study of power management today has to cover both these factors. Two common examples of distortion from electronic sources are given below.

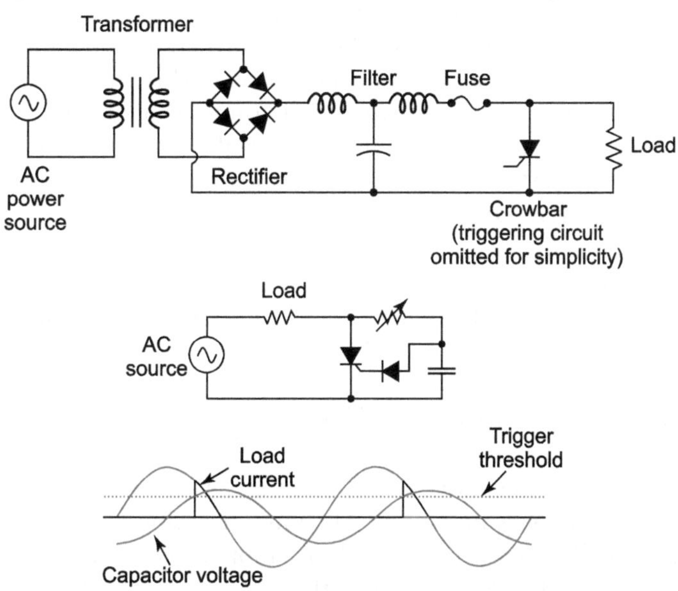

Fig. 9.1 *Two common DC and AC control circuits and capacitor voltage waveform.*

With electronic circuits, even the concept of power factor undergoes a change. Imagine that a load always draws exactly ±1 A, with the same polarity as the voltage but with a magnitude independent of the voltage, as in Fig. 9.2. With this load the voltage is 1 V RMS, and the current is 1 A RMS, giving an input of 1 VA. The average power is 0.9 W ($1.414 \times 2/\pi$). And yet, there is no negative power and the load current is in phase with the voltage. The power factor of this circuit is 0.9, yet there is no phase shift.

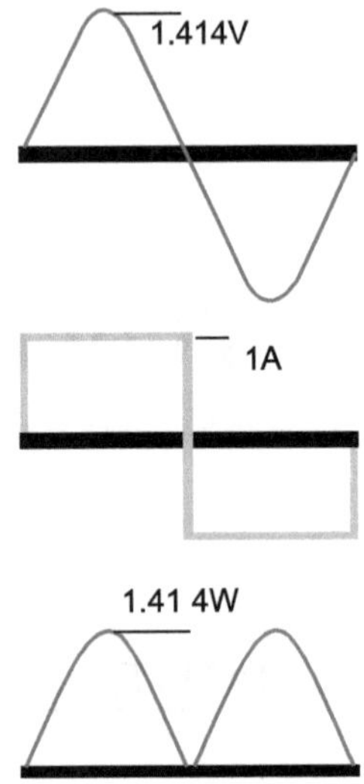

Fig. 9.2 *Power factor is not unity, even if V and I are in phase.*

The culprit is the shape of the current waveform, which is a square wave. A square wave can be shown to be the summation of a whole number of waveforms at different harmonic frequencies. In a square wave with amplitude 1 V (2 Vpp), the amplitude of the fundamental frequency is 0.9 V. Hence where harmonics are present in the load current (but not in the voltage) the harmonic currents contribute nothing to the power in the load but do reduce the power factor. A similar situation is really likely to occur in the real world. In fact, the only loads on the mains which are not likely to introduce harmonic currents are purely resistive loads such as heaters. In such situations, improvement of power factor involves reduction in such harmonics by countering their effects.

The conventional definition (PF = cos ϕ) is no longer valid for power systems of today. This definition presumes ideal sinusoidal signals for both current and voltage waveforms, which does not apply in most cases. With the presence of harmonics becoming commonplace, and to take care of situations like the above, a new definition of power factor (distortion power factor, DPF) has emerged which is applicable in such cases:

$$\textbf{Distortion Power Factor} = 1/\sqrt{(1 + THD_i^2)} = I_{1rms}/I_{rms}$$

THD_i is the total harmonic distortion of the load current. This definition assumes that the voltage stays undistorted (sinusoidal, without harmonics). This simplification is often a good approximation in practice. $I_{1,rms}$ is the fundamental component of the current and I_{rms} is the total current – both are root mean square-values. The result when multiplied with the DPF is the overall, true power factor or just power factor (PF), calculated as:

$$PF = DPF\ (I_{1rms}/I_{rms})$$

While conventional power factor correction systems, like directly connected shunt capacitors or automatic power factor control (APFC) panels measure and control DPF, new generation active control systems aim at unity PF, where current follows the sinusoidal voltage waveform as closely as possible.

9.2 POWER FACTOR INSTRUMENTS

An important realization from the above observation is that cos ϕ and PF are not synonymous. It depends on how power factor is measured. The misconception about cos ϕ is so deep rooted that many meters are actually marked 'cos ϕ' (it looks more technical than 'power factor'). Many expensive instruments do not define exactly how they measure power factor. An instrument may merely measure the phase angle using zero crossing detectors. We may trust an instrument if we know exactly what it measures. An advanced new generation instrument from companies like Fluke analyses the voltage and current waveforms. It displays harmonics of current and voltage up to 31st harmonic. These meters define two separate power factor measurements: DPF and PF. DPF is the cosine of the phase angle between the fundamental frequency components of voltage and current. It mathematically extracts the fundamental waveforms from the inputs and then measures the phase angle between them. PF is the true ratio between VA and power with all harmonics factored in.

In recent years a number of products have been offered online that claim to reduce domestic electricity bills by up to 10–20% by correcting the overall power factor of a private home. These are scams. Domestic electricity users are billed only for real power (watts) not VA because that is what domestic power meters can measure.

9.3 APFC PANELS

The simplest way of improving power factor is by connecting capacitors at load point and/or supply mains, and their manual switching as and when necessary, while keeping an eye on power factor meter reading. However, this is only possible for small and simple systems, where such control is feasible. In most medium and large industries, this control is through use of automatic check on power factor and taking self-corrective actions by switching capacitors in and out through predetermined system. Such a system is usually supplied in a control panel, complete

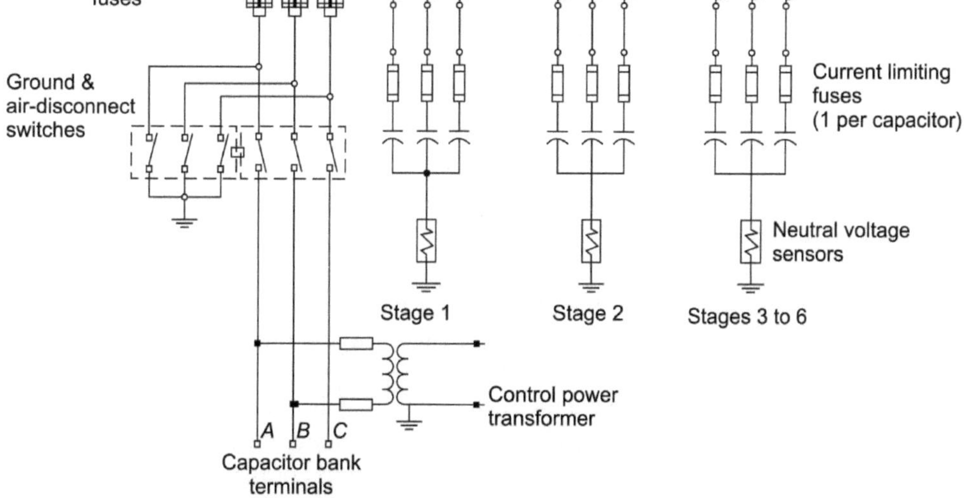

Fig. 9.3 *Typical APFC panel circuit.*

with capacitors and necessary reactors and other accessories, called 'automatic power factor control panel', or APFC for short.

A load and its KVAR are in a dynamic state. Hence matching KVAR output of a capacitor bank must also be dynamic i.e. must adjust itself instantly to its requirement, if one is to obtain a uniform set power factor all along. An APFC panel controls the load power factor by sensing various available parameters.

A power system capacitor connection system in an APFC panel is outlined in Fig. 9.3. The system contains one directly switched capacitor, and up to six KVAR capacitors/banks. The panel senses the power factor and determines capacitor requirement to add or subtract lines one to six in a finite sequence to maintain the power factor above a predetermined value. The panel is usually housed indoors, and contains all the sensing, controlling and protective devices. The general set up of an APFC panel is shown in Fig. 9.4.

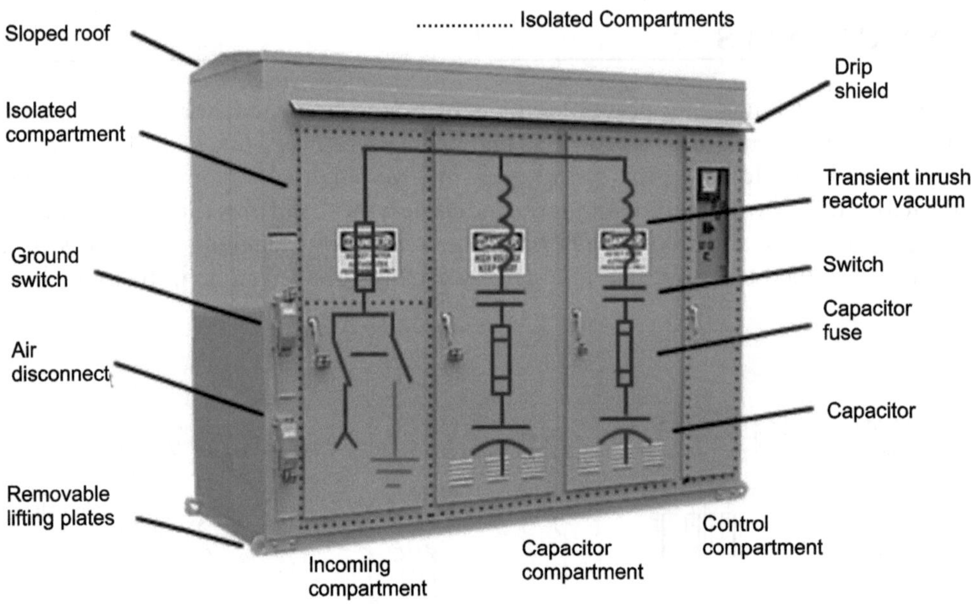

Fig. 9.4 *APFC panel components.*

[A] Sensing parameters

The first step for an APFC panel action is sensing the relevant load system parameters. Several systems are possible as outlined below.

1. **Current-sensing based APFC:** The current magnitude through a feeder or bus is sensed and fed to a relay. As this magnitude crosses a set bandwidth, the relay operates a section of a capacitor bank. This is the simplest and the cheapest relay. Its disadvantage lies in functioning with no reference to the actual load power factor – but assuming it.
2. **Power factor sensing based APFC:** This is the most widely used system. The relay senses the start of the voltage and current waveforms on a feeder and measures the time difference between them. It then converts this into power factor and compares this with a set value. Based on this difference, it operates the power contactor. Its advantage lies in its ability to

show the load power factor on an indicating meter. Its drawback is that it has no relation to the load magnitude and its KVAR requirement. This can lead to severe fluctuations, called hunting.

3. **KVAR sensing based APFC:** This is the most sensitive relay, capable of obtaining maximum benefit out of a given capacitor bank. The relay senses magnitudes of both the voltage and current waveforms and also the time or phase difference between them. It then calculates the load KVAR and computes a possible combination of sections within a capacitor bank and operates their contactors to add the required capacitor KVAR to the system. However, this may put undue stress on the contactors and related surge suppression attachments.

[B] Determining APFC panel design and size of capacitor blocks

1. Power factors for the purposes of levying penalties are based on the monthly consumptions of KVAh, kWh and KVARh as recorded on a tri-vector meter. If the basic purpose of installing capacitors is to stay safely above the penalty limit, then average power factor correction based on a 24h basis is sufficient. This helps in setting as wide a bandwidth as possible before changing a step. It prevents switching too often.

   It must be noted that KVAh and KVARh do not subtract if excessive capacitor KVAR are dumped into the system by a leading capacitive load a part of the time. It records this also as a low power factor and is subject to penalty.
2. A simple method of sizing the capacitor blocks is to divide them equally into a targeted number of steps. Besides simplicity it has an advantage of standard sizes for replacement of worn out contactors, blown fuses etc. This design is generally favoured.
3. In an ambitious method of sizing the blocks, they are designed in a binary sequence so that a large number of combinations are available for a given set of contactors etc. If the accessories are chosen properly, this can be an ideal method though slightly costlier than method (2) above.
4. It is advisable to keep as much capacitor KVAR out of the APFC control as possible, for example, the first step (load portion which is constant on a 24h basis), like continuous process industries. Capacitors may also be connected at the points of individual loads.

The remaining capacitors are then divided in a number of steps. The number of steps is kept as small as possible, by studying the load pattern. The portion that is likely to be operated often should be at the fag end. Large size contactors should be at the starting end so that they operate as few times as possible.

Modern electric power utilities face many challenges due to ever increasing complexity in their operation and structure. Voltage stability is the ability of a power system to maintain adequate voltage magnitude so that when the system nominal load is increased, the actual power transferred to that load will increase. The main cause of voltage instability is the inability of the power system to meet the demand for reactive power, when the voltage decays to a level from which it is unable to recover. This may lead to partial or full power interruption in the system.

Electrical utilities and heavy industries face a number of challenges related to reactive power. Heavy industrial applications can cause phenomena like voltage unbalance, distortion or flicker on the electrical grid. Electrical utilities may be confronted with phenomena of voltage sags, poor power factor or even voltage instability. Reactive power control can resolve these issues.

Shunt capacitors are relatively inexpensive to install and maintain. Installing shunt capacitors in the load area or at the point that they are needed will increase the voltage stability. However, shunt capacitors have the problem of poor voltage regulation and, beyond a certain level of compensation a stable operating point is difficult. The reactive power delivered by the shunt capacitor being proportional to the square of the voltage, VAR support drops during low voltage conditions thus compounding the problem.

The use of a shunt capacitor may lead to an unacceptable voltage magnitude in normal operation, and the amount of reactive power delivered is mostly dependent on the voltage magnitude. Hence, it may increase the power transfer capability but will not improve voltage stability, compared to static VAR compensators (SVC) and static synchronous compensators (STATCOM). The SVC and STATCOM can do a much better job, improving voltage stability while keeping the voltage magnitude in the acceptable region. Shunt connected controllers at distribution and transmission levels usually fall under two categories – static synchronous generators (SSG) and SVC.

9.4 POWER QUALITY MANAGEMENT SYSTEMS

(i) An SSG is defined by IEEE as a self-commutated switching power converter, supplied from an appropriate electric energy source and operated to produce a set of adjustable multiphase voltages, which may be coupled to an AC power system for the purpose of exchanging independently controllable real and reactive power. When the active energy source (like a battery bank) is dispensed with and replaced by a DC capacitor which cannot absorb or deliver real power except for short durations, the SSG becomes a STATCOM. STATCOM has no long-term energy support in the DC side and cannot exchange real power with the AC system. However it can exchange reactive power, as well as harmonic power with the AC systems.

(ii) STATCOMs are deployed at both distribution and transmission levels. STATCOM, when employed at the distribution level or at the load end for power factor improvement and voltage regulation alone, is called DSTATCOM. When it is used for harmonic filtering in addition or exclusively it is called shunt active power filter. In the transmission system, STATCOMs handle only fundamental reactive power and provide voltage support to the system and also modulate bus voltages during transient and dynamic disturbances in order to improve transient stability margins and to damp dynamic oscillations.

(iii) Flexible AC transmission system (FACTS) is a system composed of static equipment used for the AC transmission of electrical energy. It is meant to enhance controllability and increase power transfer capability of the network. It is generally a power electronics-based device. A FACTS is defined by the IEEE as "a power electronics-based system and other static equipment that provide control of one or more AC transmission system parameters to enhance controllability and increase power transfer capability."

9.5 STATIC VAR COMPENSATOR (SVC)

An SVC is an electrical device which provides fast-acting reactive power on high-voltage electricity transmission networks. The SVC uses conventional thyristors to achieve fast control of shunt

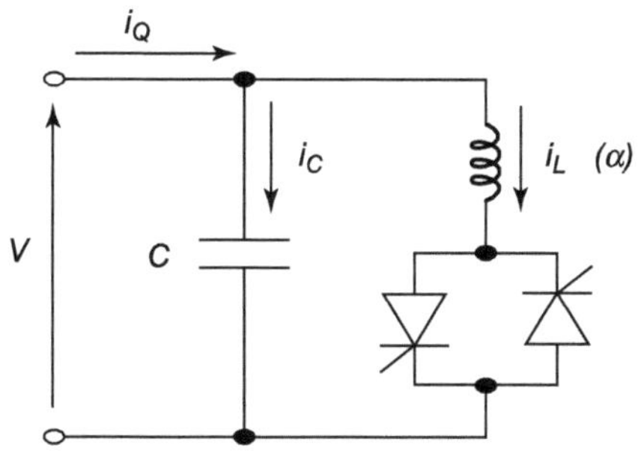

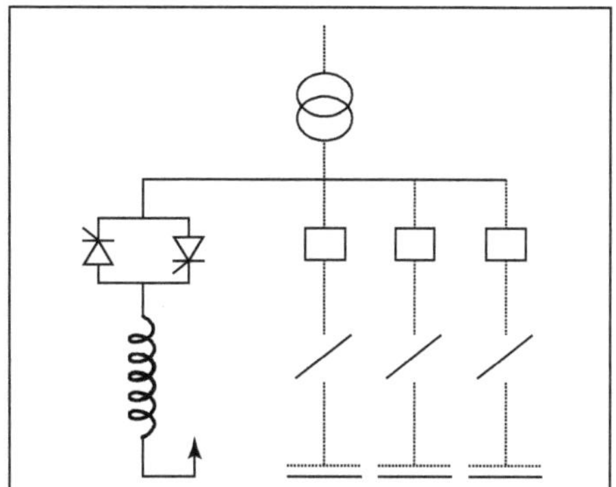

Fig. 9.5 *(a) SVC with FC and TCR.* *(b) SVC with TCR and TSC.*

connected capacitors and reactors. The configuration of the SVC, shown in Fig. 9.5(a), basically consists of a fixed capacitor (C) and a thyristor controlled reactor (L). The firing angle control of the thyristor banks determines the equivalent shunt reactance presented to the power system. A typical arrangement with three capacitor banks is shown in Fig. 9.5(b).

IEEE defines the SVC as a shunt connected static VAR generator or absorber whose output is adjusted to exchange capacitive or inductive current so as to maintain or control specific parameters of the electrical power system. Typically, the power system control variable is the terminal bus voltage. There are two popular configurations of SVC. One is a fixed capacitor (FC)

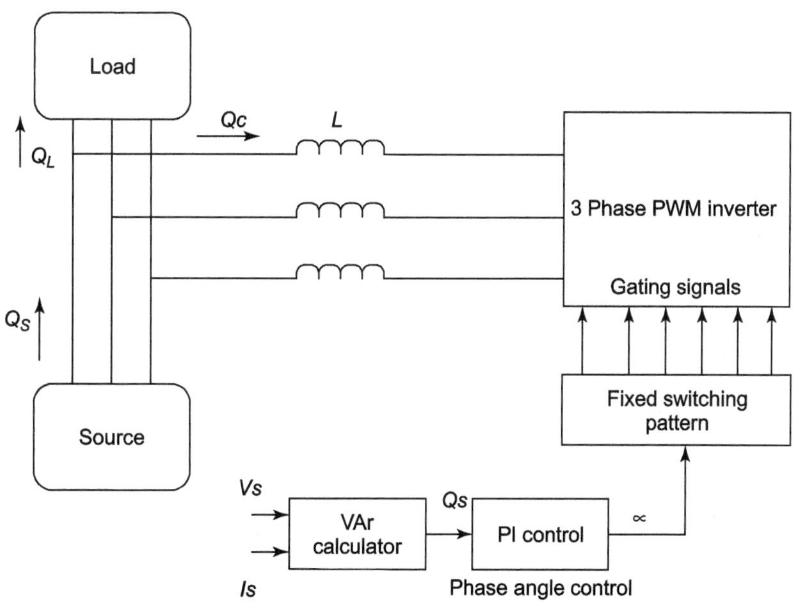

Fig. 9.6 *Block diagram of SVC system with phase angle control system.*

and thyristor controlled reactor (TCR) configuration, and the other one is a thyristor switched capacitor (TSC) and TCR configuration. SVC behaves like a fixed capacitor or an inductor. Choosing appropriate size is one of the important issues in SVC applications in voltage stability enhancement.

The SVC is an automated impedance matching device, designed to bring the system closer to unity power factor. If the power system reactive load is capacitive, reactors (usually in the form of thyristor-controlled reactors) offer inductive load, lowering the system voltage. Under inductive (lagging) conditions, capacitor banks are automatically switched in, thus providing a higher system voltage. SVCs placed near high and rapidly varying loads, such as arc furnaces or welding machines can smooth flicker voltage.

9.6 STATIC SYNCHRONOUS COMPENSATOR (STATCOM)

A STATCOM is a voltage-source converter based device, which converts a DC input voltage into an AC output voltage in order to compensate the active and reactive needs of the system. STATCOM has better characteristics than SVC when the system voltage drops substantially to force the STATCOM output to its highest – its maximum reactive power output will not be affected by the voltage magnitude. Therefore, it exhibits constant current characteristics when the voltage drops too low.

Installing a STATCOM at one or more suitable points in the network will increase the grid transfer capability through enhanced voltage stability, while maintaining a smooth voltage profile under different network conditions. These are installed near load points to take care of nearby areas, and provide additional versatility in terms of power quality improvement capabilities.

The main components of STATCOM are:

- Transformer
- Power unit
- Input reactor
- Control cabinet

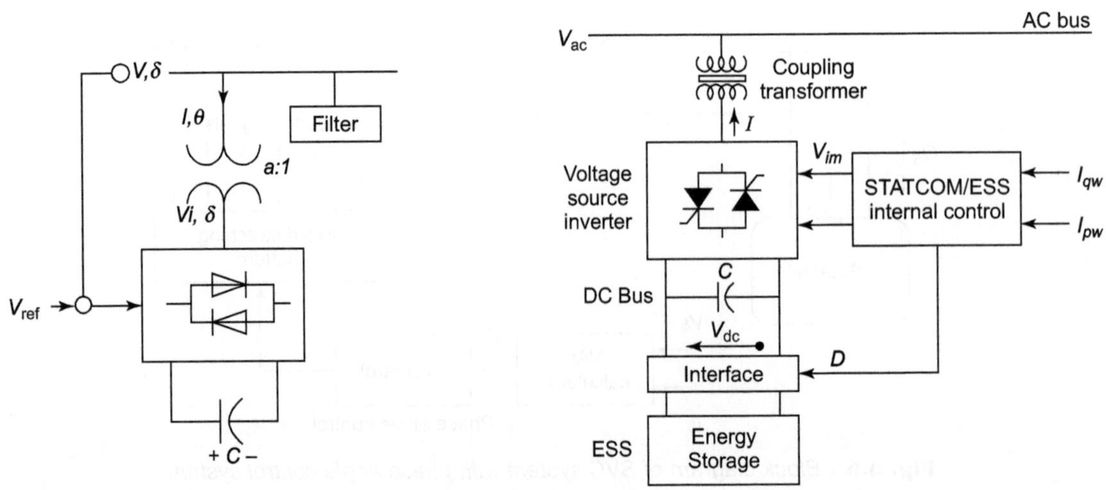

Fig. 9.7 *(a) Basic structure of STATCOM (b) STATCOM with DC link capacitor and energy storage.*

- Full digital control system
- Integrated workstation

Advantages of STATCOM

STATCOM is the most advanced VAR compensator technologically, which realizes reactive power compensation based on voltage source converter, without large volume capacitors and reactors. It uses HV Insulated Gate Bipolar Transistors (IGBTs) or Integrated Gate Commuted Thyristor (IGCT) with high switching frequency to realize reactive power regulation. Therefore, STATCOM has the following advantages compared with traditional compensation methods:

- Faster response speed
- Stronger voltage flicker restrain capability
- Wide operation range
- Low harmonic content
- Small volume

Similar to the SVC, the STATCOM also provides instantaneous and continuously variable reactive power in response to grid voltage transients, enhancing the grid voltage stability. The STATCOM operates according to voltage source principles, which together with pulsed width

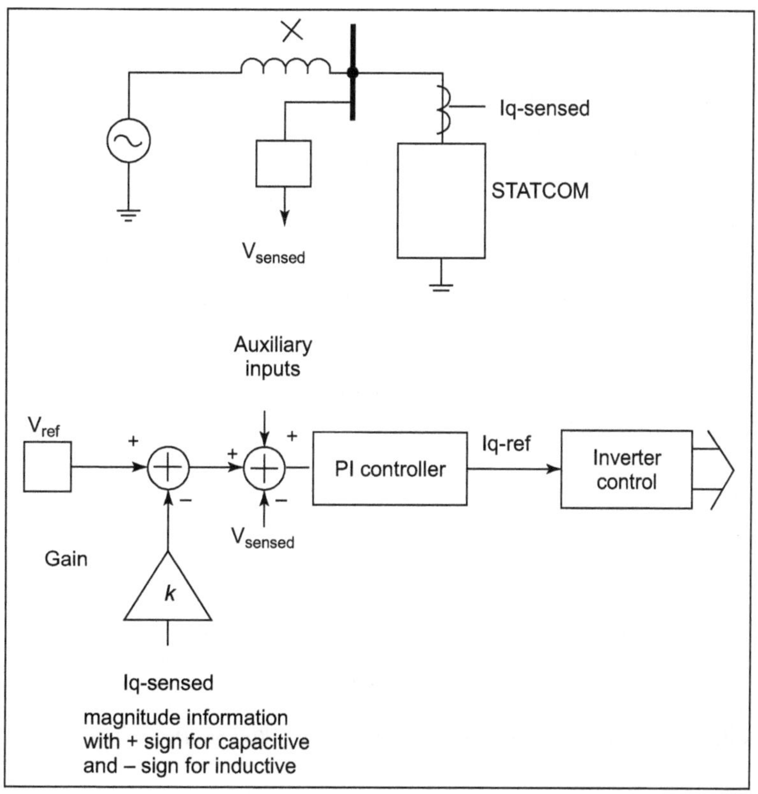

Fig. 9.8 *STATCOM control block diagram.*

modulation (PWM) switching of IGBTs gives it superior performance in terms of effective rating and response speed. The performance can be specific to active harmonic filtering and voltage flicker mitigation, but it also allows for a STATCOM to be extremely small.

The control system block diagram is shown in Fig. 9.8. The terminal voltage is sensed and a Phase Locked Loop (PLL) system is locked on to it for controlling the inverter. PLL is a control system that generates output signal where phase is related to an input reference signal. The amplitude of voltage is sensed and compared with the set reference value. The deviation controls the magnitude and polarity of the reactive current to be drawn by the STATCOM inverter from the AC system. A Proportional Integral (PI) controller is used for fast response and reduced steady state error. The inner current loop on the sensed reactive current magnitude is used to lower (in the case of capacitive compensation) the voltage reference value, thereby causing a drop in the voltage regulation. 'X' represents the system impedance at the point of connection. Auxiliary inputs are compared with the additional inputs into the control loop to make the STATCOM contribute to improvement of transient and dynamic stability.

Functions of STATCOM

- IGBT based PWM voltage source converter technology.
- Control for harmonic filtering.
- Harmonic current filtering in compliances with IEEE standards.
- Excellent dynamics with response time (less than 1 millisecond).
- Continuously monitors the harmonic current requirement and generates an adaptive waveform to match exactly the shape of the nonlinear portion of the load current.
- Injects this adaptive current into the load at the point of connection. Only fundamental (50 Hz) current is drawn from the source by the load.
- Compensates from the 2nd to the 13th harmonic in a response time of less than 100 microseconds.
- When system voltage is low, the STATCOM generates reactive power (STATCOM capacitive). When system voltage is high, it absorbs reactive power (STATCOM inductive).

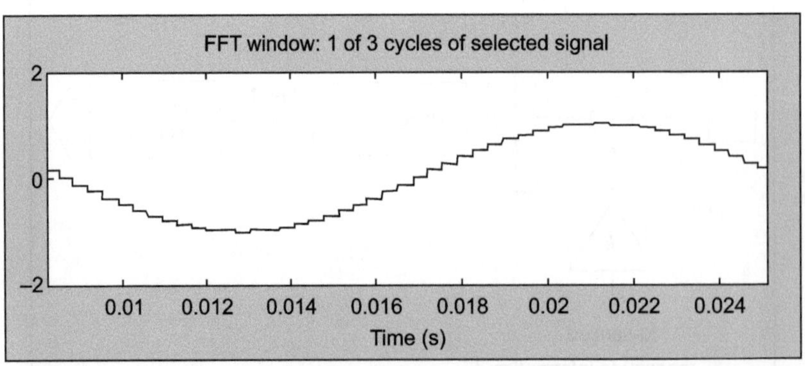

Fig. 9.9 *STATCOM maintains a near sinusoidal shape of bus voltage.*

Figure 9.9 shows the resultant waveform of bus voltage, which very nearly follows the fundamental sine wave, with negligible ripple factor. The power factor is always maintained near unity.

(a) **Shunt capacitive compensation**

Whenever an inductive load is connected to the transmission line, power factor lags because of lagging load current. To compensate, a shunt capacitor is connected, which supplies leading current.

(b) **Shunt inductive compensation**

This method is used either when charging the transmission line or, when there is very light load at the receiving end. Very small current flows through the transmission line under very low, or no load condition. Shunt capacitance in the transmission line causes voltage to rise substantially. The receiving end voltage may even become double the sending end voltage (generally in case of very long transmission lines). To compensate, shunt inductors are connected across the transmission line.

9.7 FLEXIBLE AC TRANSMISSION SYSTEM (FACTS)

The acronym **FACTS** was first introduced by Dr. Hingorani, at the joint APC/IEEE luncheon speech on April 19, 1988 with the following words: "*Allow me to introduce a new concept of a thyristor-assisted AC power system. I have called this concept 'Flexible AC Transmission System' or FACTS. It (the concept of FACTS) is a collection of thyristor based controllers, including phase shifters, advanced static VAR compensator, dynamic brake, modulator series capacitor, load tap changer, fault current limiter and perhaps others that have yet to be invented.*"

Flexibility of electric power transmission offers the ability to accommodate changes in the electric transmission system or operating conditions while maintaining sufficient steady-state and transient margins. FACTS is an alternating current transmission system incorporating power electronics-based and other static controllers to enhance controllability and increase power transfer capability. FACTS is a technology that can help improve transmission capability, increase system flexibility and control voltage on standard AC transmission lines. FACTS also improves system reliability and stability, leading to fewer electrical outages along an already jam-packed power grid.

FACTS involves the role of power electronics in generation, delivery and efficient end uses:

1. There is increasing emphasis on generating electricity using renewable sources such as wind and photovoltaic. There is also a great need for energy storage at the utility-scale, for which solutions are being investigated.
2. Reliable delivery methods increasingly use high-voltage DC transmission systems and FACTS devices such as SVC/STATCOM, thyristor controlled series capacitor (TCSC) etc. These are needed for voltage stability, which was not a serious issue in the past.
3. Power quality is a concern due to the proliferation of power electronics based loads.
4. Efficient end-use requires variable-speed drives and new types of lighting.

FACTS eliminates parallel-path flow by forcing the electricity to follow the path of least resistance, which can be changed by system engineers. This gives engineers the ability to create a path – or multiple paths – bringing flexibility to what was previously considered to be an inflexible model.

FACTS for series compensation modifies line impedance – X is decreased so as to increase the transmittable active power. The basic purpose is to provide more reactive power by the usage of capacitors. So, the concept is to connect some equipment to the transmission system and use it to control the flow of power through the system. The technology is relatively new, demonstrated FACTS availability is on order of 99.5%, and that is deemed insufficient by many utilities. Hence, it is yet to be fully incorporated into software tools for system planning and dispatching.

Software tools may be designed to select both the optimal types and the optimal locations of FACTS controllers within the system, and compare the investment benefits against reduced fuel costs, reduced system congestion, lower emissions, increased utilization of renewable sources, increased system security etc.

There are four main types of FACTS technology: TCSC, thyristor controlled phase angle regulator (TCPAR), static condenser (STATCON) and unified power flow controller (UPFC).

TCSC can provide 'dynamic control' over power lines. TCPAR is a technology that can increase control as well, but cannot itself generate additional power. STATCON units can generate additional power, but lack the technology to control it. Of these four systems, UPFC was found to be the best option from both an environmental and cost-benefit standpoint. The first of these UPFC units was installed in the summer of 1998 by American Electric Power in Kentucky.

One of the most commonly used power electronics-based building blocks is a voltage-sourced converter (VSC) in the transmission system. A basic building block of any VSC is the three-phase converter bridge. One common configuration for a three-phase bridge is shown in Fig. 9.10. The bridge has two DC terminals and three AC terminals in the midpoints of the converter legs. By controlling the states of switches in the legs, arbitrary voltage waveforms can be produced at the AC terminals.

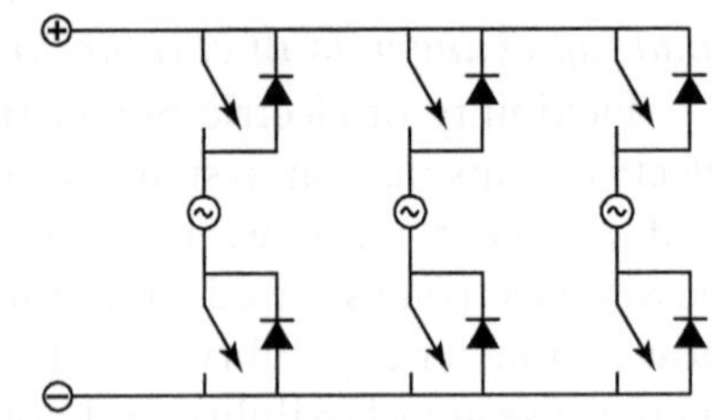

Fig. 9.10 *A three-phase converter bridge – the basic building block of a VSC.*

A VSC interfaced to a transmission system has to: (i) operate at the line frequency, and (ii) produce a balanced set of sinusoidal voltages. Therefore, a VSC coupled to the transmission system has only two control degrees of freedom – it can vary the magnitude and the phase angle of its output voltage relative to the system voltage. These degrees of freedom can be mapped into freedom to exchange active and reactive power with the transmission system. The amount of exchanged reactive power is limited only by the current capacity of the converter switches, while the active power coupled to/from the line has to be supplied from/delivered to the DC terminals, as shown symbolically in Fig. 9.11.

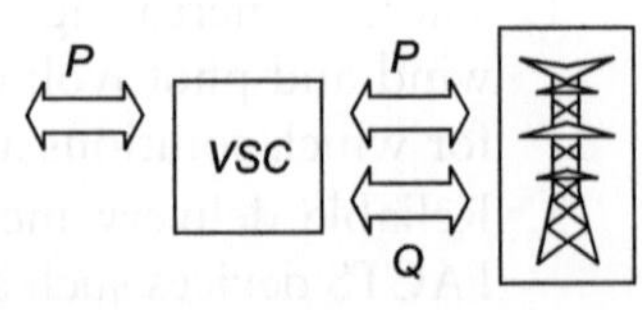

Fig. 9.11 *A VSC interfaced to a transmission line – P, Q exchange.*

Let us now look at the three exemplary FACTS controllers:

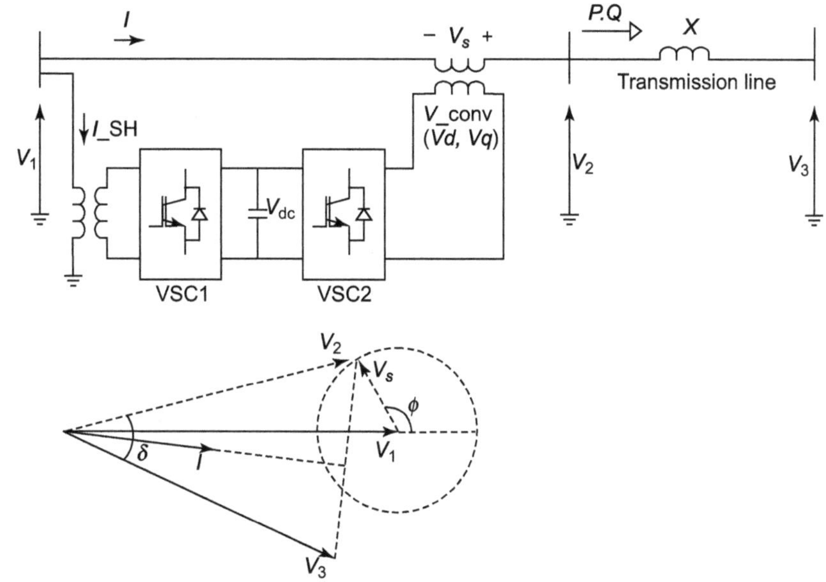

Fig. 9.12 *UPFC with STATCOM and SSSC*

(1) Unified Power Flow Controller (UPFC)

The newly developed UPFC is the most versatile member of the FACTS family using power electronics to control power flow on power grids. The UPFC uses a combination of a shunt controller (STATCOM) and a series controller (SSSC) interconnected through a common DC bus as shown in Fig. 9.12.

This FACTS topology provides much more flexibility than the SSSC for controlling the line active and reactive power, because active power can now be transferred from the shunt converter to the series converter, through the DC bus. Contrary to the SSSC where the injected voltage V_s is constrained to stay in quadrature with line current I, the injected voltage V_s can now have any angle with respect to line current. If the magnitude of injected voltage V_s is kept constant and if its phase angle with respect to V_1 is varied from 0 to 360 degrees, the locus described by the end of vector V_2 ($V_2 = V_1 + V_s$) is a circle as shown on the phasor diagram. As ϕ is varying, the phase shift between voltages V_2 and V_3 at the two line ends also varies. It follows that both the active power P and the reactive power Q transmitted at one line end can be controlled.

In addition to controlling the line active and reactive power, the UPFC provides an additional degree of freedom. Its shunt converter operating as a STATCOM controls voltage V_1 by absorbing or generating reactive power.

Both the series and shunt converters use a VSC connected on the secondary side of a coupling transformer. The VSCs use electronic devices (Gate Turn-off Thyristors, or GTOs, IGBTs or IGCTs) to synthesize a voltage from a DC voltage source. The common capacitor connected on the DC side of the VSCs acts as a DC voltage source. Two VSC technologies can be used for the VSCs:

(2) VSC using GTO-based square-wave inverters and special interconnection transformers Typically four three-level inverters are used to build a 48-step voltage waveform. Special interconnection transformers are used to neutralize harmonics contained in the square waves generated by individual inverters. In this type of VSC, the fundamental component of voltage is proportional to the voltage V_{dc}. Therefore V_{dc} has to be varied for controlling the injected voltage.

(3) VSC using IGBT-based PWM inverters: This type of inverter uses the PWM technique to synthesize a sinusoidal waveform from a DC voltage with a typical chopping frequency of a few kilohertz. Harmonics are cancelled by connecting filters at the AC side of the VSC. This type of VSC uses a fixed DC voltage V_{dc}. Voltage is varied by changing the modulation index of the PWM modulator.

DC link nominal voltage: The total value of the DC link capacitor is related to the UPFC converter ratings and to the DC link nominal voltage. The energy stored in the capacitance (in joules) divided by the converter rated power S_{nom} (in VA) is a time duration which is usually a fraction of a cycle at nominal frequency. For example, for the default parameters, ($C = 750$ μF, $V_{dc} = 40$ kV, $S_{nom} = 100$ MVA) this ratio ($\frac{1}{2} CV_{dc}^2/S_{nom}$) is 6.0 ms, which represents 0.30 cycle for a 50 Hz frequency. If the default values of the nominal power rating and DC voltage change, the capacitance value changes accordingly.

With the development of electronic control systems for power quality, it seems feasible to maintain a near unity power factor, while also ensuring distortion-free supply voltage and currents. This will go a long way towards optimum use of available generation and distribution grid capacity, releasing considerable amouns of active power for additional loads that may be connected to the existing systems. With the development of ultracapacitors, new possibilities are opening up Cto store more active power in capacitors. Further, connecting of alternative energy sources to the grid will also become feasible.

10

ELECTROLYTIC CAPACITORS

10.1 NATURE OF ELECTROLYTIC CAPACITOR

Electrolytic capacitors are a type of capacitor where an electrolyte is used to act as one electrode, or one of its plates (cathode), to get large capacitance value per unit volume. Electronic engineers often refer to them as 'electrolytics'. Their voltage ratings are much lower, tolerances are wider and ESR is higher, and loss factor one order higher than electrostatic types. Although the principle was discovered in 1886, their first major use began as late as the 1930s. Aluminium was used as electrode plates, and tantalum came much later. Niobium is also currently being used for electrolytic capacitors. These capacitors being essentially polar in nature, all electrolytics have polarity clearly marked on them.

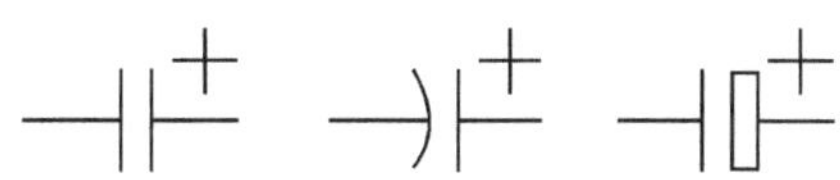

Fig. 10.1 *Common symbols for electrolytic capacitors.*

Electrolytic capacitors find use in DC power supplies, along with batteries, and in power filters. In electronic circuits, electrolytic capacitors are used in low frequency filters, timers, coupling, as energy storage devices and a host of other applications. They have high capacitance with low volumes as compared with electrostatic capacitors. Generally the voltage levels in DC circuits are low, but in several applications like photoflash camera, capacitor voltages can go to 450–600 V.

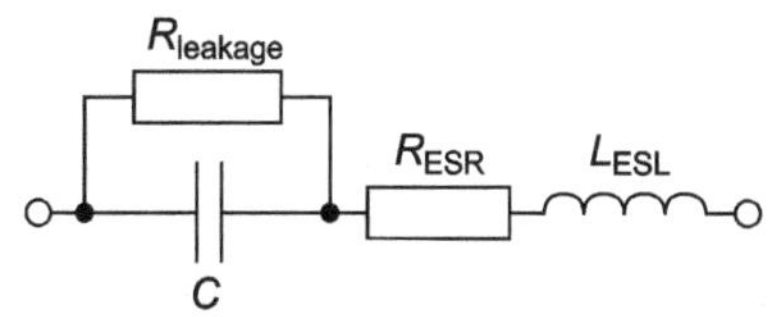

Fig. 10.2 *Simple equivalent circuit of electrolytic capacitor.*

10.2 TYPES OF ELECTROLYTIC CAPACITORS

Depending upon electrode material and the electrolyte, electrolytics can be categorized as:

- Wet aluminium electrolytic capacitors
- AC aluminium electrolytic capacitors
- Dry aluminium electrolytic capacitors

- Wet tantalum capacitors
- Dry tantalum capacitors
- Tantalum polymer capacitors
- Niobium oxide capacitors.

Dielectric constants of common dielectrics in electrolytic capacitors are as follows:

- Aluminium oxide 8.5
- Tantalum pentoxide 27
- Niobium oxide 41

10.3 BASICS OF ELECTROLYTIC CAPACITORS

The dielectric in aluminium electrolytics is a thin layer of aluminium oxide. These capacitors contain corrosive liquid and can burst if connected in reverse polarity. The oxide insulating layer will tend to deteriorate in the absence of sufficient rejuvenating voltage, and eventually the capacitor will lose its ability to withstand voltage if the voltage is not applied for a long time. In such a case, the capacitor can be reformed by connecting the voltage source through a resistor until it is charged to full voltage. Their high loss angle and bad frequency and temperature characteristics make electrolytics unsuitable for high frequency.

Electrolytics are available from low voltages below 10 V to voltages as high as 600–800V. They are available from very miniature sizes in strip packaging to large sizes used for energy storage in large power supplies. The high capacitance arises from the high dielectric constant, high breakdown field strength, the rough surface, and the extremely small, uniform thickness of anodically formed metallic oxide. Electrolytic capacitors have uniform dielectric stress and can operate at high field strength (within 80% of their breakdown strength) of the order of 1,000 V/μm for two reasons:

First, an anodization (formation) process is performed on the electrode at a fixed voltage, and the dielectric oxide layer grows all over the area to the thickness required to support that voltage. Second, the capacitor electrolyte continues the healing work of the original forming electrolyte on the electrode foil, repairing and thickening the dielectric locally as required. This healing process is driven by the capacitor's DC leakage current, which is drawn whenever a DC voltage is applied to the capacitor, that is, whenever it is in operation. In fact, electrolytic capacitors often last longer when they are in continuous, mild use than when they are only charged up briefly every year or decade.

The dielectric material of electrolytic capacitors is produced from the anode metal itself in what is known as the forming process. During this process, current flows from the anode metal – which must be a valve metal such as aluminium, niobium or tantalum – through a conductive bath of a special forming electrolyte to the bath cathode. The flow of current causes an insulating metal oxide to grow out of and into the surface of the anode. The thickness, structure and composition of this insulating layer determine its dielectric strength. The applied potential between the anode metal and the bath cathode must be above the oxide breakdown voltage before significant current will flow.

An electrolyte must have the following characteristics:

(1) It must be electrically conductive.
(2) It must have a forming property to heal any flaws on the dielectric oxide of the anode.
(3) It must be chemically stable with the anode and cathode, sealing materials, etc.
(4) It must have superior impregnation characteristics.
(5) Its vapour pressure must be low.

Power applications mostly use high voltages. AC motor start application, rectified DC high voltage sections, RFI filter, photoflash and bypass applications must withstand peak voltage up to 375 V or higher. In all cases, invariably aluminium electrolytic capacitors are used. Aluminium is the most versatile material for these capacitors as they allow large capacitance values, and also can sustain high working voltages. The dielectric oxide layer is easily formed on aluminium foil electrode.

The disadvantages of electrolytic capacitors are their non-ideal, high loss characteristics which arise from the unidirectional electrical properties of the oxide layer, resistive losses from the high electrolyte resistivity, poor frequency response due to the roughness of the surface oxide, and finite capacitor life due to breakdown and degradation of the electrolyte.

10.4 ALUMINIUM ELECTROLYTIC CAPACITORS

Aluminium electrolytic capacitors are characterized by the following components:

- Anode: aluminium – etched and formed aluminium foil/mechanically formed shapes
- Electrolyte: solid/liquid /paste chemicals – Act as de facto cathode terminal
- Kraft or manila paper as electrolyte carrier
- Dielectric: oxide layer film on anode
- Cathode: etched foil/can – serves mainly as current collector from electrolyte.
- Connection leads: tabs
- Terminal disc – This is generally rubber bonded bakelite disc.
- Pressure vent – either built up in disc, or as separate plug. Generally in small DC capacitors, the can bottom is grooved to crack open in case of failure.

Aluminium has been used in electrolytics on account of its flexibility in use, etching, electrode formation, winding, and cost. All electrolytics are essentially polar in nature, and the polarity cannot be reversed. The capacitor acts as a short circuit if polarity of applied voltage is reversed. AC electrolytics are in fact two electrolytics connected in series back to back. One capacitor works in any given polarity while the other is a short circuit.

Aluminium electrolytics do not have competition from tantalum or other capacitors in higher voltages as other electrolytics are essentially for low voltage applications. Even in low voltage, they have the advantage of lower cost per unit. There are basically two types of Aluminium electrolytic capacitor, the plain foil type and the etched foil type. The thin aluminium oxide film dielectric and high breakdown voltage give these capacitors very high capacitance values. In etched foil type the aluminium oxide on the anode and cathode foils are chemically etched to increase its surface area and permittivity. However, plain foil type has the advantage of higher AC current capacity.

Fig. 10.3 *Aluminium electrolytic capacitors of various types.*

Tolerance range of the common etched type is quite large, up to 20%. They find use in coupling, DC blocking and bypass circuits. Plain foil types are better suited as smoothing capacitors in power supplies. Typical values of capacitance range from 1 μF to 47000 μF. Aluminium electrolytics are polar in nature and reversing the applied voltage on the leads will destroy the capacitor.

10.4.1 Construction of Aluminium Electrolytics

Most large capacitors use a wound element, impregnated with liquid electrolyte, sealed in a can, with terminals connected to external tags. The element, as shown in Fig. 10.4, consists of an anode foil, a cathode foil and a separator paper (spacer) in between, saturated with liquid electrolyte. The separator is a paper (kraft or manila) in 1 to 4 layers, and is of low density, normally 0.55 to 0.65 to permit maximum retention of the electrolyte. The separator also adds to the loss factor to the extent it restricts free flow of current through electrolyte. Small DC capacitors have paste type electrolytes. The choice of electrolyte determines some of the capacitor characteristics. A mixture based on ethylene glycol and water will freeze at very low temperatures. Such 'wet' capacitors must be operated upright and tend to have lower operating voltages. A number of similar electrolytes are to be found, many of which have a 'gel' consistency. Dry electrolytic capacitors, as they are called, use electrolytes such as manganese oxide, but tend to have lower volumetric efficiency.

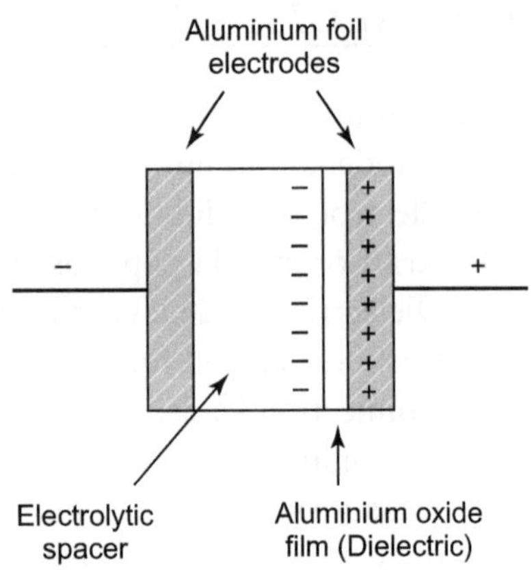

Fig. 10.4 *Al electrolytic principle.*

The current collector from the electrolyte is cathode foil or the container. Cathode foil is of thickness of 15 to 120 microns, depending on current and application, and has a large etched surface to give maximum capacitance value per unit area. Anode is etched and formed aluminium much thicker than cathode. It can be an etched and formed solid shape or plate, or a foil. In most power applications, etched and formed foil is used as this gives maximum surface area, has more flexibility in use and is easy to produce. Solid formed shapes are an outdated technology now, and all modern day aluminium electrolytics are made from foils.

Both anode and cathode foils are made out of high purity aluminium. The etching process is so typical and closely guarded that there are only a few manufacturers making the etched foil. Even the formation is a specialized technique, and most capacitor manufacturers purchase the etched and formed foil, as well as the cathode foil from these few sources.

Anode foil and a cathode foil are interleaved with electrolytic paper separator and wound into a cylindrical shape to form the capacitor element. At this stage, it has configuration of a capacitor with electrolytic capacitor paper separator and the aluminium oxide layer being the dielectric, however, the unit has hardly any capacitance. When this capacitor element is impregnated with liquid electrolyte, the anode foil and cathode foil are electrically connected. With the aluminium oxide layer formed on the anode foil acting as the sole dielectric, a capacitor with a high value of capacitance is now attainable, the electrolyte functioning as a cathode.

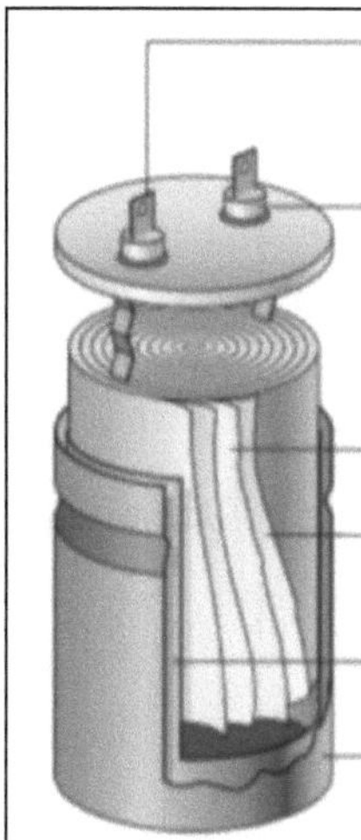

Fig. 10.5 *Al electrolytic capacitor construction.*

(a) **Etched and formed anodes**

The anode base metal is one electrode while the electrolyte acts as the other electrode. Electrolyte also helps in the formation of oxide layer and then to maintain and replenish it if it depletes. The electrolyte fills up the aluminium can, which acts as the second electrode in DC. The electrolyte may be in liquid, paste or powder form, and is a mixture of chemicals processed suitably to give desired electrical and thermal properties.

The combination of the etching and forming processes determine the formation voltage 'V_f' and the gain of the foil produced. The gain is defined as the capacitance per unit macroscopic area of the produced foil, divided by the capacitance per unit area of un-etched foil of the same formation voltage. It is also often mentioned as µF per unit area (per sq. cm. or per sq. inch). High energy density aluminium electrolytic capacitors use high-gain foil. It may be noted here that the breakdown field strength of the order of 1 Vnm^{-1} is much higher than for polymer films. The '*k*' value of 8.5 for aluminium oxide is also about three times larger than for most film dielectrics.

The anode and cathode etched foils are made from pure aluminium foil using a special technique to create hills and valleys on surface to increase the effective area many folds. Thickness and type of anode, its forming voltage, gain (µF/sq. cm.) as also the thickness and type of cathode film are decided depending upon the working voltage, capacitor duty and service conditions. The process, designed for individual foils towards intended results, makes possible high capacitance because etching of foils increases surface area more than 100 times and the oxide dielectric is less than a micrometre thick. Thus the resulting capacitor has very large plate area and the plates are extremely close together.

Anode foils are then oxidized by an electrochemical process using a choice of chemicals for the desired applications. The process of growing a thin layer of dielectric film on the anode foil surface is called 'formation'. The thickness, gain, and forming voltage are the main parameters of anode foil. Formation is done by passing the etched foil over rollers through a continuous bath of electrolyte, and applying increasing DC voltages between the bath and foil. A controlled current is passed through the foil in stages, until the voltage on foil reaches desired level, which is somewhat above the working voltage.

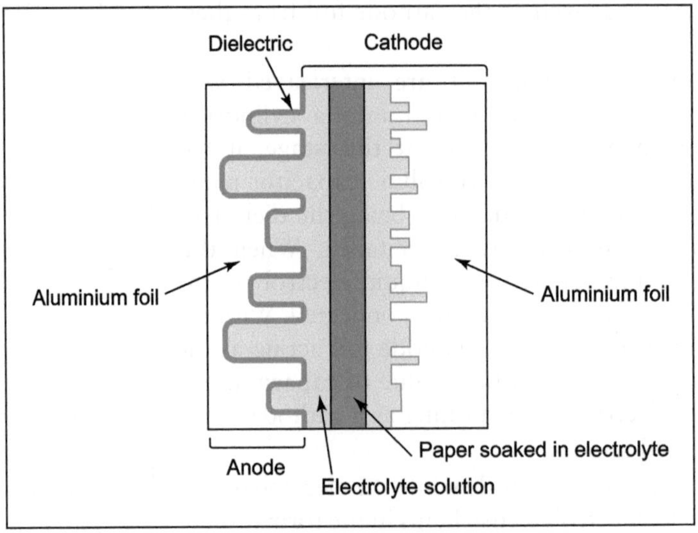

Fig. 10.6 *Dieletric system of electrolytics.*

The final formation voltage should be a little above actual working voltage of capacitor, or the peak of AC RMS voltage rating of capacitor. A voltage of 370–420 V is used for common 230 V AC motor start capacitor foils. This is to take care of surge voltage testing at 275 V AC ($275\sqrt{2}$ = 389 V, plus safety margin). Forming voltages up to 440 V or higher are also being used for high reliability capacitors working on 240 or 250 V AC supply. Various grades and thicknesses of anode foils are available for different applications. The main specification of anode foil, apart from thickness and forming voltage, is the capacitance density, viz. μF/unit area, e.g. per sq. cm. Cathode foil also is qualified in μF/unit area.

The thickness of oxide dielectric layer is about 1.4–1.5 nm/volt of formation voltage. This means a thinner oxide layer for lower voltage, which in turn means that for lower voltages, the gain of formed foil and capacitance density is significantly higher. The high currents associated with motor start capacitors necessitate an increase in foil thickness, and lower gain (more foil area for same capacitance), usually give better performance, longer AC voltage withstand time, lower power factor (lower watt loss), and better life.

The loss factor for electrolytics is relatively high, and the ESR is also considerable. In DC applications, the watt loss is not of much consequence, except when the ripple voltage and frequency is considerable. In AC circuits, however, this is a major consideration, as the heat generated during even short duty cycle of a few seconds can be damaging. The duty cycle of motor start capacitors is mainly governed by the heat generated, which can damage the capacitor itself. The duty cycle is generally specified as 15 to 20 starts of motor per hour, each start being of maximum 3 s duration. Loss factors of 1.8% to 4% are common, and in some cases may go to 7%. Every effort is made in the manufacturing process to achieve the lowest possible loss factor.

(b) Cathode

The cathode aluminium foil is generally thinner than the anode and exhibits much higher capacitance than the anode, since the cathode capacitance appears in series with the anode capacitance to yield the total capacitance. The maximum total capacitance occurs when the cathode capacitance is as large as possible. High cathode capacitance requires a very low cathode formation voltage. Cathode is etched, but not formed at all, although there is always a thin layer (around 2–3 nm) of hydrous oxide on the surface of aluminium. A thin hydrous oxide layer forms readily on aluminium with exposure to normal air in the atmosphere. Titanium passivation of cathode foil has been undertaken in recent years to offer a cathode with a capacitance approaching 200 $\mu F/cm^{-2}$. Such a high cathode capacitance is necessary only for low-voltage capacitors with high-gain anodes. Generally a cathode capacitance of fifty times the anode capacitance is sufficient. This situation yields a total capacitance which is only 2% less than the anode capacitance.

A thin foil with a surface etch is used for the cathode, giving a frequency response generally better than that of the anode, and giving capacitance so large that the total unit capacitance is not diminished. Since the cathode's voltage capability is usually only about one volt, the electrolytic capacitor unit is limited in its steady-state reverse voltage capability to about one volt. Extended reverse voltage for time intervals as short as one second can cause significant heating of the electrolyte and of the anode oxide. The current drawn during these reverse voltages can easily reach hundreds of amperes.

Note: In AC motor start capacitors, since two capacitors are connected back to back, and cathode need not have any external connection, it is sometimes totally done away with, and two anodic films are used, with paper separators between them. The electrolyte anyway works as the cathode current collector for both the capacitors so formed. Where a cathode is used, it does not have any connection, but works as thermal separator and heat conductor, and helps the heat dissipation process.

Tabs

Tabs are strips of aluminium which make contact between the conductive plates and the connection terminals in the winding. Tabs are stitched, welded or riveted on the electrode foil. Fig. 10.7 shows tabs cold welded to foils. In formed shape, the formed shape extension may

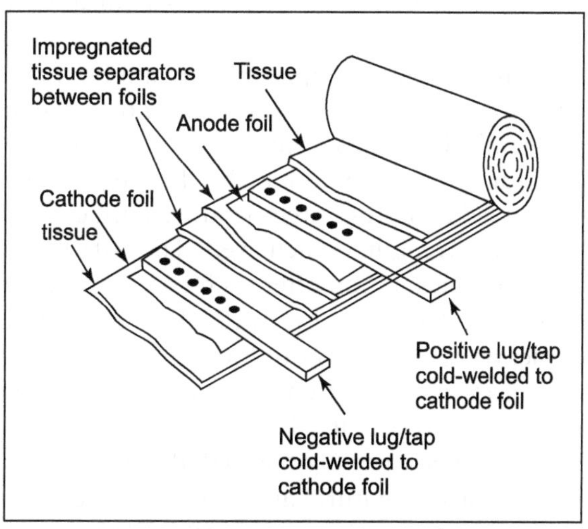

Fig. 10.7 *Tabs welded to electrode foils.*

serve as a tab. There may be several tabs connected to each of the plates. The tab paths are generally run from the capacitor section to the terminals in a fashion which keeps the inductance low and prevents tabs of the opposite polarity from coming in contact with one another or the case during movement and vibration of the capacitor unit. Tab thickness can be from 50 to 150 microns.

Tab material is not etched, but is formed to a high voltage prior to its assembly into a capacitor. There could be several tabs attached to each foil, and optimum tab placement along the foil is considered to minimize power loss due to the metal foil resistance. This leads to equal spacing from each tab to the one nearest it, and half of the inter-tab distance is provided between the outermost tabs and the foil ends. For high voltage capacitors, the tab resistance and metal foil resistance are quite small compared to the oxide and electrolyte resistance. Tabs for the cathode are not formed.

The anode has an oxide layer, and the tabs have to connect firmly to the base metal under this layer. The dielectric layer is broken in the process, and it is re-formed in the manufacturing process. The cathode tab is many times simply in physical contact with foil, and is held under winding pressure. Tabs are fitted on to the foils during the manufacturing process, either before or during winding.

(c) Separator

The separator or spacer is an absorbent material in roll form which is wound between the anode and cathode to prevent the foils from coming in contact with one another. The spacer is generally made of paper, which can be of many different types, densities, and thicknesses, depending on the voltage and effective series resistance requirements. Besides separating the anode and cathode, the spacer must wick and hold the electrolyte between the plates.

The resistance of the spacer-electrolyte combination is appreciably greater than would be accounted for by its geometry and the resistivity of the absorbed electrolyte. The electrolyte-spacer combination also impacts the capacitor's frequency response. Paper usually has a density of 0.50–0.60, and thickness depends on the voltage and applications, as also the electrolyte used. Its main function is to carry the electrolyte, and to act as separator between foils.

(d) Electrolytes

Electrolytes used for aluminium electrolytic capacitors are a mix of chemicals processed to give required properties. Most manufacturers prefer to make their own chemicals. Conventionally, for aluminium anodes, a mix of chemicals selected from glycerine, glycols, liquid ammonia, ammonium pentaborate, boric acid, di-hydrogen-ammonium orthophosphate, additives like adipic acid etc. are used. The chemicals are mixed at a pre-set temperature until desired conductivity, pH value, sparking voltage and breakdown strength are achieved. The volatile nature of chemicals makes it necessary to check and correct the solution after each batch production.

Commercially prepared electrolytes are also available in market. The electrolyte must first replace the unavoidable depletion of oxide layer of formed foil, maintain the layer in service, and keep desired properties of dielectric intact. It conducts electrical current away from electrodes, and hence must have acceptably low resistivity. The electrolyte is a major contributor to ESR and heat losses in capacitor, and has to be carefully selected and handled to produce desired results.

Electrolytes are highly corrosive in nature, and must be stored or handled in stainless steel or aluminium utensils only, as they attack most metals. They are also highly susceptible to impurities like iron and chlorine. All materials used in the manufacture, including water for cleaning need to be regularly checked for traces of these elements. These impurities severely affect the electrolyte and degrade the capacitor performance seriously. Only demineralized water is used in the manufacturing process.

Due to their corrosive nature, personnel handling the electrolytes and those nearby need to take adequate care to stay away from their fumes or physical contact. Gloves and masks are generally advised. Disposal of spent electrolytes also needs care to avoid ecological pollution.

Composition and choice of electrolyte greatly influence the various characteristics of aluminium electrolytic capacitors. Additives used in electrolytes also affect capacitor properties like low temperature performance, stability etc. The proper electrolyte is determined by the electrical ratings, operating temperatures and the application of the capacitor.

(e) **Impregnation**

The capacitor assembly is impregnated under vacuum with suitable electrolyte and the top cover is sealed by curling machine. The method of impregnation requires the winding element to be immersed into the electrolyte by either a vacuum/pressure cycle with or without applied heat, or by simple absorption. In vacuum impregnation, a vacuum is applied to the electrolyte and sections, causing hot electrolyte to be drawn into the sections, thoroughly wetting the sections. The electrolyte contains a solvent such as ethylene glycol and a solute such as ammonium borate. Should the dielectric film be damaged, the presence of the electrolyte will allow the capacitor to heal itself by forming more oxide. The vacuum soak method is limited to the use of certain separator absorbencies and certain electrolyte viscosities as well as certain types of electrolytes. Care has to be taken to limit the vacuum at low enough level so as not to evaporate electrolyte components. So a high vacuum is not desirable, and a coarse vacuum is sufficient, depending upon the electrolyte.

The greatest improvement in impregnation technique for DC electrolytics is the centrifugal impregnation method. By the employment of centrifugal force the hot electrolyte may be forced into the capacitor windings with almost unlimited pressure without any lowering of the boiling point temperature of the electrolyte. Elements are loaded in a horizontal circular stacked arrangement in a centrifuge, hot electrolyte is filled in and high speed spinning of the centrifuge forces the electrolyte into the element.

Another method of impregnating the separator material is the hand pasting procedure. This is limited to electrolytes with very high viscosities. The anode plate, the cathode plate and separator materials are cut to size and the tabs formed or folded on the plates or foils. The cathode plate is placed on a flat surface, a separator is placed on top of the foil and electrolyte is pasted or smeared on and into the separator. This is done manually with a spatula, as plaster would be applied to a surface. Anode foil is placed on the separator or plate and in turn the other separator on top of the anode foil. This second separator is then manually pasted with electrolyte. The whole laminated section is wound or folded into a concentric roll, the cathode foil forming the outside and last turn when finished. After winding, the section is more or less round but is sometimes pressed flat in a press. The rolling (winding) operation is sometimes accomplished with the aid of a hand operated winding mandrel.

There are some other methods of impregnation of electrolytes, but the ones mentioned above are most used.

(f) Sealing

The capacitor element is sealed into a can. While most cans are aluminium, phenolic outer cans are often used for motor start capacitors. The seal is not hermetic and it is usually a pressure closure made by rolling (curling) the can edge into a rubber gasket, or into rubber laminated to a phenolic board. In small capacitors moulded phenolic resin or polyphenylene sulphide may replace the rubber. Too tight a seal causes pressure build up, causing an explosion, and too loose a seal shortens the life by permitting drying out (loss of electrolyte). In motor start electrolytic capacitors, a pressure rubber vent is provided, which bursts open to throw out vapours in case of capacitor failure. This saves the capacitor from explosion.

(g) Ageing

The capacitor units are dipped in electrolyte bath and are slowly brought up a little over maximum rated voltage, but below anode formation voltage, at or near its maximum rated temperature during the ageing process (also called edge formation). The ageing process grows oxide on areas on the anode foil which have an insufficient oxide barrier. The purpose is to reform or to repair any oxide film which may have been damaged during the slitting, winding and assembly processes, thus reducing the leakage current to an acceptable low level. Heat is generated during this process, and it is often necessary to provide water cooling to keep temperature within limits.

Before being sleeved and tested the sealed capacitor is again aged by applying ageing voltage directly across the terminals, and then tested, this being the final process of the production chain. This takes little time, and no heating is necessary. A forming voltage, greater than the rated voltage, is gradually applied between capacitor terminals. The capacitor can is not insulated from the inside winding due to the electrolyte in it.

An insulating sleeve is shrunk around the can, along with an insulating disc at the base (this is single can construction). In many cases, like motor start application the assembled capacitor is further sealed in an outer can to have insulated housing, with wire leads coming out for connections (double can construction).

Normal electrolytic PCB-mount capacitor and a surface mount electrolytic capacitor construction is given in Fig. 10.8. General process of manufacture is on the same lines, but the terminals are brought out in a form suitable for surface mount application.

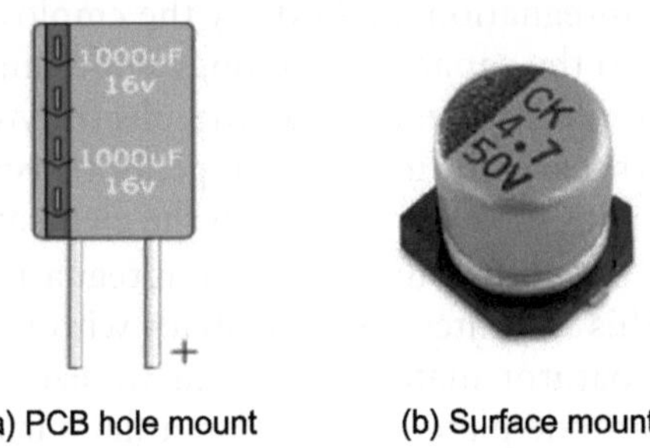

Fig. 10.8 *PCB hole mount and surface mount electrolytic capacitors.*

10.4.2 Motor Start Electrolytic Capacitors

These are basically two DC electrolytic capacitors connected back-to-back. One capacitor bears the voltage in positive half cycle, while the other acts as a short circuit. The reverse occurs in negative half cycle. The loss factor being very high, of the order of 1.8% to 7%, depending upon design and local standards governing them, the heat generation is considerable. The capacitor gets heated up very fast, which limits the time it can remain in circuit. The usual duty for this capacitor is 3 s on, and 177 s off, or 1.7% of the time cycle permissible within this overall maximum limit.

A centrifugal switch on the motor shaft cuts off the auxiliary winding circuit and with it the capacitor, once the motor achieves a predetermined speed.

A motor start capacitor of 250 µF capacitor at 230 V RMS carries a current of about 18 amp, which works out to 4140 VA, and a loss factor of 5% leads to heat loss of 200 W. If not switched off during the decided time limit, the heat generated can cause the capacitor to explode violently. To ward off this eventuality, a pressure vent is normally provided on the cover for safety, which blows off to allow the products of heat generation – air and vapours –to escape safely. The capacitor is then rendered out of service, and has to be replaced.

The external electrodes of motor start capacitor are connected to etched and formed anode foils (formed to a suitable voltage by the supplier), and interleaved by layers of electrolytic capacitor paper. The paper is of low density, its function being to retain the electrolyte in contact with the foils, as also to act as separator. The thickness, type of paper (kraft/manila), and the number of papers is decided by the desired properties and electrolyte used.

Sometimes a common cathode foil is used to physically form two internal capacitors in series, the cathode being common to both capacitor sections. The cathode is not connected to any terminal, but is a floating electrode. The cathode and the additional volume on this account serve as thermal storage and reduction of loss factor. The voltage withstand time of capacitor at rated voltage is also better in this design. The cost, though, is higher. Motor start capacitors are available in different configurations, like single can insulated body (in heat shrinkable sleeve insulation), double can aluminium can with wire, double can with terminals in insulated body etc.as shown in Fig. 10.9.

Fig. 10.9 *Motor start electrolytic capacitors.*

The voltage withstand time at rated voltage is frequently used to determine capacitor quality, and to see in-built safety limit. In many countries a time to burst the seal at rated voltage is considered satisfactory if it exceeds three minutes. Five minutes is considered by most to be among the best.

Ratings of motor start capacitors

Ratings of capacitors are in wide tolerance limits (electrolytic capacitors cannot be made to close tolerances as electrostatic ones). Motor start capacitors in India are rated at 230 V AC nominal, with surge voltage of 275 V AC (25% over-voltage). Capacitance ratings commonly in vogue are: 40–60 µF, 60–80 µF, 80–100 µF, 100–120 µF, 120–150 µF, 150–200 µF and 200–250 µF. Even much higher ratings are being made on demand. The rated power factor is below 8%. In most cases, the power factor is under 5%, with best construction giving a power factor of as low as 1.8%. The duty cycle is specified as 3 s on, followed by 177 s off, and a maximum of 20 starts per hour. This is sometimes mentioned as 1.7% duty cycle.

Testing of motor start capacitors

Motor start capacitors are subjected to the following routine tests at the final testing stage:

- Capacitance value
- Surge voltage
- Power factor
- AC high voltage test between shorted terminals and container

Type tests include the following:

1. Charge–discharge cycle at 230 V AC at rated duty cycle, viz. 3 s on, followed by 177 s off. The endurance test is carried at rated capacitor temperature, and for 10,000 operations.
2. Vent burst time: The time taken by the vent to burst is an indication of capacitor quality. A time of 3–5 min is targeted by different manufacturers: the longer the time, the better it is. At the end of the expected burst time, the rubber vent or plug should burst, giving way to gases formed inside due to overheating in this prolonged period. This is also a safety test for the capacitor, which should not explode during the test.

Metallized Motor Start Capacitors

While talking of motor start capacitors, it is pertinent to mention a new development recently which enables the use of metallized film capacitors for motor start applications. The film is very thin MPP or MPET, below 5 microns, and capacitor size is comparable to electrolytic capacitors. The cost is comparable, while the reliability is impeccable. Loss factor of these capacitors is under 0.5% for MPET, or 0.05% for MPP. The capacitors successfully withstand 250 V AC for over 30 min without failure, and do not fail up to 10 min at 275 V AC. They can be made to close tolerances of 5%, as against very wide tolerance limits for electrolytics. Failure rate of capacitors due to failure of centrifugal switch within a specified short time also goes down.

The use of metallized capacitors for motor start application opens up the possibility of making full use of enhanced voltage capacity by redesigning the motor to reduce copper in motor winding, and making the whole exercise cost effective. The capacitor being self-healing, the life is also improved. Electrolytic capacitors, by their nature, have a limited life expectancy compared with electrostatic types.

10.4.3 Energy Storage Electrolytic Capacitors

One of the main applications of aluminium electrolytic capacitors is as input capacitors for power inverters. The aluminium electrolytic capacitor provides a unique value in high-energy storage and low device impedance. Selecting the right capacitor for an application requires knowledge of all aspects of the application environment, from mechanical to thermal to electrical. These capacitors routinely offer capacitance values up to 3 F and voltage ratings from 5 V to 500 V. Some possibilities include

- 330 μF at 100 V and 6800 μF at 10 V for SMT devices,
- 100 μF at 450 V, 6800 μF at 50 V and 10,000 μF at 10 V for miniature-can styles,
- 1200 μF at 450 V and 39,000 μF at 50 V for snap-in can styles and
- 9000 μF at 450 V and 390,000 μF at 50 V for large-can, screw-terminal styles.

They are polar devices, having distinct positive and negative terminals, and are offered in an enormous variety of styles which include moulded and can-style SMT devices, axial- and radial-leaded can styles, snap-in terminals styles and large-can, screw terminal styles. Mounting can be normal style or stud mount. The following data from a manufacturer will give an idea of power available from these three configurations of capacitors:

Table 10.1 *Power Available from Electrolytic Capacitor*

Category (Terminations)	*Snap-in*	*Plug-in*	*Screw-terminal*
Application power range	0.1–30 kW	0.5–50 kW	0.5 kW–10 MW
Robustness	Moderate	Excellent	Excellent
Mounting	Circuit board	Circuit board	Circuit board or bus assembly
Ripple current per cap	< 50 A	< 50 A	< 100 A
Max temperature	105°C	105°C	105°C
Voltage range	6.3–500	6.3–500	6.3–550
Best typical life at 85°C	90k hours	> 100k hours	> 100k hours
Over-voltage withstand	Moderate	Moderate	Superior
Series inductance	Low	Moderate	Moderate

These capacitors are used in large inverter power supplies everywhere, and on most naval vessels. Flashlight cameras and professional flashlights also make use of aluminium electrolytics from 15 μF to 1500 μF and voltage ratings from 300 V to 600 V for energy storage for flash. Flashgun capacitors typically store energy from 15 joules upwards up to 600 joules or more.

10.4.4 Limitations of Electrolytic Capacitors

The disadvantage of electrolytic capacitors is the non-ideal, high loss characteristics which arise from the semi-conductive oxide properties, double-layer effects from the electrolyte-oxide charge-space region, resistive losses from the high electrolyte resistivity, poor frequency response due to the roughness of the surface oxide, and finite capacitor life due to breakdown and degradation of the electrolyte.

The anodic oxide dielectric is polar, and so are the electrolytic capacitors; that means the capacitors must be connected with the correct polarity as marked. Connecting with reverse voltage injects hydrogen ions through the oxide readily, causing high electrical conduction, heating and reduction of the anodic oxide film. Non-polar (or bi-polar) devices can be made by using two anodes instead of an anode and a cathode, or one could connect the positives or negatives of two identical devices together, then the other two terminals would form a non-polar device.

Aluminium electrolytic capacitors can withstand reverse voltages up to 1.5 V. Higher reverse voltage can cause failure by pressure build-up and rupture of the capacitor's safety vent structure. Non-polar and semi-polar devices are available that can withstand reverse voltage.

10.4.5 Safety Considerations

Pressure-Relief Vent

During operation of an aluminium electrolytic capacitor with non-solid electrolyte, gas pressure normally increases. This gas is mostly hydrogen and excess pressure is avoided by permeation of the gas through the capacitor's seal. But in cases like application of over-voltage, reverse voltage, AC voltage, capacitor failure or excess pressure can cause the capacitor to explode. To reduce the risk of explosion aluminium electrolytic capacitors are usually equipped with pressure-relief vent structures. These safety vents are intended to rupture and release the gas pressure. After rupture the capacitor has limited life because it loses the electrolyte and dries out. One should be careful not to interfere with the operation of the vent, for instance by mounting measures such as clamps, glue or potting compounds. In the case of large capacitors with internal cans secured by thermoplastic potting, they should not be mounted with the safety vents down as the potting may flow when the capacitors overheat and block the vents.

10.5 TANTALUM ELECTROLYTIC CAPACITORS

Background Information

Tantalum is a chemical element with the symbol **Ta** and atomic number 73. A rare, hard, blue-grey, lustrous metal, tantalum is highly corrosion resistant and occurs naturally in the mineral tantalite, always together with the chemically similar niobium. The chemical inertness of tantalum makes it a valuable substance for laboratory equipment and a substitute for platinum, but its main use today is in tantalum capacitors in electronic equipment.

Tantalum has been a favoured capacitor technology in space-limited designs for a long time. The tantalum capacitor is a highly reliable capacitor with solid or fluid electrolyte. Surface mount tantalum capacitors are increasingly being used in circuit designs because of their volumetric efficiency, basic reliability and process compatibility. These are more expensive than aluminium-based capacitors and generally only available in low-voltage versions, but because of their smaller size for a given capacitance and lower impedance at high frequencies, they are popular in miniature applications such as cellular telephones.

Ordinary aluminium electrolytic capacitors are rather large for many uses. In applications where size is of importance tantalum capacitors are preferred. These are much smaller than the aluminium electrolytic capacitors and instead of using a film of oxide on aluminium they use a film of oxide on tantalum. Tantalum capacitors do not normally have high working voltages, 35 V is normally the maximum, and some even have values of only a volt or so.

Tantalum electrolytic capacitors or tantalum beads are available in both wet (foil) and dry (solid) electrolytic types, with the dry or solid tantalum being the most common. Solid tantalums use manganese dioxide as their second terminal and are physically smaller than the equivalent aluminium capacitors. The dielectric properties of tantalum oxide are also much better than those of aluminium oxide giving lower leakage currents and better capacitance stability, which makes them suitable for timing applications. These capacitors can replace aluminium capacitors in many places. Also, tantalum capacitors although polarized, can tolerate being connected to a reverse voltage much more easily than the aluminium types but are rated at much lower working voltages. Typical values of capacitance range from 47 nF to 470 uF.

Tantalum capacitors can be made in much closer tolerances than aluminium electrolytics. The low leakage and high capacitance of these capacitors makes them suitable for sample and hold circuits to achieve long hold duration, and some long duration timing circuits where precision is not critical. They are a preferred choice for high reliability applications like medical electronics, computers, space equipment etc.

Tantalum capacitors are available in the following varieties:

- Tantalum foil electrolytic capacitors
- Wet tantalum capacitors: capacitors with porous anodes and liquid electrolytes
- Dry tantalum capacitors: capacitors with porous electrodes and solid electrolytes.

Tantalum foil electrodes were the first to be introduced, around 1950. They were developed as a more reliable electrolytic capacitor without the limitations of shelf life associated with aluminium electrolytics. Sulphuric acid was mainly used as the electrolyte. They can operate up to 120°C, and the leakage current depends on the purity of materials.

In all the three types, tantalum wire forms one electrode of capacitor, whereas the second electrode could be a silver coating, graphite or solder. Tantalum pentoxide, coated with MnO_2 serves as electrolyte. Tantalum in capacitors is characterized by high volumetric efficiency, which enables miniaturization, high reliability and stability over a wide temperature range (– 55°C to 125°C), attributes not matched by other types of capacitor material, such as ceramics. Frequency characteristics are also superior.

Like all electrolytic capacitors, tantalum capacitors are also polarized and they are very intolerant of being reverse biased, often exploding when placed under stress. However their small size makes them very attractive for many applications. They are available in both leaded and surface mount formats (Fig. 10.10).

(a) Hole mount leaded capacitors (b) SMD capacitors

Fig. 10.10 *Tantalum capacitors.*

The thin layer of oxide and high surface area of the porous sintered material gives this type a very high capacitance per unit volume. The cathode electrode is formed either of a liquid electrolyte connecting the outer can (wet tantalum capacitor) or of a chemically deposited semi-conductive layer of manganese dioxide (dry tantalum capacitor), which is then connected to an external wire lead. A newer development of this type (tantalum polymer capacitor) replaces the manganese dioxide with a conductive plastic polymer (polypyrrole) that reduces internal resistance and eliminates a self-ignition failure.

The conventional technology with MnO_2 was introduced 30–40 years ago as a major improvement over the wet tantalum electrolyte electrode system. Tantalum MnO_2 capacitors have established themselves as highly reliable and stable capacitors. Voltage range is typically from 2.5 to 50 V and cases sizes to better fit with specific height or space limited designs. Capacitors up to 1500 µF are on offer. The unique flexibility of the powder technology to provide thin and flat capacitors is very important for applications with critical height such as cellular phones and computers. The conventional MnO_2 technology strength is in the robustness, higher temperature and DC bias stability, voltage range up to 50 V and very good steady state reliability. The most popular applications of tantalum MnO_2 capacitors today include automotive (up to 175°C operation temperature), military, aerospace, medical and high-end applications such as servers.

10.5.1 Wet Tantalum Capacitor

The first wet tantalum capacitors were developed 40 years ago, and comprised a tantalum anode surrounded by an electrolyte inside a silver case with an epoxy end seal. This design was problematic because it could be prone to leakage of the electrolyte through the epoxy seal. It also had a limited ability to withstand any reverse voltage. The silver case material was later replaced with tantalum, which proved more stable over a range of applications.

Figure 10.11 shows the construction of a present-day tantalum capacitor. The use of a tantalum case makes it easier to construct a tantalum base-to-metal end seal that could be laser-welded to the tantalum can, making a hermetic capacitor. This eliminates the risk of fluid leakage and improves overall reliability. The use of a porous tantalum sleeve inside the case increases the area of the cathode system.

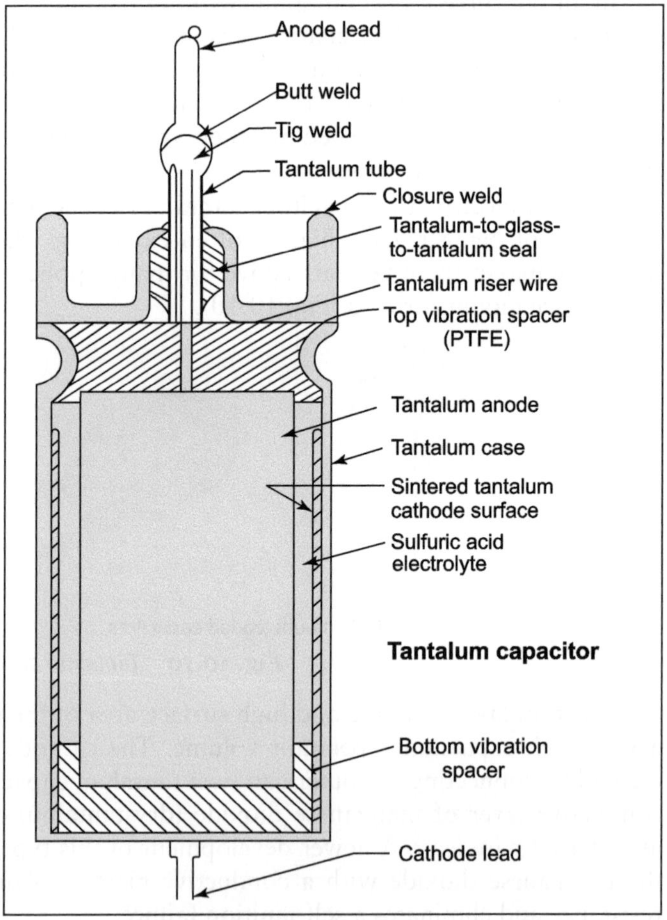

Fig. 10.11 *Wet tantalum capacitor.*

The second generation of wet tantalum capacitor technology enables higher-efficiency and higher-reliability capacitor designs, extending their applications. Traditional wet tantalum capacitors use a sleeve of pressed and sintered tantalum powder for the cathode system. Extremely fine high-purity tantalum powder is pressed into a cylindrical pellet, at the same time embedding a tantalum riser wire into the centre of the pellet. The pellet is sintered, causing tantalum particles to fuse together into a continuous structure with very high internal surface area. A tantalum pentoxide dielectric is formed over this surface by immersing the pellet in acid, making electrical contact via the riser wire and applying current and voltage. The process to make the formed pellet (often called the anode, as the riser wire forms the positive contact) is identical to that of solid tantalum anode construction.

Tantalum dielectric has excellent temperature characteristics and reliability (no wear-out mechanism), and the wet electrolyte system has many advantages over solid technology. It features a more efficient, continual self-healing mechanism, and can support higher voltage ratings (up to 125 V compared to the 50-V maximum typical for most solid capacitors) and lower leakage current. This extends the application suitability far beyond the 28 V bus. Wet tantalum capacitors also are an ideal solution, specifically where size and weight are concerns.

10.5.2 Solid Tantalum Capacitors

The reliability of the solid tantalum capacitor is heavily influenced by environmental conditions such as temperature, humidity, shock, vibration, mechanical stresses, and electric stresses, including applied voltage, current, ripple current, transient current and voltage and frequency. These factors need to be observed in service. Voltage and temperature are important parameters when estimating the field failure rate.

Structure of solid tantalum capacitors

Tantalum capacitors are manufactured from a powder of relatively pure tantalum metal. Wet tantalum capacitors contain sulphuric acid. In order to contain this corrosive material, they use a solid silver case to contain the wound foil, and the connections are made through glass-to-metal seals. The resulting capacitor is extremely effective, with low ESR, and is highly reliable.

The solid tantalum capacitor is made by a different approach, creating a sponge of high purity tantalum powder around a tantalum wire by pressing tantalum powder into a small slug (pellet) around the lead (riser). The riser wire ultimately becomes the anode connection to the capacitor. This pellet/wire combination is subsequently sintered under vacuum at high temperature (typically 1200–1800°C) to produce a mechanically strong pure and highly porous pellet having large surface area.

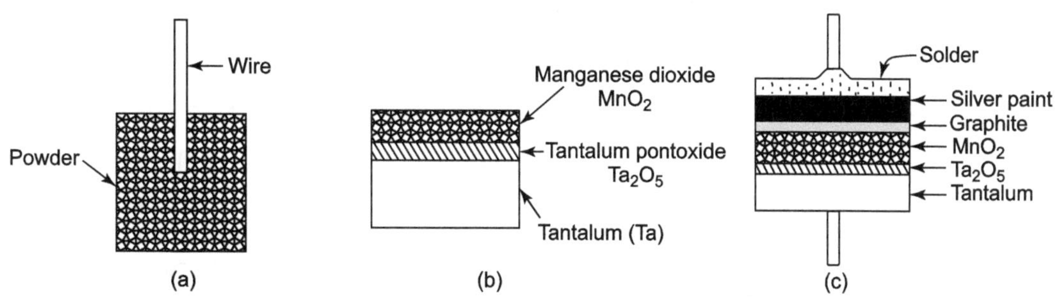

Fig. 10.12 *(a) Pressed and sintered Ta (b) Tantalum pellet (c) All tantalum layers*

The dielectric is then formed over all the tantalum particle surfaces by the electrochemical process of oxidation. The 'pellet' is submerged into a very weak solution of acid (e.g. phosphoric acid) at a high temperature, and voltage is gradually increased with controlled current to create an oxide layer of the correct thickness. Figures 10.12 (a) and (b) clarify the construction of pellets. The total dielectric thickness is determined by the final voltage applied during the forming process. During the chemical reaction, tantalum oxide (Ta_2O_5) is formed throughout the tantalum mass.

The next stage is to produce the cathode plate, making contact with every part of the top surface of the oxide. This is carried out by dipping the pellet into manganese nitrate, and then baking it at 250°C to produce the dioxide coat. The process is repeated several times, so that it is fully penetrated and a thick coating built on the outside surface; this 'manganese anode' is then coated with graphite, and finally a metal layer such as silver as in Fig. 12 (c). The cathode is a combination of silver, graphite and manganese dioxide. The carbon is coated with a conductive material (silver) to facilitate connection to the external cathode termination. Capacitors can be subdivided into four basic groups: chip/resin dipped/moulded/axial leaded.

The encapsulation of a solid tantalum capacitor can be done in several ways. The original design included soldering inside a metal can and sealed with a glass-to-metal hermetic seal. Potting with an epoxy resin inside of a moulded plastic shell was the next step, followed by transfer moulding with epoxy, and then dipping in liquid epoxy resin. The final process in vogue today is the tantalum chip, which has been encapsulated in epoxy and has several terminal designs to provide protection against the rigors of directly soldering onto ceramic or glass epoxy substrates. A cutaway view of SMD type solid tantalum capacitor is shown in Fig. 10.13.

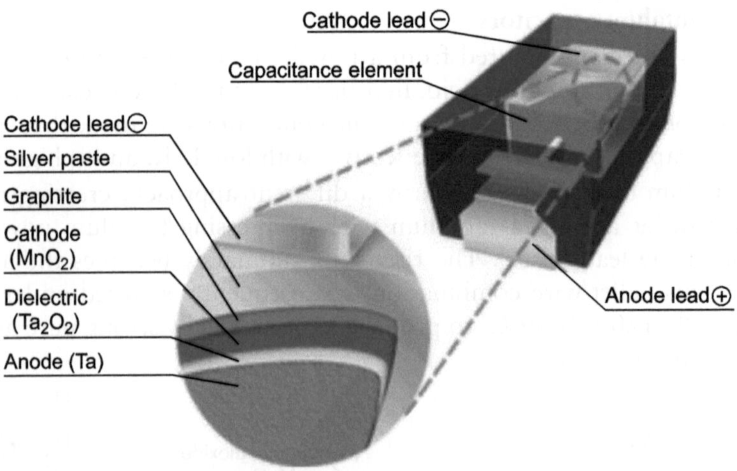

Fig. 10.13 *Cutaway view of surface mount tantalum capacitor* (Source: NEC data sheet).

The reliability of the solid tantalum capacitor is heavily influenced by environmental conditions such as temperature, humidity, shock, vibration, mechanical stresses and electric stresses, including applied voltage, current, ripple current, transient current and voltage and frequency. These factors need to be observed in service. Voltage and temperature are important parameters when estimating the field failure rate.

Characteristics of tantalum capacitors

Tantalum capacitors have very low electrical leakage (high leakage resistance), and can retain a charge for a long duration. These capacitors are not particularly tolerant of heavy charge and discharge currents. They are tolerant of hot operating environments up to 125°C, unlike aluminium electrolytic capacitors. However, de-rating of voltage has to be applied for non-solid capacitors with increasing temperature, as shown in Fig. 10.14. These capacitors are relatively expensive. They cannot tolerate heavy charge and discharge currents. Being polar in nature, correct polarity must be observed, otherwise the dielectric oxide layer will break down, reducing the resistance of the device and causing it to fail. Because of the solid nature of the tantalum capacitor construction, there is no known wear-out mechanism.

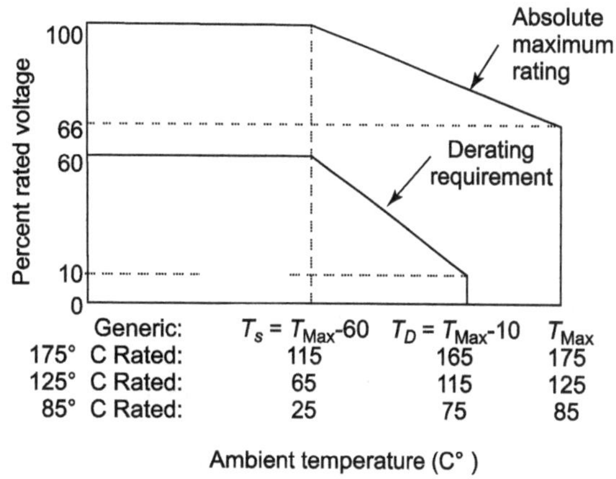

Fig. 10.14 *De-rating of non-solid tantalum capacitors with temperature.*

Tantalum electrolytic capacitors are less prone to 'drying out' than aluminium capacitors, which tend to decrease in capacitance particularly when used in hot environments. Tantalum capacitors maintain their designed capacitance under such conditions over decades. Low-voltage tantalum capacitors are commonly used in large numbers for power supply filtering on computer motherboards and in peripherals due to their small size and long-term reliability.

Tantalum electrolytic capacitors exploit the tendency of tantalum to form a protective oxide surface layer. Tantalum powder is pressed into a pellet shape, as one plate of the capacitor, the oxide as the dielectric, and an electrolytic solution or conductive solid serves as the other plate. The dielectric layer being very thin (thinner than the similar layer in an aluminium electrolytic capacitor), a high capacitance can be achieved in a small volume. Because of the size and weight advantages, tantalum capacitors are attractive for portable telephones, personal computers and automotive electronics. Most ranges of solid tantalum capacitors in miniature chip form are rated from 0.1 μF to 100 μF, tolerances down to ±5%, and working voltages up to 35 V. Tantalum capacitors make highly efficient use of PCB area and component volume for high values of capacitance.

Tantalum SMD capacitors

Tantalum SMD capacitors are widely used as they provide capacitance higher than those made from ceramics. Solid tantalum technology is used in these SMD capacitors. They remained the mainstay of SMD capacitors in early phases of SMDs due to their superior thermal properties making them ideal for SMD reflow soldering. With technological developments in recent years, Aluminium electrolytics are available in SMDs. Still, Tantalum electrolytics are preferred at many places because of other advantages.

10.6 POLYMER TANTALUM CAPACITOR

While the first patented tantalum capacitor was claimed to have ESR of roughly 2.0 ohms, a similar capacitor today has ESR of about 0.1 ohms. Today's digital electronics frequently require capacitors to have still lower ESR in the low milliohms. A substantial part of the ESR of a tantalum capacitor comes from its solid electrolyte material MnO_2. While MnO_2 is substantially more conductive than almost all wet electrolytes, especially at low temperatures, search was on for higher conductivity materials to replace MnO_2. This led to the development of the conductive polymer polyethylene-dioxy-thiophene (PEDT) which has up to 100 times MnO_2's conductivity and has generally acceptable compatibility with tantalum pentoxide, the tantalum capacitor's dielectric.

Conductive polymer electrolyte made possible remarkable improvements in capacitor ESR. For lower-voltage capacitors, improved dielectric strength and long-term reliability are also observed. Also, substantially more capacitance stability with frequency is observed. But the limitations to the technology are marginal material stability at elevated temperatures, lower self-healing capability, and reduced reliability at high rated voltages. Currently, single-anode capacitors with 100 kHz ESR under 7 mΩ have been manufactured and multiple-anode tantalum polymer capacitors have been made with ESR under 3 mΩ.

The details of these capacitors are shrouded in mystery and patents; it appears that this type of capacitor replaces the manganese dioxide electrode by a solid conductive polymer. The reason for this is that conventional tantalum capacitors tend to fail in short-circuit mode. Tantalums with manganese dioxide cathodes have been known to ignite explosively when shorting and passing high current, whereas the polymer tantalum type does not. A claim is also made that healing may occur by one or two potential mechanisms.

- The polymer next to the dielectric fault site may overheat and vaporize, disconnecting the fault site from the circuit.
- The polymer may oxidize into a more resistant material that plugs the defect site in the dielectric and reduces the flow of current.
- Another benefit of the polymer tantalum capacitor is capacitance stability over a wider frequency range compared with manganese dioxide types. Among the limitations is lower maximum rated voltage for these parts, currently at 10 V DC, and higher leakage currents than traditional capacitors.

10.7 NIOBIUM OXIDE CAPACITORS

Niobium oxide capacitors are a recent development, and these are used in place of tantalum due to certain advantages. Niobium (atomic number 41) is a sister metal to tantalum, and it has a lot of similar features. Niobium is more abundant in nature as compared with tantalum, and costs are low. Niobium oxide has been found to have better features for production of capacitors, and can be manufactured on the same machines. It exhibits a metal like conductivity and can be produced with simpler and higher yielding powder manufacturing techniques. Niobium oxide has the advantage of high ignition resistance and safety through its efficient self-arresting failure mechanism, and capacitors made from this material have better steady state reliability. They are better suited for reflow soldering. These capacitors may slowly drive out tantalum capacitors in many applications.

Niobium capacitors have largely identical performance properties to tantalum capacitors. While tantalum capacitors are suitable for voltages up to 60 V, the cheaper niobium capacitors are able to handle voltages only up to 10 V. Technology of niobium oxide capacitors is based on the conventional tantalum capacitors with MnO_2 electrode. The electrode originally made from tantalum powder is replaced by powder from niobium oxide (NbO). The MnO_2 self-healing process works efficiently in the case of NbO capacitor also. NbO capacitors have no wear-out mechanism and due to its self-healing and self-arresting mechanisms it provides the highest level of safety and reliability.

Technology for niobium oxide (NbO) capacitors is based on conventional tantalum capacitors with MnO_2 electrode. The first electrode, originally made from tantalum powder, is replaced by niobium oxide powder (NbO). The MnO_2 self-healing process also works efficiently in the case of the NbO capacitor. NbO has a very high ignition energy compared to tantalum and a much lower risk of combustion.

Although aluminium capacitors are cheap, they have very high resistances because of their liquid cathode. Tantalum and niobium capacitors are more like microscopic sponges. To obtain this structure, the grains of powder, which are only a few thousandths of a millimetre in size, are compressed around a tantalum wire to form beads measuring about a millimetre. They are then fused together to produce an anode with an all-conductive structure and a large specific surface area. This is given a wafer-thin, non-conductive oxide layer in an acid bath under electrical voltage. This layer insulates the anode. The cavities left in the sponge-like structure are filled with manganese dioxide or increasingly with conductive polymers.

Advantages of NbO capacitors

- Manufactured using existing tantalum processing plant
- Non-burn characteristics – failed devices guaranteed not to burn up to rated voltage, designed to eliminate smoke.
- Halogen free and compatible with lead-free directives (EC)
- Reliability 2 – 5 times better than Al electrolytic capacitors – up to 500 K hours. MTBF up to 50 times higher
- Sizes much smaller and thinner than Al capacitors
- Reduced voltage de-rating
- Temperature up to 125°C
- Rated voltage under 10 V

The three types of tantalum-equivalent capacitors have characteristics summarized in Table 10.2 below for reference.

Table 10.2 *Properties of Tantalum-Equivalent Solid Electrolyte Capacitor Systems*

Characteristic	*Tantalum*	*Tantalum-polymer*	*Niobium oxide*
Anode	Tantalum	Tantalum	Niobium oxide
Electrode	Manganese dioxide	Conductive polymer	Manganese dioxide
ESR (D case typical)	35 to 100 mΩ	15 to 30mΩ	50 to 100 mΩ
Cap tolerance	10% or 20%	20%	20%
DC leakage current	0.01 CV	0.1 CV	0.02 CV
Basic reliability	1%/1000 hours at 85°C	1%/1000 hours at 85°C	0.2%/1000 hours at 85°C

The figures above are given for the sake of comparison between different electrolytic capacitors. Niobium is a sister metal to tantalum and it has a lot of similar features. It is more abundant in nature as compared with tantalum, and costs are low. Niobium oxide has been found to have better features for production of capacitors. It exhibits a metal like conductivity and can be produced with simpler and higher yielding powder manufacturing techniques. Niobium oxide has the advantage of high ignition resistance and safety through its efficient self-arresting failure mechanism, and a capacitor of this material has better steady-state reliability.

Fig. 10.15

11

CERAMIC CAPACITORS

11.1 BACKGROUNDS

Ceramic capacitors are the workhorses of the capacitor world these days. These capacitors are used extensively owing to a combination of their cost and performance. A wide variety of dielectrics are used, but as the name suggests, they are all ceramic in nature. Ceramic capacitors have also been used for many years, starting from valve or tube circuits dating from the 1930s. The vast majority of ceramic capacitors that are used today are in the form of surface mount technology devices. Millions of these ceramic capacitors are used every day in every form of mass produced electronics equipment.

Ceramic capacitors are available in a variety of formats ranging from leaded components to surface mount technology (SMT) varieties. These are available in all the common formats. Ceramic capacitors or disc capacitors, as they are generally called, are made by coating two sides of a small porcelain or ceramic disc with silver, which are then stacked together to make a capacitor. For very low capacitance values a single ceramic disc of about 3–6 mm is used. Ceramics having a high dielectric constant (High-K) enable relatively high capacitances to be obtained in a small physical size. They exhibit large nonlinear changes in capacitance against temperature and are non-polarized devices.

Ceramic capacitors are normally used for radio frequency and some audio applications. They are also used for XY capacitor applications. Capacitance ranges from as low as a few pF to around 0.1 µF. In view of their wide range and suitability for RF applications they are used for coupling and decoupling applications. These capacitor s are the most commonly used type, being cheap and reliable. Their loss factor is particularly low, although dependent on the type of ceramic in use. Stability and tolerance is not as good as silver mica types, but their cost is much less. In view of their constructional properties, these capacitors are widely used both in leaded and SMD types.

Low breakdown voltage of ceramics means that low-K ceramics with the good electrical properties have poor volumetric efficiency and are usually found only in small values. High-K ceramics have poor electrical properties, are highly dependent on temperature, voltage, and frequency, and a significant ageing rate. Since ceramics have no self-healing mechanism, a high level of quality control is necessary during manufacture. Ceramics are most cost-effective in small sizes.

The actual performance of ceramic capacitors is highly dependent upon the type of ceramic dielectric used. Very high values are available; parameters such as the temperature coefficient and tolerance need to be checked. Ceramic capacitor values range from as low as a few picofarads to around 0.1 microfarads. In view of the wide range and suitability for RF applications they are used for coupling and decoupling applications, where they are by far the most commonly used type, being cheap and reliable and the loss factor is particularly low although this is dependent on the exact dielectric in use. Ceramic dielectrics are made from a variety of forms of ceramics. The exact formulas of the different ceramics used in ceramic capacitors vary from one manufacturer to another but common compounds include titanium dioxide, strontium titanate and barium titanate.

A ceramic capacitor is a non-polar device, most common being the 'disc capacitor'. This was used extensively in radio receivers and other devices from about 1930 through the 1950s, and in transistor equipment from the 1950s through the 1980s. Presently ceramic disc capacitors are widely used in electronic equipment. They are available in various shapes and styles, some of which are shown in Fig. 11.1, and include:

- Tubular shape (not popular now)
- Multilayer rectangular block, surface mount
- Bare leadless disc, sits in a slot in the PCB and is soldered in place, used for UHF applications disc, resin coated, with through-hole leads
- Disc, resin coated, with through-hole leads

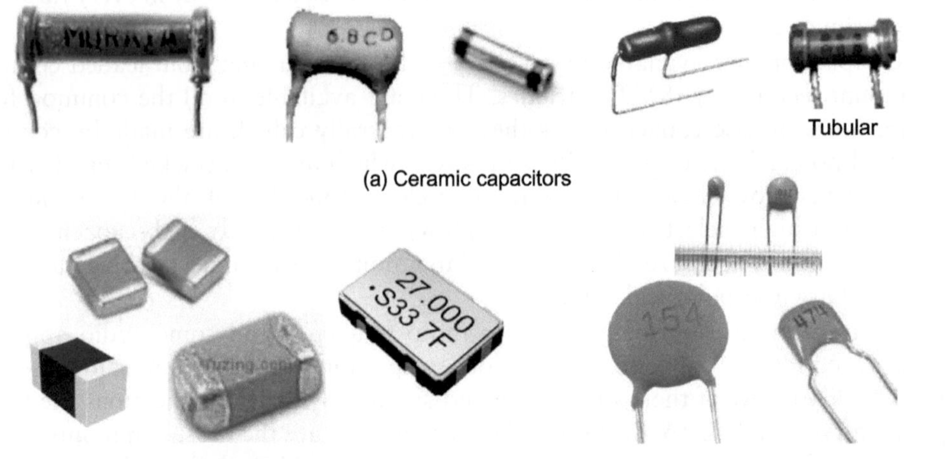

Fig. 11.1 *Ceramic capacitor types.*

A number of dielectrics are used in ceramic capacitors. For low values a dielectric type 'C0G' is normally used. This has the lowest dielectric constant but gives the highest stability and lowest loss. Where higher values are required in a given size, a dielectric with a higher dielectric constant must be used. Types with nomenclatures X7R and for higher values, Z5U are used, however their stability and loss are not as good as those of the capacitors made with C0G dielectric.

11.2 CLASSES OF CERAMIC DIELECTRICS

The ceramic dielectrics are made of a variety of titanates, zirconates and oxides (and many other things as well). Common ingredients include titanium dioxide, barium titanate and strontium titanate. The EIA (Electronic Industries Association, USA) classifies ceramics into four classes (1–4), and into types within those classes. Lower class number indicates better overall characteristics, but larger size for a given capacitance. The following are the types of dielectrics used in ceramic capacitors, depending upon stability and temperature requirements. Class 4 has almost disappeared, and we shall discuss the remaining three classes.

C0G or NPO: Ultra stable Class I dielectric, with negligible dependence of electrical properties on temperature, voltage, frequency and time. This dielectric is used in circuitry requiring very stable performance. Values are typically 1 pF to 0.1 μF, tolerance ± 5%. High tolerance and good temperature performance is their main advantage. They are larger in size and more expensive. A typical Class I capacitor will have a temperature coefficient of 30 ppm°C^{-1}. This will typically be fairly linear with temperature. These also find use in high Q filters.

The common compounds used as dielectrics are magnesium titanate (for positive temperature coefficient), or calcium titanate (for capacitors with negative temperature coefficient). Using combinations, it is possible to obtain a dielectric constant of between 5 and 150. Temperature coefficients vary between +40 and –5000 ppm°C^{-1}.

Class I capacitors also offer the best performance with respect to dissipation factor; a typical figure may be 0.15%. It is also possible to obtain very high accuracy (~1%) class 1 capacitors (5% or 10% tolerance is usual). The highest accuracy Class I capacitors are designated C0G or NP0.

X7R: Stable Class II dielectric, with predictable change in properties with temperature, voltage, frequency and time. These offer better performance with respect to volumetric efficiency, but at the cost of lower accuracy and stability. These are normally used for decoupling, coupling and bypass applications, where accuracy is not of prime importance. A typical Class II capacitor may vary capacitance by as much as 15% from – 50°C to +85°C temperature range, and may have a dissipation factor of 2.5%. Tolerance may vary from ±10% down to +20% to –80%. These figures would not present a problem when used as blocking, decoupling, bypassing and frequency discriminating elements. This dielectric is ferroelectric, and provides higher capacitance than Class I. These are typically 100 pF to 2.2 μF, tolerance ±10%. These are good for non-critical applications like coupling, timing, biasing etc.

Z5U/Y5V: General purpose Class III dielectrics with higher dielectric constant and greater variation of properties with temperature and test conditions. Very high capacitance per unit volume is attainable for general purpose applications where stability is not important. Values are typically 1 nF to 10 μF, tolerance ±20%. They are also not normally able to withstand high voltages. The dielectric used is often barium titanate that has a dielectric constant of up to about 1250. These are good for bypass, coupling applications.

Their advantage lies in low price and small size. Y5V and Z5U are general purpose EIA Class III dielectrics, with + 22% to – 56% (Z5U) and + 22% to – 82% (Y5V) variations over temperature range respectively and very high capacitance density. These are very stable with time, typically ageing less than 2% per decade. General purpose chips are used in bypass and decoupling functions and other applications where capacitance change over the operating temperature range is not critical. Class III ceramic capacitors are typically used in decoupling or in other power supply applications where accuracy is not an issue. However they must not be used in applications where spikes are present as these may damage the capacitor if they exceed the rated voltage.

Figure 11.2 brings out the long-term stability performance of the three types of dielectrics. C0G is fully stable over time, while X7R and Z5U undergo depreciation in value, Z5U showing as much as 15% downward drift over 1000 hours.

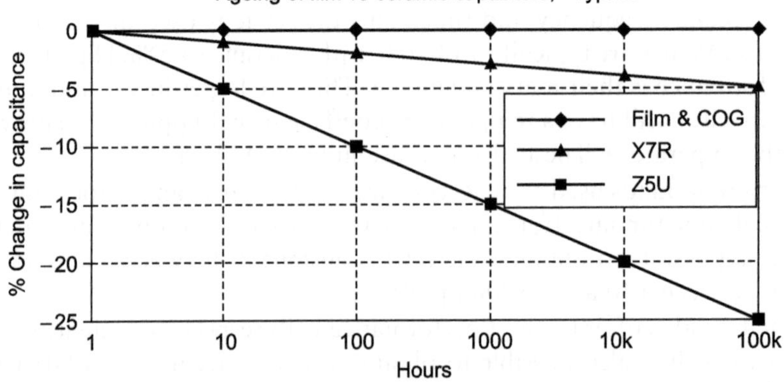

Fig. 11.2 *Effect of ageing on ceramic capacitors* (Source: Cornel Dubilier).

Apart from these, ceramic chip capacitors are available with 1% accuracy, values up to about 1 μF, typically made from lead zirconate titanate (PZT) ferroelectric ceramic.

These classes correspond roughly to low *K*, medium *K* and high *K*. None of the classes can be said to be 'better' than any others – the relative performance depends on application. Class I capacitors are physically larger than Class III capacitors, and for bypassing and other non-filtering applications, accuracy, stability and loss factor may be unimportant, while cost and volumetric efficiency do matter. Hence these are primarily used in filtering and in RF applications. Class III capacitors are typically used in power supply applications. As ceramic technology has improved, ceramic capacitors are now commonly available in values of up to 100 μF, and they are increasingly competing with electrolytic capacitors, where ceramics offer much better electrical performance at prices which continue to fall with improvement in technology.

[A] Ceramic disc capacitor

One form of ceramic capacitor looks almost exactly like the classical model of a parallel plate capacitor. A square or circular-shaped ceramic dielectric is prepared and coated with conductors on each flat face. If the value of K is known for the dielectric, the area of the conductors and the thickness of the dielectric can be measured, and the capacitance can be calculated directly.

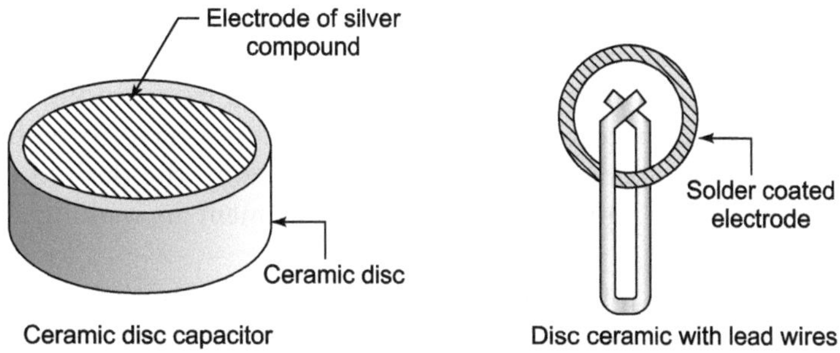

Fig. 11.3 *Ceramic disc capacitor construction.*

A circular shape is called a 'ceramic disc' capacitor. The dielectric is made from finely powdered materials, barium titanate being the most common. Disc elements are pressed in dies and then fired at high temperature to produce a very dense structure. Single-plate elements are usually cut from larger sheets of fired ceramic material. Electrodes for both discs and single-plates are formed from a compound containing powdered silver, powdered glass and an organic binder. This material is screen-printed onto the discs or sheets from which the single plates are cut. Another firing step removes the binder and melts the glass, binding the silver glass matrix to the ceramic surfaces.

The outer surface is easily solderable, and wires are usually attached as seen in the figure in a radial configuration. The springy hairpin shaped wires hold the ceramic elements while the assembly is dipped in solder. The lower end of the hairpin shaped wire is cut off later. This process is mechanized, and dipped discs are among the cheapest capacitors available.

[B] Monolithic/MLCC ceramic capacitors

A much more sophisticated design is the monolithic ceramic capacitor. It offers much higher capacitance per unit volume. A monolithic capacitor consists of thin dielectric layers interleaved with staggered metal-film electrodes, and has leads connected to alternate projecting ends of the electrodes – the assembly is compressed and sintered to form a solid monolithic block. The ceramic material acts both as dielectric and as encapsulation of the basic element. Electrodes are buried within the ceramic and exit only on the ends. The ends are surrounded with silver-glass compounds, which form the terminations.

The construction of ceramic capacitors has changed dramatically over the last few decades. Earlier, the common form was that of a single-layer ceramic disk. A metal layer was screened and fired on each side, and wires soldered on, to make up to 1000 V/0.01 μF capacitors. With the decline of the vacuum tube, their demand has declined. Today engineers require maximum capacitance in a smaller volume. This gave rise to 'multilayer ceramic capacitor' (MLCC). MLCCs are made by screening many very thin alternating layers of unfired ceramic and silver-palladium alloy electrodes, and firing the result to make a finished capacitor.

The manufacturing process for monolithic/multilayer ceramics is much more complicated and sophisticated than that needed for discs or single plates. The powdered ceramic materials are mixed with a binder and cast on moving belts into thin flexible sheets which are wound onto reels and stored. The sheets are then printed with electrode patterns. The ink used in this printing is pigmented with finely divided precious metals, usually chosen from among platinum, palladium and gold. Precious metals are necessary because the electrodes have to be fired above 1000°C along with the ceramic in an oxidizing atmosphere to develop the desired ceramic properties. The precious metal electrodes represent a major cost element in making monolithic ceramic capacitors.

After the ink is dried, pieces of sheet are stacked above one another, each piece representing one dielectric layer. The electrode patterns are printed so that alternate electrodes exit from opposite ends. Finally, cover layers which do not bear electrodes are placed on top and bottom. The assembly is compressed and fired, when the ceramic sinters into one homogeneous structure, from which the capacitors get the name 'monolithic.'

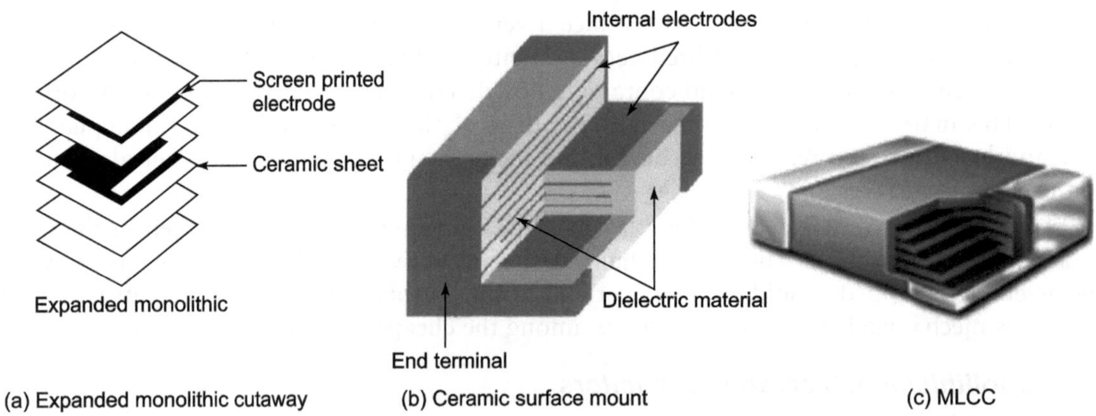

Fig. 11.4 *Monolithic multilayer capacitor.*

Twenty or 30 electrodes are very common in commercial practice and 60 or 80 might be used to obtain larger values of capacitance. The diagram above shows three electrodes. In a multilayer ceramic capacitor, $C = \varepsilon_0\varepsilon_k A(N - 1)/d$, where N = number of stacked plates. SMD ceramics are in extensive use, and their numbers go to almost 100 billion every year. SMD capacitors are in the shape of a rectangular block or a cuboid. Inner electrodes are connected to two terminations, either by a 65:35 silver/palladium (AgPd) alloy, or silver dipped in a barrier of plated nickel, and finally covered by plated tin (NiSn).

When DC bias is applied on a circuit, the degree of capacitance ageing varies depending on the level of DC bias voltage. Different grades of monolithic capacitors ceramics show wide variations when a DC bias is applied across them. Figure 11.4 indicates DC bias characteristics of monolithic ceramic capacitors at normal temperature. The main component of temperature compensation type capacitors (C0G, U2J characteristics, etc.) does not vary due to DC bias. Conversely, the capacitance of a high dielectric constant type capacitor (X5R, X7R, Z5U, Y5V characteristics, etc.) decreases due to DC bias, especially with Y5V characteristics. It will be noted that C0G ceramics are most stable, and DC bias has no effect on their value.

Ageing phenomenon is a basic characteristic of high dielectric constant type ($BaTiO_3$) ceramic capacitors, and the degree of change in capacitance from ageing varies depending on the type of ceramic material used. Here also C0G ceramics are absolutely stable. Therefore, when using high dielectric constant type ceramic capacitors, change in capacitance from the ageing phenomenon should be taken into consideration, and especially when the stability of capacitance is important, it should be verified on the actual circuit.

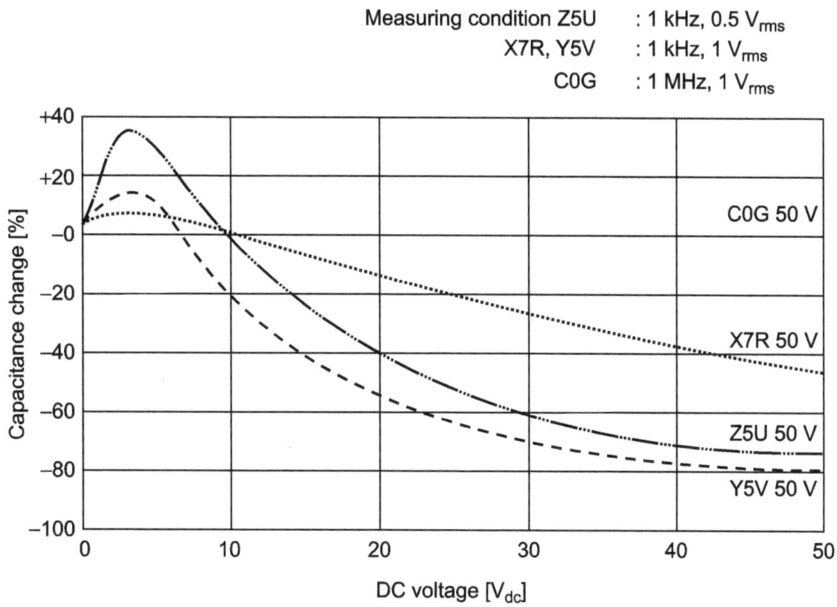

Fig. 11.5 *Effect of DC bias on MLCC value.*

[C] Tubular ceramic capacitors

In the early 1980s the filter connector, in its infancy, used exclusively tubular type capacitors. These capacitors served the needs of the industry well at that time. However, low yields and an array of quality problems suggested that the tubular capacitor was no longer sufficient for the systems it was designed into. Therefore, in the late 1980s the monolithic planar array was born into existence. This new technology incorporated the monolithic chip capacitor technology and adapted it to a multiline configuration. This gave both the ability to achieve higher capacitance per line as well as higher withstanding voltages. The two technologies are vastly different in their design and capabilities.

Wall thickness of the tube capacitor is 0.050″ (≃ 0.4 mm) minus the web dimension of the ground plane minus the wall thickness of the ferrite. Typically it is around 0.015″ thick. This limited thickness has to be designed to withstand the voltage rating of the system, achieve the desired capacitance and be strong enough for system vibration.

A ceramic tubular capacitor is a length of ceramic tube, having inner and outer surfaces painted with silver ink to form its plates. Tubular capacitors use the inside and outside surfaces of a ceramic tube, to form a coaxial capacitor. The wall thickness of the tube is dictated by the pin-to-pin spacing of the connector, the metal ground plate used to ground the capacitor, and the size of the ferrite in a Pi section filter. They have replaced ceramic disk capacitors in surface-mounted circuits to save board space and permit automatic placement. They are protected with a coat of protective resin.

The systems of today typically require much higher capacitance values and/or require higher voltage ratings. The fighter aircraft today has several requirements that exceed 2000 V DC and vibration requirements are the highest in industry. The 0.015″ tubular capacitor cannot handle these high vibration requirements and there is no space to increase either the capacitance or the voltage rating. The dielectric material in the capacitor typically is X7R type material to achieve the highest capacitance with the least change in capacitance over the temperature range. The tube has electrodes (which are stacked together in parallel to increase capacitance) running parallel to the contact. This in combination with the pin-to-pin spacing limits the capacitance to about 7000 pF at 200 V DC.

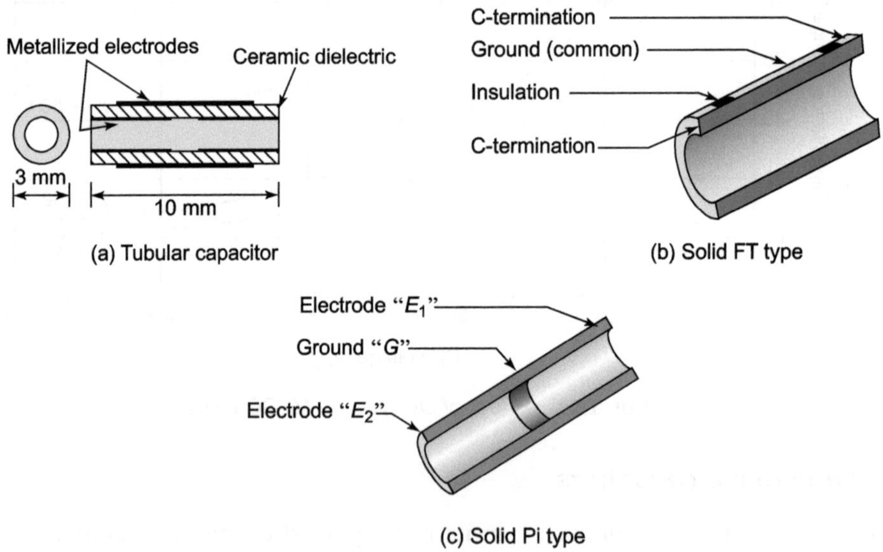

Fig. 11.6 *Tubular ceramic capacitors.*

The beneficial properties of tubular ceramic capacitors include:

- They are ideal for multi-pin connector applications
- High ratio of capacitance to volume
- Low inductance, non-polar
- Impervious to moisture and contamination
- – 55°C to + 125°C operation

Wide variety of tubular ceramic capacitors are available in small size, lightweight, non-polar and offer high dielectric strength. Operating temperatures of –55°C to +125°C are achieved with no voltage de-rating. All capacitors are fired to produce true monolithic structures, which are impervious to moisture and contamination. Outer terminations feature a nickel barrier and a final metal layer, typically silver. The tubes may be circular or square for surface mount.

[D] Leadless ceramic capacitors

The three kinds of leadless ceramic capacitors for direct soldering are the silvered disc, past and trapezoidal types. These capacitors are the most appropriate choice for TV tuners, FM tuners, high frequency bypass circuit in satellite-TV/radio transmitters and receivers and low voltage DC motor circuit, because of the minimized residual inductance due to the fact that the electrodes are directly soldered to the chassis or the copper leaf on the PCB.

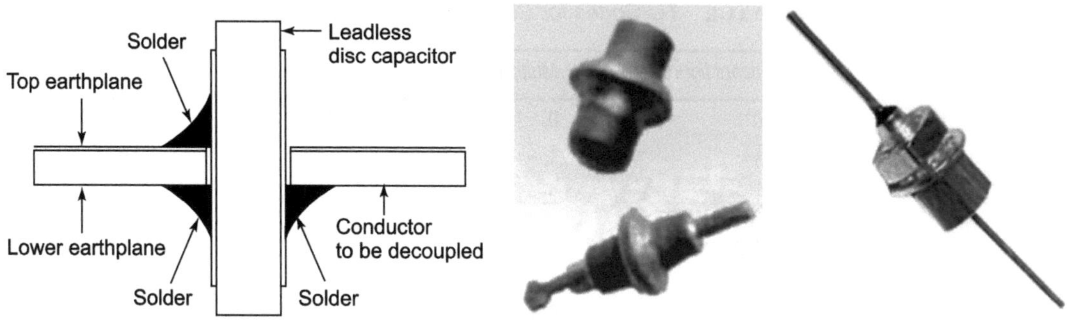

Fig. 11.7 *Leadless ceramic feed through capacitors.*

[E] SMD ceramic capacitors

SMD component capacitors of ceramic dielectrics are used today in large numbers. Millions of these ceramic capacitors are used every day in every form of mass produced electronics equipment.

Care must be taken when soldering these capacitors. If heat is applied for too long, then the terminations can be damaged. Fortunately modern versions are far more robust than much older capacitors which used to suffer from metallization if heat was applied for too long. Despite this, care should be taken, especially if these components are being soldered manually. Normally production methods using infrared reflow with carefully controlled heat profiles are recommended.

11.3 VALUES AND TC CODES

Ceramic capacitors are also known by codes indicative of their drift and tolerance, as given by a letter and two digits, which indicate their type.

There is a three-digit code printed on a ceramic capacitor specifying its value. The first two digits are the two significant figures and the third digit is a base 10 multiplier. The value is given in pF. A letter suffix indicates the tolerance, as per Table 11.1:

Table 11.1 *Ceramic Capacitor Tolerance Codes*

C	±0.25 pF	M	±20%
D	±0.5 pF	P	+100 — 0%
J	±5%	Y	—20 + 50%
K	±10%	Z	—20 + 80%

Example: A label of '104K' indicates 10 × 10^4 pF = 100 nF ±10%

Temperature-compensated capacitors use a different EIA (Electronic Industries Association, U.S.A.) code. Here, the first letter gives the significant figure of the change in capacitance over temperature in ppm°C^{-1}. The second character gives the multiplier. The third character gives the maximum error from that in ppm/°C^{-1}. All ratings are from 25°C to 85°C:

Class I capacitors are known by codes as per Table 11.2 below:

Table 11.2 *TC Codes for Class I Capacitors as per EIA*

Significant figure	*Multiplier*	*Tolerance*
C: 0.0	0: −1	G: ±30
B: 0.3	1: −10	H: ±60
L: 0.8	2: −100	J: ±120
A: 0.9	3: −1000	K: ±250
M: 1.0	4: +1	L: ±500
P: 1.5	6: +10	M: ±1000
R: 2.2	7: +100	N: ±2500
S: 3.3	8: +1000	
T: 4.7		
V: 5.6		
U: 7.5		

For instance, a C0G will have 0 drift, with an error of ± 30 ppm°C^{-1}, while a P3K will have −1500 ppm°C^{-1} drift, with a maximum error of ± 250 ppm°C^{-1}. Table 11.3 below shows the breakdown of the ETA three-digit '*TC*' codes for Class II ceramics:

Table 11.3 *TC Codes for Class II Ceramic Capacitors*

Low temperature limit	High temperature limit	Maximum allowable capacitance change from +25°C (0 VDC)
X = – 55°C	4 = + 65°C	F = ±7.5%
Y = – 30°C	5 = + 85°C	P = ±10%
Z = + 10°C	6 = + 105°C	R = ±15%
	7 = + 125°C	S = ±22%
	8 = + 150°C (Special)	T = +22%/–33%
	9 = + 200 °C	U = +22%/–56%
		V = +22%/–82%

X 7 R = ±15% ΔC over–55°C ~ + 125°C

Common '*TC*' designations include:

X5R = ± 15% change over – 55°C ~+ 85°C Standard Tolerance: K = ±10%

X7R = ± 15% change over – 55°C ~+ 125°C Standard Tolerance: K = ±10%
Y5V = + 22%/– 82% change over – 30°C ~+ 85°C Standard Tolerance: Z = –20%/+80%
Z5U = + 22%/– 56% change over – 10°C ~+ 85°C Standard Tolerance: M = ± 20%
The exception to the above system is ultra-stable 'TC': **C0G = NP0**
NP0 = 0 ± 30PPM/°C over the range of – 55°C ~ + 125°C ...Standard Tolerance: J = ± 5%
NP0 stands for negative positive zero [originated from military standards].

High frequency use

Ceramic capacitors are suitable for moderate to high-frequency work (into hundreds of megahertz range, or even into the low gigahertz), as modern ceramic caps are fairly non-inductive compared to the other major classes of capacitors (film and electrolytic). Capacitor technologies with higher self-resonant frequencies tend to be expensive (typically, mica or glass capacitors). Sample self-resonant frequencies for one set of C0G and one set of X7R ceramic capacitors are as per Table 11.4 below:

Table 11.4 *Frequency Limits for Ceramic Capacitors*

	10 pF	*100 pF*	*1 nF*	*10 nF*	*100 nF*	*1 μF*
C0G (Class 1)	1550 MHz	460 MHz	160 MHz	55 MHz		
X7R (Class 2)			190 MHz	56 MHz	22 MHz	10 MHz

12

MICA CAPACITORS

12.1 INTRODUCTION

Mica's thermal, electrical, and chemical properties make for excellent capacitors. Capacitance changes within the usable temperature range from ± 500 ppm°C^{-1} to 50 ppm°C^{-1}, depending on the construction technique. Mica capacitors exhibit very little voltage dependence, with dC/dv less than 0.1%. They exhibit high Q, or conversely small power factors (range 0.0001 1/N 0.0004) that are quite independent of frequency. This, combined with low inductance designs, makes mica capacitors ideal for high frequency and RF applications. Specification sheets of mica capacitors commonly show parameters plotted into the gigahertz range.

Although more expensive than other dielectrics, mica is an ideal form of dielectric for very high performance capacitors such as silver mica capacitors. These capacitors are available from lower voltages to high voltage ranging up to 70 kV DC. Values may vary from 20 pF to 10 μF. Mica also finds use in trimmer capacitors. Mica capacitors are extensively used in airborne, space, radar applications and lasers where low inductance is necessary. Mica capacitors are still indispensable in high-power RF transmitter applications and snubbers, for which special rectangular, cylindrical and button-style cases are used. Nowadays mica is finding increasing use in equipment that encounters very high temperatures like rockets, missiles and jet engine ignition system. Another niche is high-voltage applications that are a result of mica's high dielectric breakdown and corona resistance.

Mica capacitors, which are now obsolete in many applications, were in use in the early 20th century. They consisted of sheets of mica and copper foil sandwiched together and clamped. However, mica capacitors are currently used in some low power RF designs, and pulse (snubber) applications, but advances in ceramic capacitor performance have slowly eroded mica's traditional edge in these areas over the years. Perhaps more importantly, mica capacitors tend to be bulky – a result of the relatively low dielectric constant. For example, a 300 pF dipped mica capacitor may be as much as 16 times larger (in volume) than a good 300 pF MLC (NPO) capacitor.

Mica capacitors have good stability because their temperature coefficient is small. On account of their excellent frequency characteristics, they are used for resonance circuits, and high frequency filters. Also, they have good insulation, and so can be utilized in high voltage circuits. They were often used for vacuum tube equipment like radio transmitters. Mica capacitors do not have high values of capacitance, and can be relatively expensive.

Figure 12.1 shows 'dipped mica capacitors'. These can handle up to 500 V. The capacitors from the left: Capacitance: 47 pF (printed with 470J) [width 7 mm, the height 5 mm, thickness 4 mm] Capacitance: 220 pF (printed with 221J) [width 10 mm, height 6 mm, thickness 4 mm], Capacitance: 1000 pF (printed with 102J) [width 14 mm, height 9 mm, thickness 4 mm].

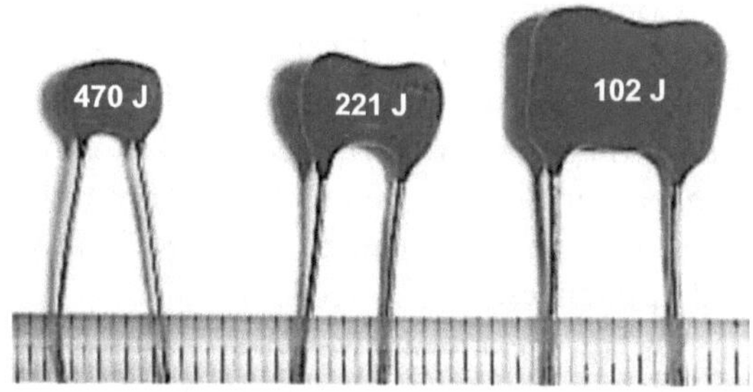

Fig. 12.1 *Dipped mica capacitors.*

These capacitors have no polarity. Silver mica capacitors are not as widely used these days as they used to be. However these electronic components can still be obtained and are used where stability of value is of the utmost importance and where low loss is required. In view of this, one of their major uses is within the tuned elements of circuits like oscillators, or within filters. They are stable up to several megahertz. Values are normally in the range from a few pF up to two or possibly three thousand pF.

For this type of capacitor the silver electrodes are plated directly on to the mica dielectric. Again several layers are used to achieve the required capacitance. Wires for the connections are added and then the whole assembly is encapsulated.

Fig. 12.2 *Commercial mica capacitor.*

Mica capacitors are available in the following types:

- Silver mica capacitor
- Stacked mica capacitor
- Dipped mica capacitors
- Moulded/potted mica capacitors
- Trimmers

Some common varieties are shown in Fig. 12.2 above.

12.2 MICA DIELECTRIC

The mica dielectric properties govern the performance of silver mica capacitors. It was also one of the first dielectric materials to be used for capacitors in the early days of wireless because of its combination of stability and general physical and mechanical attributes. Mica is invaluable in electrical industry because of its unique combination of physical, chemical and thermal properties, low power loss factor, dielectric constant and dielectric strength.

Common specifications are 1000 V and even 1500 V per mm of thickness without puncturing, and mica provides a high factor of safety at these figures. Properties like low loss factor and dielectric constant make mica ideal for use in capacitors. No other natural substance has been found to possess properties equal to those of mica. Muscovite mica can be split into flexible and transparent films as thin as 0.006 mm, which gives added advantage in making built-up mica tapes and films that can be punched in any shape and size to make capacitors for instruments and appliances.

Of all the known varieties of mica only muscovite and phlogopite are of commercial importance and valued in the electrical industry. Muscovite finds the largest use while phlogopite has a limited application. Phlogopite does not possess the splitability and flexibility of muscovite. On the other hand phlogopite is superior to muscovite in heat resistance. Muscovite can withstand temperatures up to 700°C, and phlogopite up to about 1000°C. Phlogopite is, therefore, preferred where a high temperature is required. Other mica types have no use except for lepidolite which is a source of lithium.

For silver mica capacitors the silver electrodes are plated directly on to the mica dielectric, although originally thin sheets of silver foil were placed between the mica dielectric. Again several layers are used to achieve the required capacitance. Wires for the connections are added and then the whole silver mica capacitor assembly is encapsulated to provide protection.

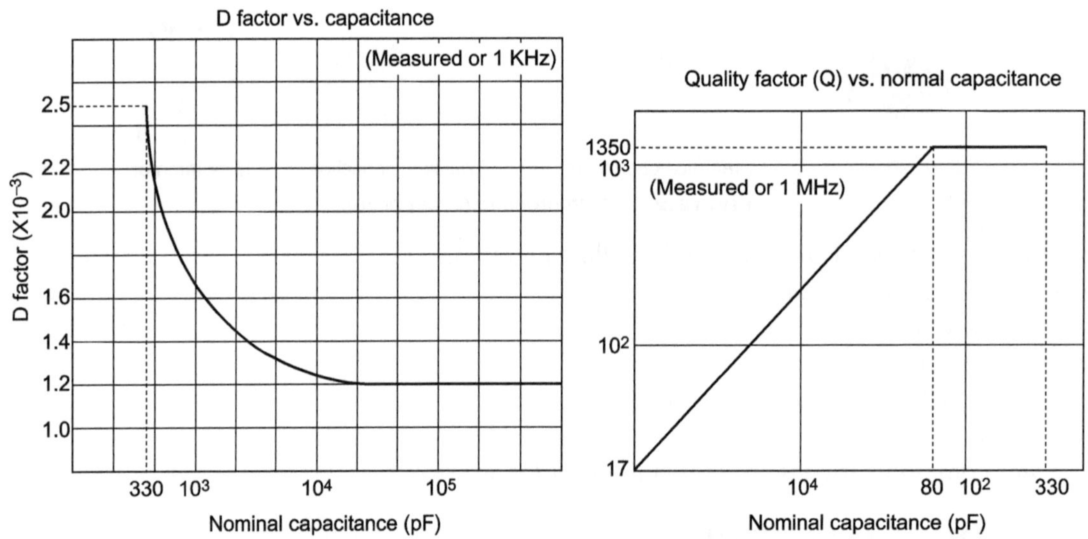

Fig. 12.3 *Relation between (a) dissipation D and (b) Q factor and capacitor value.*

Figure 12.3 (a) shows dissipation factor of mica in relation to its value 1 kHz. It can be seen D shows little variation from 2.5×10^{-3} at 330 pF to 1.2×10^{-3} up to 10,000 pF and is absolutely stable thereafter. Quality factor Q,(b) measured at 1 MHz shows increase up to 80 pF, and thereafter it remains very stable.

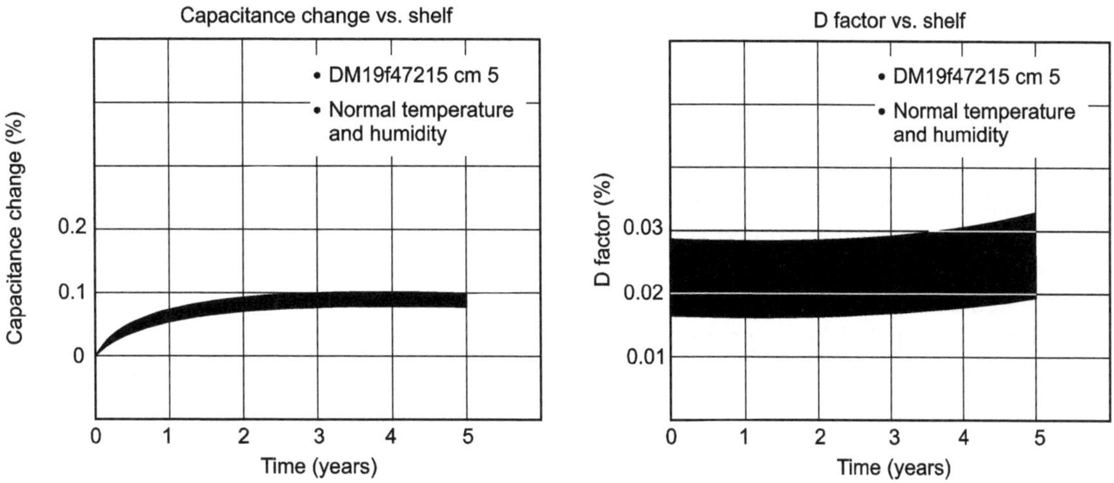

Fig. 12.4 *Capacitance and D factor of mica are stable under shelf storage.*

Figure 12.4 shows high capacitance and tan delta stability over five years. Mica is thus a very stable dielectric over many years, and can be relied upon for long service life. The temperature coefficient of capacitance of mica is also very low, in some grades as low as 70 ppm°C^{-1}. An endurance test on mica capacitors over 10,000 hours shows practically no variation in value or D. Frequency stability for both D and capacitance is impeccable, as seen from Fig. 12.5 below:

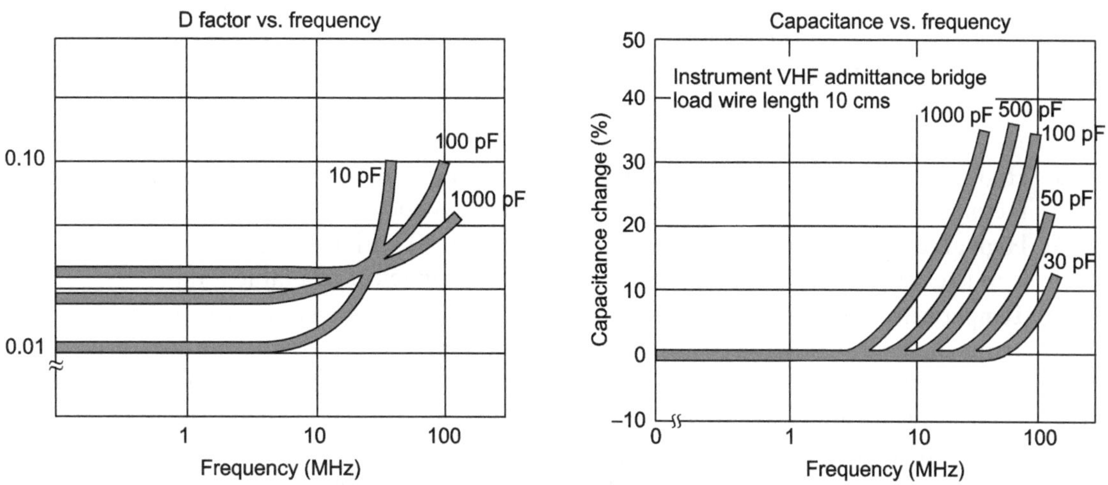

Fig. 12.5 *Relation of D and capacitance value of mica capacitor with frequency* (Source: Jo Kwang).

12.3 TYPES OF MICA DIELECTRICS

[A] Mica plates

This is the basic form of mica. Mica plates are punched from original flakes in sizes required, interleaved with metal (usually tin), stacked together, and electrodes connected in parallel. Multiple layers of parallel printed silvered mica plates interleaved with foil, usually of tin/lead, are mostly used for the manufacture of the present day stacked units. The capacitance per plate is calculated and a pre-determined number of plates are stacked together to arrive at the required capacitances for the finished capacitors. Thick pieces of plain mica are used as backing on either side to reinforce the units. The finished units are then dipped in varnish and subjected to pressure with heat to make the units compact.

[B] Silver mica/silvered mica

The reason for the continued use of silver mica capacitors is their ability to offer very high levels of performance, better in many areas than any other type of capacitor. Silver mica capacitors are high precision, high stability and high reliability capacitors. They are available in small values, and mostly used at high frequencies.

Silver mica capacitors, also known as silvered mica capacitors, replaced the earlier foil type mica capacitors. Instead of being clamped with foils these use sheets of mica coated on both sides with deposited metal. The assembly is dipped in epoxy. They have the advantage of greater stability, since there are no capacitive air gaps that can change dimension. Airtight enclosure removes the risk of oxidation or corrosion of plates or connections. They also have greater value per volume; since there are no air gaps between plates and mica, the conducting surfaces can be thinner. And no clamping mechanism is needed.

Properties of the silver mica capacitor are summarized below:

- **High Q:** Silver mica capacitors have very high levels of Q and conversely small power factors. These are both almost independent of frequency.
- **High accuracy:** Silver mica capacitors can be obtained with tolerance figures of ±1%. This is much better than virtually every other form of capacitor available today.
- **Temperature coefficient:** The temperature coefficient of silver mica capacitors is much better than most other types of capacitor. The temperature coefficient is positive and is normally in the region of 35–75 ppm°C^{-1}, with +50 ppm°C^{-1} being an average value.
- **Value range:** Values for silver mica capacitors are normally in the range from a few picofarads up to two or possibly three thousand picofarads.
- **Low capacitance variation with voltage:** Silver mica capacitors exhibit very little voltage dependence.
- **High Q:** Very high levels of Q and conversely small power factors. These are almost independent of frequency.
- Rated voltages are in the range of 100 – 1000 V for standard dipped mica capacitors. Rated voltages for RF transmitting capacitors are up to 10 kV. Given the excellent electrical performance of mica capacitors in general, these are typically marketed with close tolerances (1% or better).

[C] Reconstituted mica paper

Mica paper is a premium dielectric material made from Muscovite mica. Muscovite mica is processed by removing all soluble contaminants. The resultant flakes are then mixed with purified water to form slurry, which is then processed, producing a continuous sheet of uniform thickness. The National Electrical Manufacturers Association defines mica paper as flexible, continuous, and uniform layers of mica reconstituted into a paper-like, electrical insulating material composed entirely of small, thin, overlapping flakes or platelets, which have sufficient strength to be self-supporting and capable of being wound into roll form for commercial use. Capacitor grade mica paper does not contain binders, adhesives, foreign matter, or colouring agents, and is substantially free of any substance that will adversely affect its performance. Mica paper thickness usually lies between 12.7 µm and 50.8 µm.

- Mica paper exhibits low dissipation factor over a wide temperature range.
- Good insulation resistance over a wide temperature range. Mica paper capacitors are ideal for circuit applications that require extremely low leakage.
- Nearly constant capacitance over a wide temperature range compared to other popular film, paper-film and ceramic type dielectric capacitors. Especially important in applications employing tuned circuitry.
- Excellent resistance to corona deterioration. Mica paper is known for its ability to resist the effects of partial discharge. Mica paper very closely approximates the characteristics of mica splittings in this respect. In applications where corona or partial discharge is generated, mica paper capacitors offer excellent reliability.
- Excellent resistance to radiation, making mica paper capacitors well suited for use in nuclear, space and other applications involving exposure to gamma rays, x-rays and neutrons.
- High energy storage per unit volume. At voltages in excess of 1.0 kV, energy densities of 0.50 – 1.50 joules per cubic inch are common. Higher energy density means a smaller capacitor which delivers the excellent characteristics of mica paper.

High reliability and the ability of mica paper capacitors to operate at high temperatures and high voltage stresses for extended periods of time make these capacitors particularly suited for applications in ignition systems, lasers, power transmission, high-voltage power supplies and transmitters, high-voltage filters, radars, space crafts, nuclear equipment, medical instruments and a host of others. Mica paper capacitors may be used up to as high as 150 kV, and surge currents up to 100 kA.

12.4 CONSTRUCTION OF MICA CAPACITORS

[A] Stacked mica foil sections

Perfect mica sheets are selected and stacked between tin foils. The individual pieces of stacked units are cut into sections with a saw cutting machine or a gang cutting machine. The cut units are literally the capacitors themselves, apart from having the terminals and the coatings. All the processes for testing of the various electrical properties are carried out at this stage on the unit.

[B] Sintered chips

In this type, multiple layers of over-the-edge printed mica are stacked together. These units do not have any foil but use the silver itself for the purpose of termination. The units are edge bonded using high content silver paste on which the terminals are directly soldered.

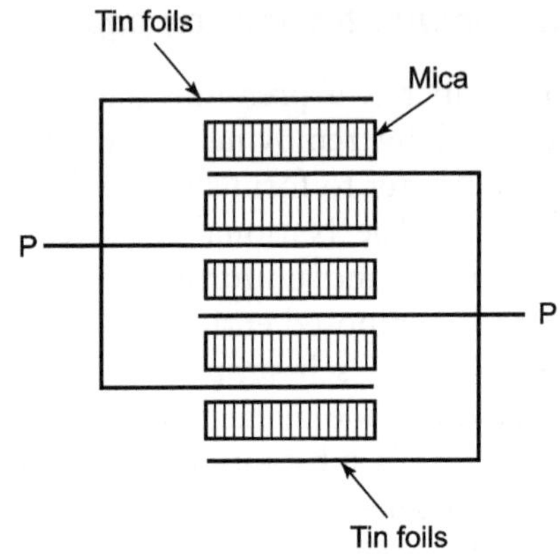

Fig. 12.6 *Stacked mica capacitor*

[C] Stacked silver mica capacitors

Cut films of mica are thoroughly cleaned with the help of de-greasing agents and are then silvered on both sides by a screen-printing process. There is perfect adhesion between mica and silver which results in optimum performance of the plates. Variations on thickness of the silver deposit affect the dissipation factor of the plate, while the overlapping print area of the silver (electrode) and the thickness of the mica (dielectric) control the capacitance of the plate. Individual plates are used in the manufacturing of single film mica capacitors, whereas multiple pieces are used for all other types. These plates are printed for multiple patterns to reduce the manufacturing cost involved therein and in the onward processes.

For silver mica capacitors the silver electrodes are nowadays plated directly on to the mica dielectric, although originally thin sheets of silver foil were placed between the mica dielectric. Several layers are used to achieve the required capacitance. Wires for the connections are added and then the whole silver mica capacitor assembly is encapsulated to provide protection. Today a ceramic encapsulation is used, although early versions, used in some valve or vacuum tube radios had a form of wax encapsulation. Wax was effective in protecting the capacitor from moisture, but when warmed, the wax melted, and often these capacitors had little wax on them from the warm environment of a vacuum tube or valve radio.

[D] Dipped mica capacitors

The clamped units having radial terminals are powder coated with electronic grade epoxy. These units are hung in a device, brought to a heating station, heated at about 240°C and then immersed in epoxy powder. The process is repeated a number of times to achieve the desired dimensions and complete encapsulation of the product.

[E] Reconstituted mica paper capacitors (RMPC)

The dielectric material used in the design and construction of mica paper capacitors is reconstituted mica paper, which is impregnated with a polymer resin, i.e. polyester, epoxy or silicone. In combination with aluminum foil electrodes, the mica paper dielectric can be wound on conventional automatic or semi-automatic capacitor winding machines. Flag tabs are inserted during the winding process, or the foils may be extended and metalized, depending on the application. The process is amenable to fast production, and the capacitors are popular for high voltage and specialized applications. Different winding techniques such as straight winding with embedded foil and series winding with embedded or extended foil can improve performance or reduce costs. For example, straight winding RMPCs for low-voltage applications (less than 8 kV DC) are the most economical option due to low labour and material costs. But series winding for high-voltage applications (greater than 8 kV DC) can also cut cost because fewer bare sections are required. And many times, requirements are so complicated that several capacitors must be connected in parallel, series, or a combination of the two to get just the right capacitance and voltage ratings.

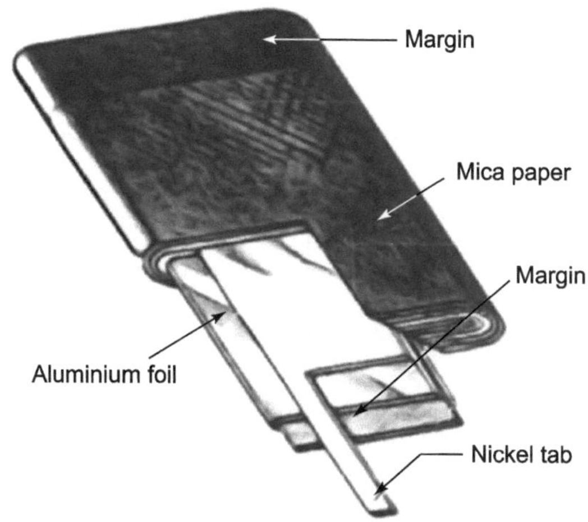

Fig. 12.7 *Mica paper capacitor element.*

[F] Moulded/potted mica capacitors

Moulded mica capacitors, or mica paper capacitors (mica paper windings in a void free assembly) are moulded in epoxy resins. These are capable of withstanding severe environments. Depending on the electrical, mechanical and environmental requirements, various epoxies are available to meet individual applications. Moulded capacitors can be made in practically any shape. Wide varieties of mounting configurations are required. A wide variety of terminations including but not limited to wire leads, threaded inserts and studs, turret terminals, machined brass are available.

(a) Moulded mica paper capacitors (b) Multi element capacitors (c) HV DC mica capacitors

Fig. 12.8 *Moulded mica capacitors.*

In Fig. 12.8 above, (b) is a combination of capacitors from 2 μF 800 V to 0.1 μF 12 kV DC, with capacitors in configuration, while (c) shows one configuration available for capacitors up to 100 pF to 1 μF and up to 15 kV DC.

12.5 COLOUR CODES FOR MICA CAPACITORS (TC CODES)

Both digital and colour codes are in use for mica capacitors, and codes are similar in general to standard codes for other components.

(a) Digital code: The following format as per Table 12.1 is used for digital coding of values, tolerance and voltage ratings, as also to indicate temperature range, temperature coefficient and drift properties and frequency range.

Table 12.1 *Digital Codes for Mica Capacitors*

DM 5 — F — D — 181 — J — O — 3

- DM 5: STYLE
- F: CHARACTERISTIC
- D: WORKING VOLTAGE
  - Y: 50VDC
  - A:100VDC
  - C: 300VDC
  - D: 500VDC
- 181: CAPACITANCE
- J: CAPACITANCE TOLERANCE
  - D: ± 0.5 pF
  - K: ± 10%
  - J: ± 5%
  - G: ± 2%
  - F: ± 1%
  - E: ± 0.5%
- O: TEMPERATURE RANGE
  - O:–55°C to +125°C
  - O:–55°C to +150°C
- 3: VIBRATION GRADE FREQUENCY 10Hz TO 2 KHz

CHARACTERISTIC

LETTER	TEMPERATURE COEFFICIENT [PP M10C]	CAPACITANCE DRIFFT
C	–200 to +200	± (0.5% + 0.1 pF)
D	–100 to +100	± (0.3% + 0.1 pF)
E	–20 to +100	± (0.1% + 0.1 pF)
F	0 to +70	± (0.05% + 0.1 pF)

(b) Dot colour codes: Most small size mica capacitors are marked with dot colour codes. Six dots are commonly used, as shown in Fig. 12.9. The dots also have arrows on them to indicate the direction of read ing. To read the code, capacitor should be held with arrows pointing from left to right. The first dot indicates the material of dielectric, which in case of mica, is always white.

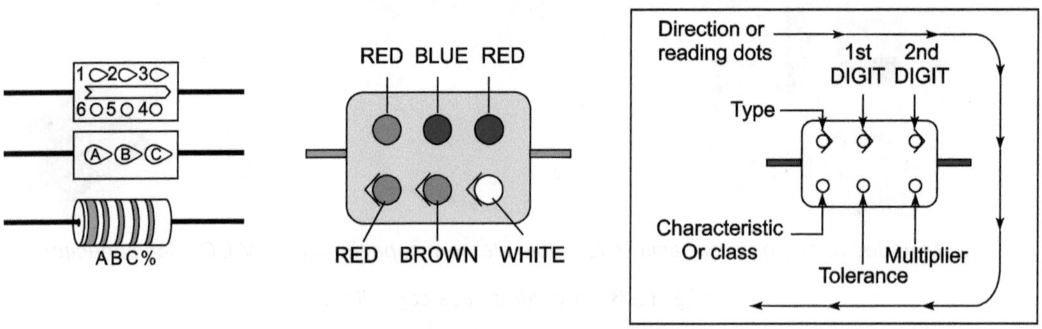

Fig. 12.9 *6–Dot colour markings for mica and moulded mica paper capacitors.*

The colour codes are read from the guidelines as per Table 12.2. Detailed colour codes are given in Appendix J.

Table 12.2 *Colour Codes for Capacitors*

Colour	*Digit A*	*Digit B*	*Multiplier D*	*Tolerance T > 10 pf*	*Tolerance T < 10 pf*	*Temp. coeff. TC*	*Working voltage V*
Black	0	0	× 1	± 20%	± 2.0 pF		
Brown	1	1	× 10	± 1%	± 0.1 pF	– 33 – 10-6	
Red	2	2	× 100	± 2%	± 0.25 pF	$-75 - 10^{-6}$	250 V
Orange	3	3	× 1000	± 3%		$-150 - 10^{-6}$	
Yellow	4	4	× 10k	+ 100%, – 0%		$-220 - 10^{-6}$	400 V
Green	5	5	× 100k	± 5%	± 0.5 pF	$-330 - 10^{-6}$	100 V
Blue	6	6	× 1m			$-470 - 10^{-6}$	630 V
Violet	7	7				$-750 - 10^{-6}$	
Grey	8	8	× 0.01	+ 80%, – 20%			
White	9	9	× 0.1	± 10%			

13

ULTRACAPACITORS: THE FUTURE OF ENERGY STORAGE

Over the past few years, a lot of research and investment is being directed into a new type of capacitors, known as electrochemical capacitors, whose basic function would be storage and supply of electrical energy. Particularly, double layer variant of these, known alternatively as electrochemical double layer capacitors (EDLC), digitized energy storage devices (DESD), Ultracapacitors and supercapacitors have been the subject of intense research and hold out good promise as energy storage devices of future. They are getting all the more importance as these capacitors offer an eco-friendly alternative to batteries for energy storage, and can work with batteries for backup power peaks.

A number of manufacturers in the USA, Europe, China, Korea and Russia are already manufacturing ultracapacitors for various applications. Some of the leading names include Evans Electric, Maxwell, Nesscap, Panasonic, Capex, Ioxus, ELNA, EEstor, Heter and ESMA.

Depending on the required energy densities and lifecycle, plus overall efficiency, ultracapacitors are expected to play a very important role in automotive, aerospace and military applications. For automotive operations, ultracapacitors help extend the lifecycle of the vehicle batteries subjected to a very aggressive power profile. In consumer products such as copier machines, ultracapacitors are already utilized as a source of energy to shorten the warm-up time of copiers hence eliminating the need to increase the power rating of their utility connection. Ultracapacitors are considered as a part of a hybrid energy system (HES) on trucks and heavy vehicles, aiming at reducing emissions and improving fuel economy.

13.1 ELECTROCHEMICAL CAPACITORS

Electrochemical capacitors (EC capacitors) do not use separate dielectric material. The dielectric is formed as an extremely thin layer at the contact surface of two different electrode materials. One of the electrodes is a highly porous solid, and a liquid electrolyte flowing through it increases the effective contact area tremendously. This, coupled with extremely low barrier thickness allows very high capacitance values, even a couple of thousand farads, to be accommodated in capacitors. They combine the high energy potential of batteries with high energy transfer rate and fast recharging capabilities of capacitors.

The surface area in EC capacitor is limited solely by the porosity in the activated carbon electrode. EC capacitors are also made using two solid porous electrodes, and an electrolyte forms dielectric layers at both electrodes. These are called double layer capacitors. Nanotechnology has made possible materials with extremely high porosity, creating huge potential in the development of these capacitors. Figure 13.1 shows the basic components of button type EC capacitor.

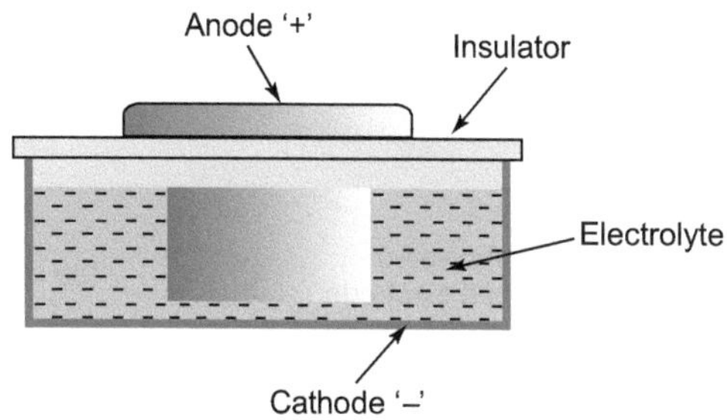

Fig. 13.1 *EC capacitor schematic.*

Consider the basic equation,

$C = \epsilon_0 \times kA/d$ ($\epsilon_0 = 8.85 \times 10^{-12}$ F/m)

If A is $1000m^2$, d in nanometers, it follows that C will be of very high order compared with conventional capacitors. Commercially available materials have porosity enough to give 1000 – 3000 m^2/g so it is possible to pack very high values, even several thousand farads, in a capacitor.

The voltage rating of an electrochemical capacitor is limited only by the decomposition potential of its electrolyte. For an aqueous electrolyte like KOH, this limits the usable cell voltage to 0.8 – 1.6 V. Use of non-aqueous electrolyte may increase this to 3 – 5 V. To get higher voltages, series parallel stacks of capacitors are used, each rated about 1.3 V. For example, 11 capacitors in series (11 × 13 = 14.3 V) may be used to get a 14 V capacitor.

Standard Oil Company of Ohio is credited with the invention of electrochemical capacitor (as it is known today), in 1962. Nippon Electric Company (NEC) introduced 'Supercapacitors' under their license in 1970. Initially used as backup power devices for clock chips and CMOS computer memories, many other applications have evolved over the years. The capacitor was made into a bipolar device, and mass production started in 1980. Panasonic developed 'Goldcap' capacitors in 1978, primarily to replace coin cell batteries in memory backup applications at that time. Spiral wound capacitors were developed in the 1990s, when much larger capacitors made their appearance. The first vehicle application capacitor by Panasonic was made in 1999 (2000 F 2.3V), using this construction.

13.2 ELECTROCHEMICAL DOUBLE LAYER CAPACITORS (EDLC)

The EDLC (or ultracapacitor, as it is commonly known) is constructed on the principle shown in Fig. 13.2. Generally two active porous carbon electrodes are used, with electrolyte filling the remaining space, and a separator used between these two sections. Two layers of dielectric are thus formed, a capacitor at each surface (C1 and C2), and hence the name 'Double Layer Capacitor' or EDLC. The resultant capacitance C is the series combination of these two (1/C = 1/ C1 + 1/ C2), with a corresponding increase in operating voltage.

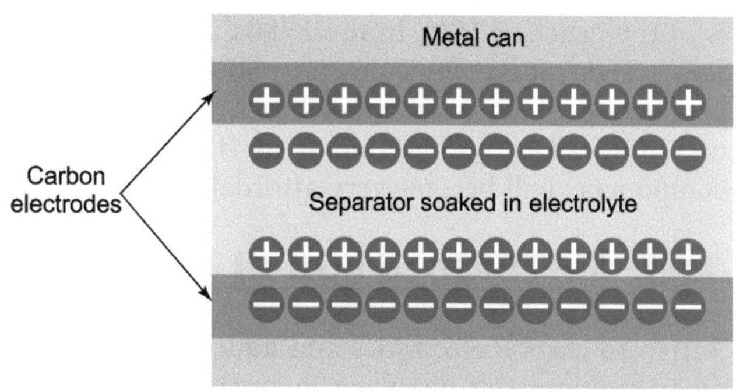

Fig. 13.2 *EDLC basic construction.*

The carbon electrode (porous collector) of EDLC is made from activated carbon particles painted or rolled on a metal foil collector. Figure 13.3 shows porous electrode foil in black while the separator is light grey.

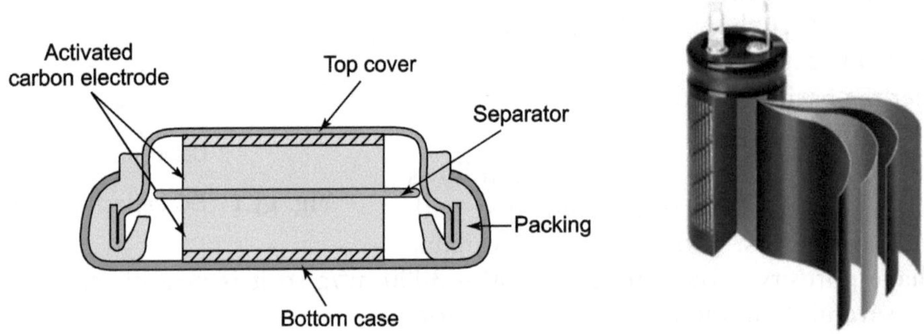

Fig. 13.3 *Button capacitor and wound EDLC.*

Energy stored in ultracapacitors is very high. A conventional capacitor has energy density of 0.15 watt-hour (Wh) per kg, whereas EDLCs may store up to 15 Wh. (1 Wh = 3600 watt-sec = 3600 Joules). The energy storage of these capacitors is their biggest asset, and the stored energy levels are comparable to that of a battery. This fact led Evans Capacitor Company to name one of its ultracapacitor series as 'Capattery'. Efforts are on to increase the density, when it will store nearly half the energy compared to a similarly sized Li-ion rechargeable battery.

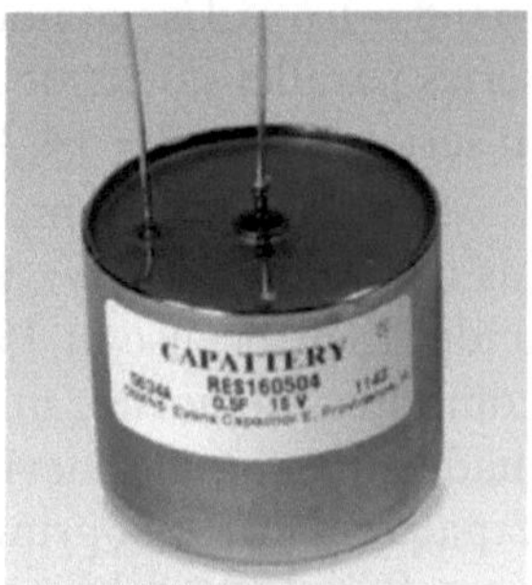

Fig. 13.4 *Capattery.*

Li-ion capacitors already store up to 21–25 Wh/kg. Targets are pointing to the possibilities of approaching 40 Wh/kg or more in the near future. In fact, one manufacturer JOEL, Japan, is hopeful of increasing this level up to a maximum of 75 Wh/kg. By comparison, Ni-MH battery packs around 45 Wh/kg. While present energy densities are much below that of batteries, their power delivering capabilities of up to 2000 W/kg are ten times those of batteries.

The ratio of surface area to charge-separation distance (dielectric thickness) in ultracapacitor is of the order of 10^{12}. In the 1980s, ultracapacitor of 470 farads 2.3 V would cost $2.00 per farad. Today it has come down to $0.10 per farad, and costs continue to decrease rapidly. Mass production of ultracapacitors has helped reduce the costs, and it is expected by some these may well fall by another factor of 10 or 20, to a level of 1 or 0.5 cents per farad, and these components will become very affordable.

13.3 MATERIALS FOR ULTRACAPACITORS

Activated carbon electrodes and aqueous electrolytes like sulphuric acid, potassium hydroxide or ammonium salt dissolved in organic solvent (e.g. propylene carbonate) are used for cells with operating voltages of 1.2–2.7 V. These basic cells of EDLC are connected is series parallel combination to increase the voltage ratings for different applications. Activated carbons have specific capacitance of 50–100 F/g, which creates the possibility of several thousand farads in a single cell. Capacitance values of few millifarads to several thousand farads, voltages from 2.5 V to almost 400 V (in banks) are available. Units used for capacitance values are farads or kilo farads.

(a) Carbon nanotubes (CNT) (b) Graphene (c) Aerogel

Fig. 13.5 *New materials for ultracapacitors.*

The advent of nanotechnology and resultant improvements in materials has given a big impetus to these capacitors. The base material most commonly used in new generation EDLCs is CNT or carbon nanotubes (Fig. 13.5(a)), a collection of tiny nanotube structures that form a highly porous solid. The nanotubes are typically of diameters the order of 5 nm, or 1/10,000 times the diameter of a human hair. These are spaced about 5 nm apart, and may be about 100 μm long. An electrolyte fills up this vacant space completely to form the other electrode. The effective electrode area is several thousand times greater than conventional electrolytic construction, allowing very high values. Capacitors up to 100,000 farads have been made in a single unit. Energy stored in capacitors is expressed as ½ CV^2, and one may calculate the tremendous amount of energy that can be stored in ultracapacitors.

Another material developed and in use by some manufacturers is graphene, a one-atom thick sheet layered structure (Fig. 13.5(b)). This structure, apart from giving maximum surface area, ensures full penetration of electrolytes in the atomic structure of the electrode, giving maximum capacitance per unit area.

A third class of new material, carbon aerogel (Fig. 13.5(c)), has been developed for ultracapacitors. The materials, so porous and lightweight that they are called 'solid smoke', 'blue smoke' or 'frozen smoke', are 90–99.8% air, with densities ranging as low as 1.9 mg per cc, are feather light, but can support up to 8000 times their weight. Their surface areas can be as high as 3000 square meters or more per gm. One square inch of aerogel will have more surface area than a football field. They are almost transparent and are in the Guinness Book of World Records as the best insulators and the lowest density solids.

New materials for anodes and experiments with various electrolytes, including those similar to ones used in Li-ion batteries are helping improve energy densities beyond what has been possible till now. One such electrode material is ruthenium oxide (RuO_2), having specific capacitance of 650 F/g, on which research is in progress. Manganese dioxide and barium titanate are other materials being targeted for electrodes. Electrolytes are also undergoing experiments for better results.

Their use in ultracapacitors has opened a new chapter and increased energy storage manifold. These can be produced as thin films. The energy produced in a carbon aerogel capacitor is 325 kJ/kg, about 70% of that by a lithium polymer battery. Aerogel can also be made using materials other than carbon, and trials are in progress on such materials.

13.4 SYMMETRICAL AND ASYMMETRICAL CAPACITORS

[A] Symmetrical capacitors

These have both electrodes of same material, mainly carbon. They have cell voltage limitations of 1.2 V for aqueous electrolytes and 3.5 V for non-aqueous capacitors. Carbon electrodes of these ELDCs are made of high porosity carbon having active area of the order of 1000 to 2000 square meters per gm. in particulate, plastic bonded or film form, in contact with a current collector. Carbon may be pasted on aluminium foil and wound on high-speed machines. While winding pressure is enough for cylindrical capacitors, high pressures are applied to cells for keeping continuous contact of electrodes with current collectors.

[B] Asymmetrical capacitors

These designs use different materials for the two electrodes, e.g. nickel hydroxide and carbon, and their cell voltage can be higher. Higher cell voltages lead to lesser capacitors in series for same voltage. The positive electrode may be made of sintered pellet of high capacitance density (tantalum powder) and Ta_2O_5 as the dielectric film formed over it. RuO_2 film forms the negative electrode. Asymmetrical capacitors can be discharged only to a certain minimum voltage (like a battery), to about half the rated voltage. However, they store 4 to 5 times the energy density, or even more, as compared to symmetrical type. Lithium ion capacitors (LIC) are being developed as a separate class of asymmetrical capacitors with high energy density, and their outlook seems promising.

[C] Li-ion capacitors

A lithium-ion capacitor (LIC) is a hybrid type of capacitor. Activated carbon is used as cathode. The anode of the LIC consists of carbon material pre-doped with lithium ions. This pre-doping process lowers the potential of the anode and allows a high output voltage. The electrolyte used in an LIC is a lithium-ion salt solution. A separator material is used between electrodes to avoid direct contact. Typically, output voltages for LICs are in the range of 3.8–4.0 V. As a consequence, LICs have a high energy density. Furthermore, the capacity of the anode is several orders of magnitude larger than the capacity of the cathode. As a result, the change of the anode potential during charge and discharge is much smaller than the change in the cathode potential.

Li-ion capacitors have created a place for themselves, as they offer about four times the energy density of a conventional EDLC. LICs feature a self-discharge of less than 5% after 3 months, and less than a 10% drop in capacity after 100,000 cycles. Though much costlier than EDLC, they are used where small size or high energy density is needed.

13.5 ULTRACAPACITORS VS. BATTERIES

Energy storage in batteries is in the form of chemical energy. Charging and discharging of a battery are electrochemical processes. There are maximum and minimum rates at which they can be charged or discharged. Electrochemical processes are by nature dependent on ambient temperature. These factors require larger batteries to get enough starting current for automobiles, and also cause starting problems in cold weather. Capacitors, when used with batteries in cold atmospheres, provide heavy discharge current for this purpose, and the vehicle can run smoothly in all weathers. Even in normal conditions, ultracapacitors reduce battery size requirements for vehicles. They can be used to give large pulses of energy in most applications. Capacitors may undergo few hundred start-stop cycles every day without any problems.

In an ultracapacitor, energy is received and stored as electrical energy, and is directly available for use without any time gap. It can work in a much wider temperature range than batteries, and even as low as – 55°C. The change in capacitance even at –40°C is below 30%, and the capacitor remains functional at both extremes of temperature.

The rate at which an ultracapacitor can be charged or discharged depends purely on the current that can be permitted, depending upon cables and accessories. Very heavy currents are available for starting an engine or diesel generators. Ultracapacitors are used in a big way as complementary battery backup, in order to reduce battery size, as also to take over full battery function in engine start applications. (Battery sizes need to be large enough for starting torque in vehicles, though smaller sizes are enough while running.) Although batteries excel in energy storage, EC capacitors excel in supplying high-power levels. The power density of EC capacitors can reach up to several kW/kg. However, batteries may only reach levels of 0.1 – 0.5 kW/kg.

The Ragone (pronounced "Ra-go-ni") Chart (Fig. 13.6) shows power and energy densities of capacitors, supercapacitors, batteries and fuel cells and brings out sharp differences between these energy storage devices. Power is 'rate of delivery or discharge of energy'. Conventional capacitors have the maximum power, but are very low in energy levels. Batteries, on the other hand, have large amounts of stored energy, but because of chemical reaction process for delivering energy, cannot deliver the energy very fast, being limited by electrochemical energy conversion rate. Fuel cells are fast coming up as an alternative to batteries with even more energy packed, but are slower than batteries in delivery rate. A combination of battery/fuel cells and ultracapacitors can work very effectively to store energy and power automobiles or in many other applications.

The service life of a battery is rather limited, from a few hundred to a few thousand charge discharge cycles. ELDC by nature are extremely long lasting, and charge discharge cycles

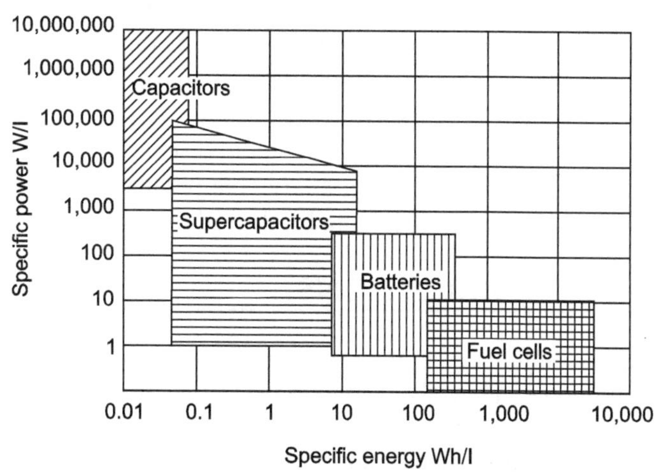

Fig. 13.6 *Ragone chart showing energy density and power density/litre.* (Source: www.cap–xx.com)

up to 500,000 are available. Thus the life expectancy could go to 10–15 years and beyond. A battery can be very unreliable, as there is no way to assess how much life is left in it. When one battery fails, the entire battery bank fails. In contrast, a voltmeter connected across the capacitor gives direct indication of balance charge (or usable energy) remaining on capacitor. Table 13.1 brings out the comparison in basic energy storage characteristic of ultracapacitors and battery.

Table 13.1 *Comparison of Ultracapacitor and Battery*

Parameter	*Ultracapacitor*	*Battery*
Charging time	1–30 s	1–6 hrs
Discharging time	0–30 s	0.3–3 hrs
Wh/Kg	4–20	20–100
Power density (W/kg)	1000–2000	20–50
Charge/discharge eff.	0.90–0.98	0.7–0.85
Recycle life	100,000	200–4000
Temp. °C	–30 to +70	–10 to +60

An ultracapacitor offers number of advantages over a battery:

- Much higher energy efficiency. Batteries need 30% more energy to charge than to store.
- An ultracapacitors is much lighter. Its weight is 1/5th that of a battery.
- They can deliver 10 times greater power.
- Ultracapacitor life is almost unlimited. Batteries last under 2 years.
- Regeneration charging is made possible with ultracapacitors.
- Fast charge/discharge – quick response as there is no electrochemical process involved.
- In conjunction with batteries, they complement batteries to reduce battery size and weight.
- Ultracapacitors are perfect rechargeable energy sources.
- As battery backup, they prolong battery life, by avoiding deep discharging.
- Ultracapacitors can deliver high currents and power.
- Life of over 100,000 charge-discharge cycles, which means a useful life of 10 – 15 years or more.
- Same efficiency for charge and discharge >98%.
- They can work over a large temperature range from –30 to 70°C.
- All constituents being biodegradable, they are totally pollutant-free.

13.6 APPLICATIONS

Today millions of EDLCs provide backup power for the memory used in microcomputers and cell phones. They also supply brief bursts of energy to numerous consumer products containing batteries. In a camera, for example, an ultracapacitor can extend battery life by powering functions, like zooming in for a close-up, or a flashlight.

Ultracapacitors are ideally suited for missions where short duration power support is required, like installations that need up to 60 seconds support during changeover to a secondary power source like gensets, or to perform an orderly shutdown. Fuel cell vehicles have been made with ultracapacitors in supporting role, for starting the vehicle from rest, or at stop signals for restarting, since fuel cells have slow response time. Fuel cell size can be reduced by this pairing.

Potential applications for LICs are in the fields of wind power generation systems, uninterruptible power source systems (UPS), voltage sag compensation, photovoltaic power generation, energy recovery systems in industrial machinery and transportation systems. Low self-discharge and high power densities are their major advantages.

(a) **Low power applications of ultracapacitors**

- Ratings up to 4 farads 5.5–6.3V
- Coin (button) type/cylindrical
- Small high-frequency devices reduce battery size
- Useful for memory function – in laptops, mobile phones, radio tuners, electronic diaries, PDAs, pagers, data loggers, vending machines
- Useful for peak power assistance in applications like digital cameras
- Can be used as energy storage for motor starters, solar panels and other applications

It may soon be possible to recharge mobile phones within seconds, and they may give much longer service life.

(b) **Automotive and traction industry can use ultracapacitors in the following areas:**

- Regenerative braking
- Electric drive, hybrid vehicles
- Power steering, seat belt restraints
- Air conditioning
- Battery backup, cold start
- Accessories – radio, horn, etc.
- Subsystems to improve fuel efficiency and reliability

Fifty to sixty million passenger vehicles are being produced worldwide annually – so the possible number of ultracapacitors for these applications can run into billions.

(c) **Ultracapacitors in vehicles**

Cars are being designed with hydrogen fuel, battery power, solar power or fuel cells, with ultracapacitors as backup power sources. Vehicles have also been made to run exclusively on ultracapacitors as power source. Braking energy is also almost fully recovered in ultracapacitors. China started ultracapacitor-powered buses (Capabus) on commercial scale on two routes in Shanghai in early 2005–06 which stop at alternate stops for charging (Fig. 13.7). The bus, a 41-seater, plus driver, gets charged in two minutes at bus stops via pantograph raised from its top. Many plug-in vehicle models by different makers are being designed to run on

Fig. 13.7 *Capabus*

battery or fuel cells, with ultracapacitors taking over engine starting, regenerative braking and other functions. Batteries serve as the energy storage, while the capacitor supplies high power for short duration.

On city roads, large fuel savings result from hybridization of vehicles. Honda motors, Zenn and a few others are developing car models exclusively powered by ultracapacitors. Toyota Supra-HV-R car, powered by ultracapacitor alone, won Japan's 24-hour endurance race in 2007.

(d) Traction applications

Traction applications include material handling or utility vehicles like forklifts, golf carts, fixed route buses etc. ELDCs are being increasingly used in everyday field, and are supporting or even competing with batteries as a source of stored energy. Figure 13.8 shows one such ultracapacitor from Maxwell, the size of a coffee table, comprising 144 nos. of 150 F each in series parallel combination, to make a working voltage of 395 V. Trucks, forklifts and delivery vans are being powered by them. NASA is considering their use for reusable satellite launch vehicles.

Fig. 13.8 *Ultracapacitor for car.*

Cold start is expected to dictate the size of battery, but not for ultracapacitors. If the size, weight, maintenance costs, replacement and downtime costs are considered, it makes sense to use ultracapacitors by themselves or as backup with batteries in many applications. EDLCs are much lighter than batteries, and weight is an important criterion in vehicle design. Table 13.2 gives some typical energy requirements where ultracapacitors are currently deployed:

Table 13.2 *Energy Requirement for Car Operations*

Event	*Energy needed*
Start–stop launch assist	15–25 Wh
4% grade @35 mph × 1 min	170 Wh (Regen)
2% grade @35 mph × 1 min	70 Wh (Regen)
Accessory 1 kW × 1 min	16.7 Wh

Korean Railway Experience

Korea used EDLCs at substations in 2004–05 on two routes of its railway traction systems for braking power regeneration. Ninety per cent of energy used is for traction, and 10% for auxiliary power. Up to 40% tractive power capable of returning to supply can be regenerated during braking. Their experience has been very positive. Measured regeneration on one substation was 15,464 kWh per day, and 20,155 kWh at the other. Their suggestion for 1500 kWh 1500 V DC system unit at a substation included 4896 ELDCs 2.7 V/5000 F @5.69 Wh per kg and power density 5.12 kW per kg, with a set of converters etc. at a total cost of 1400 million KRW ($1.13

mn). They calculated the cost recovery period about 2.1 years. The energy and power densities today being higher, and costs lower, economics will work out still better.

Stationary Applications

Ultracapacitors are being used in LED lighting systems, portable equipment, data backup storage systems and laptop computer supplies. Mobile phones, clock chips, UPS systems, small DC motors, hand held flashlights and a host of applications use ultracapacitors as primary power supply. Diesel generators are being started using ultracapacitors. An ultracapacitor powered household inverter can get charged in just a few minutes when supply comes on even for a short time, and inverters can run the whole day normally even during severe power cuts.

Industrial applications

- Power quality improvement
- Frequency regulation (fast discharge and low maintenance)
- Managing peak loads and surges
- Energy storage devices for solar and wind energy.
- Standby and backup energy provider (Fig. 13.9 shows a UPS of 100 kW backup)
- Help to maintain grid stability
- As a power source for hand tools and machinery
- Ultracapacitors may in near future become as commonplace as batteries are today in all applications. They also find use in pulsed power application in avionics and military devices. They can be charged and discharged literally in microseconds. The same technology could be used to store energy from alternate energy sources like wind or solar power, fuel cells or wave, which depend on energy storage devices for their operation. Solar panels or windmills may feed electrical energy directly into ultracapacitor cells or banks, to be used by inverters to feed household electrical installation.

System in Fig. 13.9 has 9 EDLC modules in series as storage device. Its operating voltage is 600 V and gives a support power not less than 100 kW for 10 s. The system contains converter, control system and voltage leveller in capacitor cells.

Fig. 13.9 *100 kW UPS from ESMA.*

Transmission line stability

It is possible to increase the stability of a transmission system by adding energy storage. This serves to dampen oscillation through the successive generation and absorption of real (as opposed to reactive) power. Very large ultracapacitor based UPS systems may be used for this purpose.

There is also transient stability – the stability required after a utility event (loss of substation or a major line). During a transient event, achieving stability requires a substantial capability to absorb energy quickly. This is somewhat analogous to 'dynamic braking' because generator turbines must be slowed. A typical specification is 100 MW with 500 MJ (< 5 s).

Power quality and uninterruptible power

Power outages can result in costly interruptions for large commercial and industrial customers. Energy storage has been used for decades in power quality applications at distribution and utilization voltages. The next logical step is power quality for wide areas at the transmission level. Examples of this already exist. Dynamic voltage restorer (DVR) technology can provide energy storage equivalent to ½ s of peak load. Similar principles can be applied to large loads (>20 MW) served from low-voltage transmissions. A typical specification is 1–5 MW and 10–50 MJ. A typical model for peak power application is explained in Fig. 13.10.

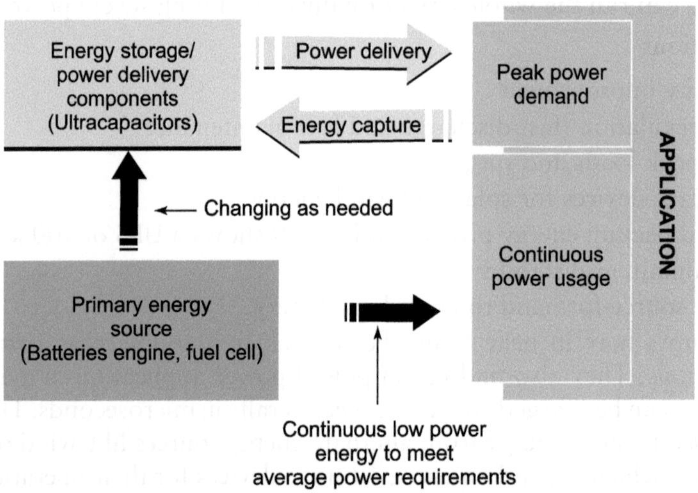

Fig. 13.10 *Peak power management using ultracapacitor water desalination/purification.*

Research on ultracapacitors is in an advanced stage for desalination and purification of water by high surface carbon capacitive deionization (CDI) technology. Desalination by CDI technology uses ultracapacitors to collect ions at electrodes, which do not interact chemically at 2 V level, but just get attracted electrostatically to electrodes and stay there. The ions get dispersed in flowing water during recharging when the external voltage is removed. The charge on capacitor cell so removed can be transferred to second uncharged cell, which functions alternatively. Desalination by this process does not involve appreciable energy consumption.

The carbon developed by a company in California is claimed to be the cheapest in the world, about 1/10th the cost of other materials. They claim energy densities of 7.5 kW per kg of carbon with surface area of 2000 to 2500 sq. meters per gm. Ultracapacitors using this carbon are said to be the lightest, and ideal for mobile and traction applications. An ultracapacitor of an incredible 435,000 farads has been developed in a size of 2 ft. × 2 ft. and less than 4 ft. length. Use of these capacitors in desalination may be programmed to absorb power from grid in off-peak hours, and return to grid during peak hours. In the USA, peak power rate is $115 per MWh, and the off-peak rate is $18 per MWh. The differential rate of peak and off-peak rates in fact generates revenue for the operator while desalining.

The cost of desalination compares favourably towards ultracapacitor use:

Table 13.3 *Comparison of Energy and Costs for Water Purification*

Technology	kWh per 1000 L	Energy cost Rs. per 1000 L
Distillation	53,000	442,000
RO	9,300	74,400
CDI	340	2720

Ultracapacitor sizes

Compared to conventional capacitors, ultracapacitor dimensions are very small. A button cell capacitor of 1 farad 5.5 V may measure 18.8 mm in diameter and 4.7 mm in thickness, and will weigh just 3.2 g. A 120 F 2.8 V capacitor has typical dimensions of 25 mm diameter and 45 mm length, whereas 3000 F 2.8 V is accommodated in 60.5 mm diameter and 138 mm length. The energy density in most commercially available capacitors varies from 5.2 to 5.8 Wh/kg. Sizes will become substantially smaller in the near future with newer developments taking place and energy densities going up.

Fig. 13.11 *Some button cell capacitors (coin capacitors).*

Ultracapacitors are often connected in banks to store bulk energy and deliver large power requirements of cars, trucks or gensets. ESMA of Russia specifies ultracapacitor ratings in terms of energy storage capacity in kJ (30 to 3000 kJ), maximum operating voltage (from 16 V to 52 V), operating voltage window (downward up to 25% of maximum voltage) and maximum power up to 49 kW, with different models for specific applications.

Fig. 13.12 *Flat and stackable ultracapacitors.*

Large capacitors and their banks are used in many places to give large values of capacitors, according to the energy and voltage requirements. Figure 13.13 shows some capacitors and banks used for the purpose.

Fig.13.13 *Large size ultracapacitors and banks.*

13.7 ULTRACAPACITOR MEASUREMENT

Measurement of ultracapacitor value is done starting from the basic equation I = Cdv/dt. A constant current is drawn from a charged ultracapacitor for measured potential drop, and the time for the same is noted (or vice versa), and capacitance calculated. For example, if 1 amp current is drawn from the capacitor to drop the potential drops by 1 V, say from 2.5 V to 1.5 V, the capacitance in farads is equal to the seconds measured. The voltage measured is the sum total of open circuit capacitor voltage, plus the IR drop due to ESR. This allows us to calculate the ESR.

Ultracapacitor cells are connected in series parallel combination to achieve the power requirements of the user equipment. The individual capacitors cannot be of exactly same values, and hence the voltage will get divided unevenly, which is harmful to capacitors. Hence balancing circuits are used to even out the voltages across capacitors. Individual voltages need to be well within the operating limit of capacitor ratings. ELDCs do not have large safety margins for over-voltages, and manufacturers specify maximum voltage on capacitors as not to exceed by more than 10% of rated voltage.

13.8 ULTRACAPACITOR BANKS

Ultracapacitors are usually rated 2.3–2.7 V, and need to be connected in series parallel combination to get the required capacitance and voltage. The large tolerance of capacitor, coupled with variation in leakage currents, creates uneven voltages across individual capacitors while charging, and may exceed the rated voltage of a unit. A mechanism has therefore, to be in place to balance the voltages between them. This can be done in different ways:

1. Leakage resistance is connected across each capacitor, which carries at least ten times the capacitor leakage current. This reduces the effect of capacitor leakage current variation. This mode is the cheapest, and is available when the resultant power loss is permissible.
2. Zener diodes may be connected across individual capacitors (with series resistors). In case capacitor voltage exceeds a limit, the zener starts conducting,

Fig.13.14 *Electronic balancing circuits in ultracapacitor bank. The bank has 18 capacitors in series.*

and will not allow the charging current to flow through it. This involves less loss than the resistor method.

3. Electronic balancing circuits are interconnected across capacitors, which compare voltages across them, and keep all voltages within tolerance limits and properly balanced. Provision is also made to cut off overall charging voltage of bank, once a required limit is reached. Figure 13.14 is a typical example of such a balancing circuit.

13.9 LIFE OF ULTRACAPACITORS

Ultracapacitors have a long life, and field failures are insignificant, so long as rated voltages are not exceeded. The failure considered over time is when the parameters deviate beyond specified

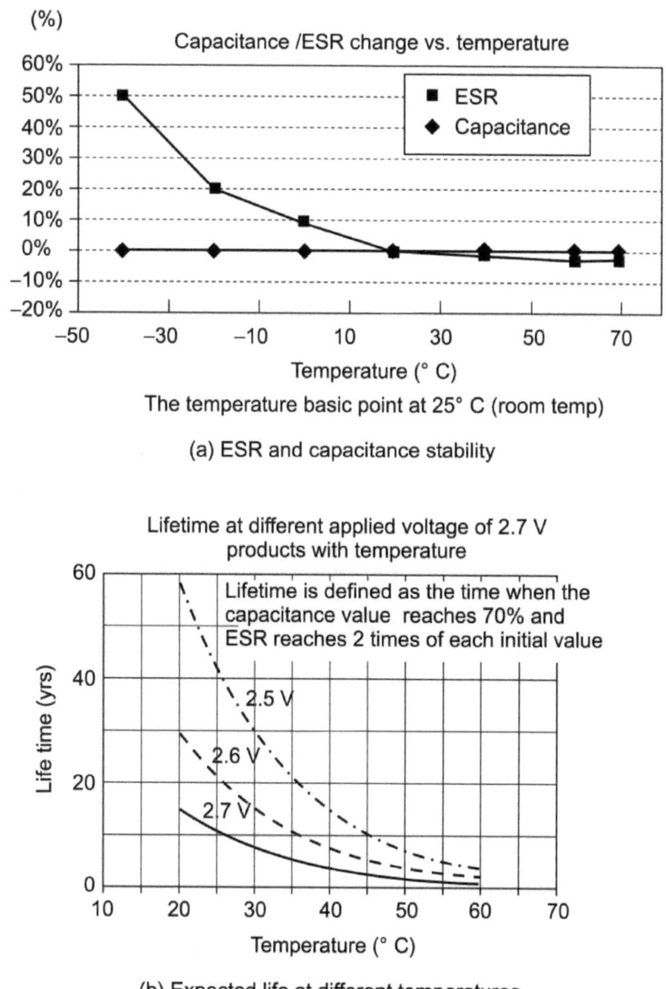

Fig. 13.15 *Long-term stability of ultracapacitors.*

limits, viz. drop in capacitance value below 30%, or increase in ESR over 100%. Life depends upon the ambient temperature and the actual voltage across the capacitor.

Figure 13.15(a) shows the high stability of capacitance from –40 to 70°C, while ESR is stable above 15°C. The capacitor is still workable at as these low temperatures, where batteries simply cannot work. The service life of ultracapacitors can reach twenty years and beyond at most places, and depends on the temperature and applied voltage (Fig. 13.15 (b)). This means they can work without maintenance for the full lifetime of equipment in which they are fitted.

Charging voltage and temperature affects the performance of EDLCs in a big way. For example, a capacitor will drop its value by 30% and its ESR increase 140% over 22000 hours at 45°C when charged at 2.7 V, whereas at 2.5 V value will drop only 15% and ESR rise by 40% over the same period. The same changes in value and ESR can be seen at both voltages after 88,000 hours at 25 at 25°C, and at 5500 hours at 65°C.

13.10 PSEUDOCAPACITORS

Pseudocapacitors have been developed with still higher energy densities by use of metal oxide electrodes (in place of carbon) and using electrochemical reaction similar to battery technology for energy storage, and have improved energy density. They need only 60% of the volume of EDLC capacitors for same capacitance. This also means a pseudocapacitor holds 80% more energy for the same volume. Evans Capacitor Company makes hybrid capacitors with high cell voltages of 10 V or more using tantalum oxide anode, RuO_2 cathode and sulphuric acid electrolyte. This is claimed to have high specific power and energy, and low ESR. NESSCAP Ltd. has developed a credit card size product with 30 W of power, apart from pseudocapacitors up to 300 F.

13.11 PRESENT TECHNOLOGY STATUS AND FUTURE SCENARIO

Table 13.4 *Power Densities of Ultracapacitors and Batteries*

Device	Energy density
Existing commercial ultracapacitors	0.5–10 Wh per kg
MIT LEES Experimental	30–60
EEstor experimental	200–300 – target 450
Lead acid battery	30–40
Li-ion battery	120
Gasoline	12000

The ultracapacitors are produced in button cell form, cylindrical and rectangular shapes, and are banked for higher capacities and voltages. India does not have a manufacturer as yet, but it may not be before long that ultracapacitors will be manufactured here.

Most production and research work is being done in the USA, Russia, Europe, Korea and Japan. Additionally, many others are operating under license in China, Japan and Korea. A Texas based company has filed a patent for 'technology for replacement of electrochemical battery', and car manufacturers are gearing up for the technology in their electric cars. The car is expected to run 250 miles on a 5-minute charge – if so fast a charging is possible (charging time may be 4–5 hours on a household plug). Several companies have also shown interest in ultracapacitors in defence equipment and vehicles. They hope to develop an energy storage device 10 times lighter than a Li-ion battery, which will charge in minutes, and at half the cost. According to one company, they are developing an energy storage unit from ceramics, which will charge in 5 minutes and run 500 miles on $9 worth of electricity. If this claim comes to reality, it could make IC engines obsolete.

There are claims that it is possible to develop a ceramic ultracapacitor with a barium titanate dielectric, which can store 280 Wh per kg, which is more than double that of Li-ion battery. They cite a permittivity of 18,500 for this dielectric.

Battery manufacturers are not far behind, and Li-ion batteries are being developed with higher and higher power capacity and energy storage, and it seems both technologies will grow in parallel and will complement each other. However, as per claims of some manufacturers, ultracapacitors may overtake batteries in not too distant a future as main storage of bulk energy for most power applications. It is certain that ultracapacitors, with other energy sources, or by themselves, are changing the scenario of energy storage applications in the world.

Advances in lighting industry, alternative energy sources like solar and wind, have created the necessity to store the electricity produced when available, and use as and when necessary. Batteries are being presently used for both solar and wind energy, but their limited life spans and ecological factors are pushing for a better alternative. Ultracapacitors are being looked upon as a viable alternative. The alternative energy sources can be fully utilized once this option is available. Hundreds of thousands of ultracapacitors are being manufactured by a number of companies. Worldwide ultracapacitor market is projected to rise between 15% and 25% annually and may be still higher as costs come down, and usage increases.

Ultracapacitors and batteries represent complementary pieces of grid storage system where ultracapacitors are used to take care of small time disruptions in power and batteries store current for long-term requirement and peak power management. These together represent a large attractive growth sector.

14

AUTO IGNITION AND CDI CAPACITORS

14.1 AUTO IGNITION CONDENSER

The conventional breaker point type ignition systems have been in use since the early 1900s. The automotive ignition system has two basic functions: it must control the spark and timing of the spark plug firing to match varying engine requirements, and it must help induce voltage across ignition coil to a point where it will overcome the resistance offered by the spark plug gap and fire the plug.

An automotive ignition system is divided into two parts: the primary and secondary circuits. The primary circuit carries low voltage, operates on battery current and has primary of ignition coil, the breaker points and the ignition switch. The secondary circuit consists of the secondary windings in the coil, the high-tension lead between the distributor and the coil, the distributor system, and the spark plugs. The distributor switches the primary current on and off and distributes the current to the proper spark plug each time a spark is needed. The distributor is a stationary housing surrounding a rotating shaft. The shaft is driven by the engine's camshaft through the distributor drive gears. A cam near the top of the distributor shaft operates the contact points mounted within the distributor housing.

The system produces a high voltage of the order of 15–17 kV to give a spark across a spark plug gap to ignite the fuel mix in a petrol engine. The capacitor serves to protect the ignition switch from sparking due to high frequency of current interruptions of considerable order, while also helping increase the secondary high voltage across the spark plug.

Under normal operating conditions, the points are closed and power from the battery is fed through a resistor or resistance wire to the primary circuit of the coil. It is then grounded through the ignition points in the distributor. Energizing current in primary winding induces a very large, intense magnetic field. This magnetic field remains as long as current flows and the points remain closed.

As the distributor cam rotates, the points are pushed apart, breaking the primary circuit. Interrupting the flow of primary current causes the magnetic field to collapse. This causes the secondary of coil to produce a very high voltage (since there are many more turns in the secondary windings); the voltage from the primary windings is magnified considerably up to 40,000 V.

This interruption of the primary current is accomplished by mechanically opening the primary circuit through the points in a synchronized sequence to send high voltage pulses through a rotary switch (distributer) to the sparkplugs. One of the drawbacks of this process is that the interruption of current in the primary coil generates an inductive back-voltage which tends to cause sparking across the points. The system needs a sizable capacitor (condenser) across the contacts so that the voltage surge charges the capacitor and saves the point from damage due to destructive sparking.

Fig. 14.1 *Auto condensers.*

Ignition capacitors or auto condensers (Fig. 14.1) are generally rated 0.18 – 0.33 μF with voltage withstand capacity of 900 – 2000 V DC depending upon design. These capacitors face a rigorous duty as they charge and discharge at very high frequency continuously. (IC engine rotational speed can be anywhere from 6000 to 12,000 RPM.) Paper or PP Film extended foil construction has been used all along for these capacitors, and they also incorporate a spring inside to make them withstand severe vibrations encountered. The working temperature could go as high as 100°C.

The ignition capacitor case must be of steel, and contains a spring contact to take care of vibrations. The high rate of repetitive peak currents and the working temperature requirements necessitates the use of extended foil paper/foil or PP/foil impregnated capacitors, and metallized capacitors are unsuitable. Currents are collected by metal discs at top and bottom. The can is

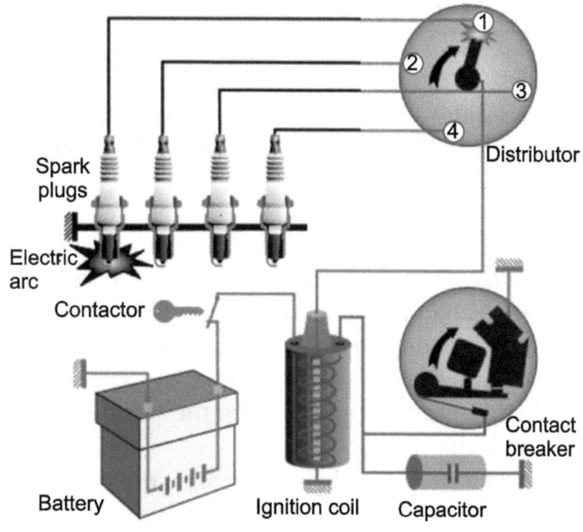

Fig. 14.2 *Ignition circuit schematic.*

one terminal, while the lead coming out through the top sealing and fixed to the top metal disc serves as the other external electrode. The sealing is generally rubber, or Bakelite sheet with epoxy resin sealing.

A general schematic diagram is shown in Fig. 14.2. Voltages across the capacitor can reach 700–750 V every time the contacts open. Considering the engine RPM, reaching 6000 or much higher, depending on design of engine, this puts a severe duty on the condenser. Its condition is much more severe considering that the assembly is mounted right near the engine, and the temperature may reach as high as 100°C. The vehicle movement and engine working cause high vibration levels, under which the capacitor must work. The capacitor-point combination has to work flawlessly for years, and failure of any one induces the failure of the other. A condenser which is defective or improperly grounded will not absorb the shock from the fast-moving stream of electricity when the points open and the current can force its way across the point gap, causing pitting and burning.

This capacitor has been the mainstay of all two and four wheelers, as well as all petrol driven vehicles thus far, and used in a configuration as shown in Fig. 14.3. It is almost out of use now, being replaced by ignition systems based on Capacitor Discharge Ignition (CDI) or electronic ignition.

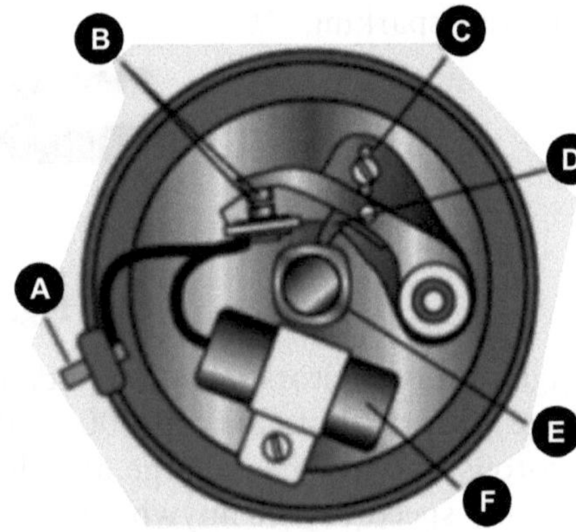

A. To coil
B. Point
C. Points dwell adjustment
D. Point riding on cam
E. Rotating cam
F. Condenser

Fig. 14.3 *Condenser-point assembly.*

14.2 CDI: AN INTRODUCTION

In modern two wheelers, such as mopeds or lawn mowers, the ignition system design is based exclusively on CDI. In automobiles CDI was in the past only used in the replacement module market. Today, due to new standards of pollution control, the CDI system is becoming one of the most efficient choices available.

CDI principle

The spark necessary to ignite the air/petrol mixture in the combustion chamber is produced by the CDI module. A high voltage coil produces, after rectification, a positive voltage between 100 and 400 V. The high voltage for an automobile CDI is supplied by a DC/DC converter. This stage produces generally 400 V from the 12 V battery voltage. A capacitor between 0.68 and 2.2 μF stores the charge from the high voltage supply. During the second phase of the ignition cycle the capacitor is discharged through the ignition circuit.

A switch transfers the energy stored in the capacitor to the primary of the ignition coil. This function is carried out by a SCR or a triac. The switch is generally linked to a diode for the reverse current. A sensor synchronizes the spark with the engine rotation. For a small engine, the sensor detects a bump at each engine revolution. For car modules the sensor system gives a pulse for each cylinder ignition point.

An SCR gate operates at correct lead angle for all RPMs. The ignition coil, a step-up transformer, delivers high voltage to the spark plug, when the capacitor discharges into its primary. The voltage can be between 5 and 20 kV depending on the working conditions. The spark plug is the final element of the ignition chain. High engine efficiency and complete gas combustion are linked to a good spark quality. Generally a minimum of 20 million joules is necessary at the spark plug.

CDI ignition is most widely used today on automotive engines. A CDI module has a capacitor which sends a short high voltage (about 250+V) pulse through the coil. The coil acts as a transformer and multiplies this voltage to step up the voltage about 100:1 (Fig. 14.4). A typical 250 V CDI module output is stepped up to over 25,000 V output from the coil. The CDI output voltage of course can be higher. The huge advantage of CDI is the higher coil output and 'hotter' spark. The spark duration is much shorter (about 10 – 12 microseconds) and accurate.

The main advantage of CDI unit is that it has no moving parts. Hence control and reliability is very superior to any mechanical system. The term 'engine tune-up' is meaningless with CDI systems, and except for replacing plugs and inspecting wiring, there is nothing else to do. CDI coils have low impedance (0.5–1 Ω) and can reach high RPM. This type of ignition makes high power and short sparks.

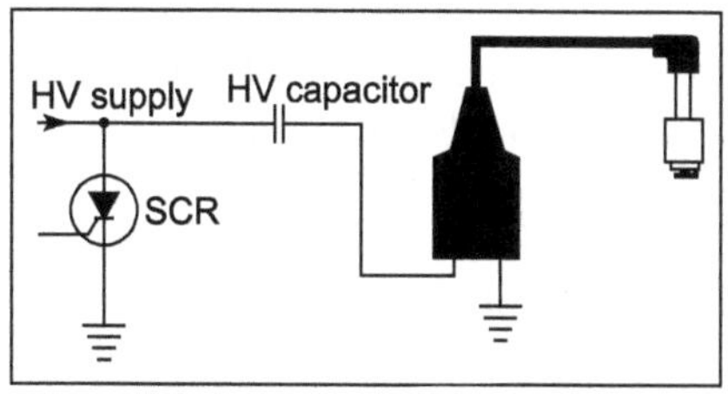

Fig. 14.4 *CDI ignition.*

CDI ignition makes the spark by discharging a capacitor loaded with high voltage (200 – 450 V) into the ignition coil by using a SCR (thyristor). Solid-state components are particularly sensitive to thermal stress, vibration, moisture and power surges. Thus, designs have to incorporate factor to improve CDI reliability. These include things like:

- Improved cooling and heat sinks.
- Epoxy resin or epoxy-rubber encasing components so they can't get wet.
- Using separate ignition modules for each plug (so a single failure won't kill the whole engine). Heavier duty components that can withstand the heat, vibration and duty cycles.

14.3 CDI UNITS AND CIRCUITS

CDI is being used for most of the modern auto/motorcycle applications. This is simply due to the ability to fine-tune all aspects of the combustion process electronically. Where simplicity and reliability is a factor, induction systems have an advantage. The longer spark duration of induction systems gives a better chance that combustion WILL take place.

CDI typically has very short spark duration near 10 – 12 microseconds. Instead of one big spark, a shower of short duration sparks is flooded across the combustion stroke. This makes for a much more efficient burn. Using this technique newer CDI can achieve longer spark duration times (near 250 μsec). This is particularly better for starting lean mixtures (which are hard to ignite), and in high compression situations. Figure 14.5 shows a typical CDI unit and its circuit diagram.

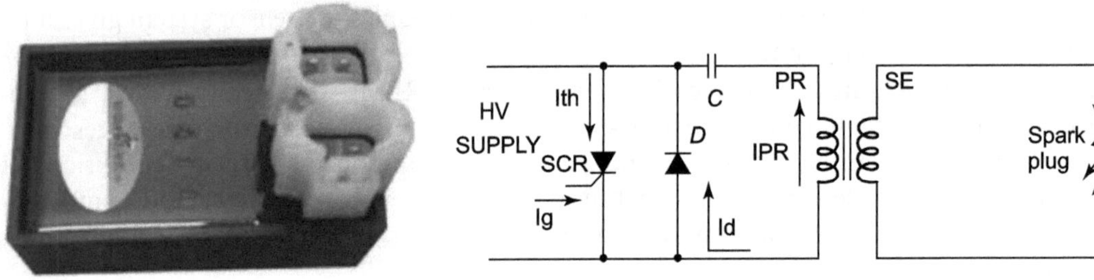

(a) (b)

Fig. 14.5 *(a) CDI unit and (b) circuit diagram.*

The principal advantage of a CDI system is the ability to present a superior spark to the air/fuel mixture inside the combustion chamber, thus maximizing burn efficiency. The easiest way to get a bigger spark is to increase the spark plug gap size and increases the voltage to ionize the air/fuel mixture. A CDI system provides the higher voltage required by the increased spark plug gap size, thus providing a very intense spark.

Fig. 14.6 *CDI capacitors.*

CDI capacitors are usually rated 400 V DC and at temperature category 85°C or 100°C. These are usually MPP or sometimes MPET construction, with tolerance of ± 10% or closer. Climatic category of 40/85/56 (MPP) or 55/125/56 (MPET) is preferred. Capacitance values are from 0.68 to 2.2 μF for different designs. Dissipation factor is under 3% for MPET and 0.05% for MPP at 100 kHz. Time constant (IR in M Ω × μF) is 10,000 sec and above. Applicable international standards are IEC 384-1 for MPP and IEC 384 – 2 for MPET. CDI capacitors could be axial lead or box type (Fig. 14.6) or of dip coated construction.

Capacitors for CDI have to be made from special grade MPP/MPET/MPEN film, with lower surface resistivity on the base as well as the heavy edge, to take care of repetitive surge currents and discharge currents. Long life expectancy under these extreme conditions, as well as at high operative temperatures necessitates careful selection of materials, and also due care in the production process. End contacts also play a vital part. After all this, it is still imperative to keep the currents under wraps by proper circuit design, and recondition the capacitors as quickly as possible between successive operations. Capacitors should be suitable for high frequency and high dv/dt conditions encountered in service.

Fig. 14.7

15

ELECTRONIC GRADE CAPACITORS

15.1 INTRODUCTION

Capacitors are needed everywhere in the electronics industry. Whether in mobile phones, game, consoles, computers, digital cameras or cars, capacitors are the second most common passive components in electronic circuits after resistors. An idea of their use is clear from their average consumption figures in Table 15.1 in some common applications:

Table 15.1 *Capacitors in Modern Applications*

Device	*No. of capacitors per unit*
Mobile phone	260
Digital camera	310
Game console	315
Computer	700
Car	1700

Capacitors perform various functions in circuits like timers, filter, decoupling, bypass applications, tuning circuits, biasing circuits, RFI suppression and a host of others. They come in different varieties, sizes, ratings, types, dielectrics, configurations and terminations to suit individual applications and environments. There are fixed and variable capacitors in different dielectrics, electrolytic capacitors with aluminium, titanium and niobium electrodes, ceramic and mica capacitors of different grades to choose from. Mounting of capacitors can be done by clamps, brackets, direct soldering onto the PCB through radial or axial leads, while SMD capacitors are directly soldered on to PCB through reflow soldering.

Most electronic capacitors in electronic circuits are used on PCB mountings. These capacitors have their leads coming out and spaced at standard lengths, usually from 5 mm to 27.5 mm (called pitch), for ease in mounting. The leads may come out in axial direction (straight out from both ends), or radial direction (perpendicular to winding), as shown in Fig. 15.1. Small capacitors are often supplied as strip packaging, which facilitates their use for large-scale production lines. They may come in box enclosed, round can type, resin dipped construction, or open configuration (heat shrunk and sealed). Large capacitors are made in round or rectangular cans and may be fixed with clamps. SMD components are supplied in strip packs in a form which is usable on automatic feeding on PCBs by robotic machines.

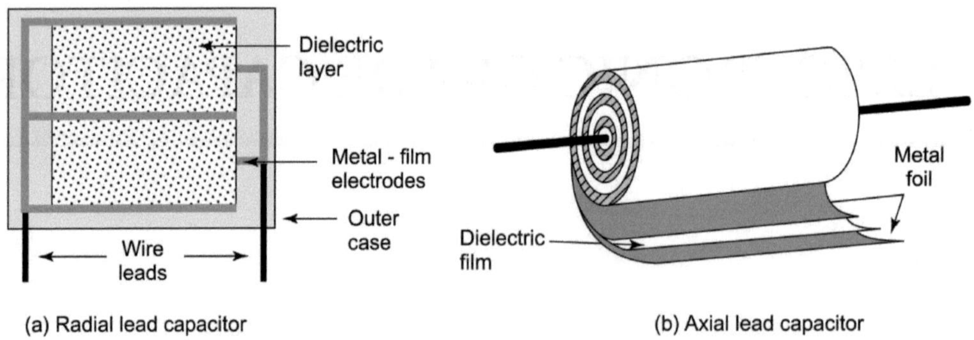

Fig. 15.1 *Main PCB mount electronic capacitor constructions.*

While mounting these capacitors, care must be taken for proper mounting or clamping so as not to subject them to excessive vibrations. They should be located away from the source of heat. Vibration and temperature are two major reasons for component failures in electronic circuits.

One particular concern in AC applications is the dielectric's loss characteristics. Capacitor losses are typically represented by the dissipation factor (DF), in units of percentage and equivalent series resistance (ESR), expressed in mΩ. The DF and ESR performance of a capacitor with temperature and frequency are critical when making capacitor selections for AC applications. If these values are too high at the application temperature or at the operating frequency, it will result in reduced capacitor life, or in the extreme case, catastrophic failure.

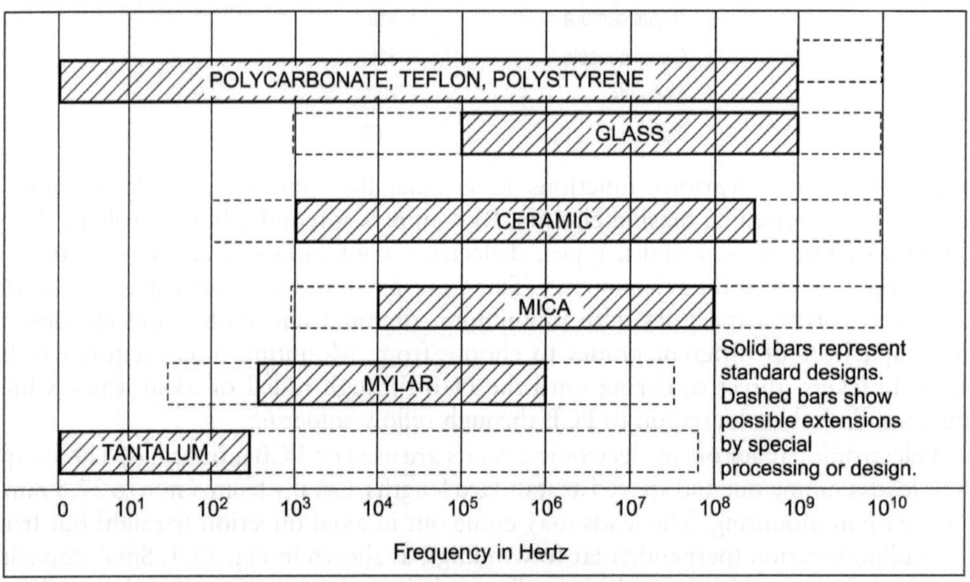

Fig. 15.2 *Dielectrics and their frequency range.*

Different types of capacitors using different types of dielectrics are only able to operate within a particular frequency range. Fig. 15.2 below shows the different styles of dielectrics and their operational frequency range.

15.2 NON-INDUCTIVE AND INDUCTIVE CAPACITORS

(a) **Non-inductive capacitor:** This is a capacitor constructed such that it has practically no inductance; foil layers are staggered during winding, so an entire layer of foil projects at either end for contact-making purposes; all currents then flow laterally rather than spirally around the capacitor.

Extended foils are preferred for high frequency response over metallized ones where ripple currents, pulse or surge currents are high. Extended foil construction gives good frequency response over a large spectrum, and inductance and resonance frequency is limited only by lead wires and other factors. Leads are soldered or welded to extended foils before encapsulation or sealing of capacitors. These capacitors are characterized by high dv/dt response, lower losses and higher current capabilities.

(b) **Inductive winding:** In this type of capacitor, dielectric film extends on both sides of the foil, is simpler to wind and is generally cheaper. Tabs are brought out from electrodes to form external connections. These capacitors have a resonance frequency above which it may become inductive when coil inductance becomes predominant. Frequency response gets poorer at higher frequencies as inductance becomes comparable to capacitive reactance. Losses increase as frequency goes up.

15.3 FILM/FOIL AND METALLIZED FILM CAPACITORS

Plastic film capacitors can be divided into three basic construction types: film/foil capacitors, metallized film capacitors and mixed technology capacitors. Film/foil capacitors consist of two metal foil electrodes made of aluminum foil separated by a piece of plastic film. The plastic film can be polyester, polypropylene or polycarbonate. There are other types of plastic films but these films are used in specialized applications. The thickness of the plastic film typically ranges from 2 to 20 μm, while the aluminum foil thicknesses range from 5 to 9 μm.

(a) **Film/foil capacitors:** A film/foil capacitor is made by alternating two pieces of aluminum foil with two layers of plastic film. These interleaved layers are wound around a spindle in a manner that prevents the metal layers from touching. Film/foil capacitors can be wound in two different ways – inductive (insert tab) and non-inductive (extended foil). Both extended foil (non-inductive) capacitors and inductive capacitors are in use. Long-term stability of film/foil capacitors is excellent, the failure at the end of useful life is generally in short circuit mode. Failure may cause blowing of fuse or explosion of capacitor.

Polystyrene Film Capacitors are exclusively made in film/foil construction.

(b) **Metallized film capacitors:** These are commonplace today for most applications, and are non-inductive by nature. Self-healing is their main advantage, and they offer many advantages of extended foil capacitors. But dv/dt rating is generally lower than film/foil type. Capacitance stability is not as good, as a result of self-healing properties. Lower cost, better frequency response and self-healing properties has made these popular in the electronics industry. Special films with extra heavy edge and increased metallization thickness have led to much higher dv/dt characteristics and lower losses comparable to extended foil types. This has made possible their use in some snubbers, commutation capacitors and pulsed power applications.

Self-healing property gives the benefit of long life to these capacitors. Field failure rates are low, and failure is usually in open-circuit mode. This comes with a sacrifice of long-term capacitance stability, as there is an infinitesimal loss of capacitance at every self-healing event, which ultimately becomes noticeable, followed with an increase in loss factor. Nonetheless, their advantages make them ideal for most noncritical electronic applications, and they are found almost everywhere in the electronics industry.

(c) **Mixed technology capacitors:** A mixed technology capacitor is a combination of film/foil and metallized film capacitor types. Metal foil is one electrode, while one metallized film is used, so that the second electrode is metallized coating. The film displays self-healing properties due to metallized film, but the effect of self-healing is limited to one electrode only, resulting in smaller capacitance loss, and a better stability compared to metallized capacitors. Normally these are high voltage capacitors with internal series connections. Mixed technology capacitors have the characteristics of both capacitor types, such as high pulse carrying capabilities and self-healing properties.

15.4 PLASTIC FILM CAPACITORS

Polypropylene and polyester films have been dealt with in detail in chapters 2 and 3. We shall proceed to discuss other films used in electronic capacitors.

[A] Polystyrene capacitors

These capacitors use polystyrene film as the dielectric. This type of capacitor is not for use in high-frequency circuits, because they are wound like a coil and are inductive. They are suited well in filter circuits or timing circuits which run at several hundred kHz or less. The components in Fig. 15.3 show the inner foil electrode. The colour of the electrode in transparent film capacitor is the colour of metal used for foil. Aluminum foil gives silver colour, copper gives red and tin is a silver coloured electrode. The capacitor on the left has a height of 10 mm, is 5 mm thick, and is rated 100 pF, the one in the middle has a height of 10 mm, 5.7 mm thickness, and is rated 1000 pF, and that on the right has a height of 24 mm, is 10 mm thick, and is rated 10,000 pF. Styroflex capacitors, as these are called, have no polarity. These capacitors are highly stable. As mentioned in an earlier chapter, polystyrene is no longer made and has been replaced by PP and COG ceramics.

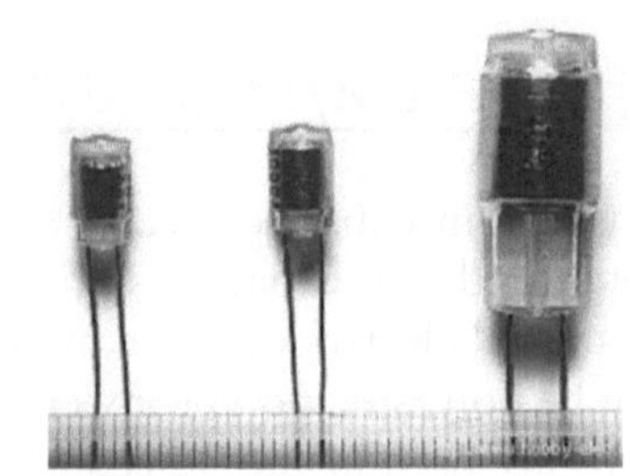

Fig. 15.3 *Polystyrene capacitors.*

[B] Polycarbonate film capacitor

Polycarbonate capacitors have been used in a wide variety of applications because of the superior performance offered. Typically they are used in applications where precision capacitors are needed (<5%). They are generally used in electronics circuits such as filters, as well as for timing and precision coupling applications.

Polycarbonate capacitors can also be used for AC applications. They are sometimes found in switching power supplies. Although the dissipation factor is low, the current must be restricted to prevent them from overheating, although they can tolerate temperature better than many other types of capacitor.

[C] Polycarbonate capacitor replacements

With polycarbonate capacitors being less widely available these days since the Bayer Corporation ceased production of polycarbonate for use as a dielectric, a number of alternative types of capacitor have been sought, especially for use in some military applications. A variety of types can be used as almost direct replacements:

- Polyethylene naphthalate (PEN)
- Polyphenylene sulphide (PPS)
- Polyimide (PI)
- Polytetrafluoroethylene (PTFE)

Of these, polyphenylene sulphide, PPS is being widely used in many areas as an almost direct replacement.

[D] Polyphenylene sulphide (PPS)

Discussed in 2.6.1 (E), PPS has a superior, overall capacitance stability with temperature over the whole range. The ESR performance at room temperature over the frequency range from 100 to 100,000 Hz for polyphenylene sulphide is superior to polycarbonate. A PPS device may run hotter than a PC device without any problem in some AC applications. PPS, unlike PC, can operate without degradation at capacitor temperatures exceeding +125°C in DC applications.

[E] PEN film This has been referred earlier in Sec. 2.6.1(F).

Other films mentioned above (PI and PTFI) are in the development stage, and not common as of now. Polyimide (PI) film appears very promising, with working temperature of up to 250°C, and it has properties similar to polyester films. The film thickness available is up to 3 microns. Its high cost relegates it to use in high temperature applications.

15.5 GLASS CAPACITORS

Glass dielectric capacitors are viewed as specialist capacitors, and in view of their properties, they offer real advantages in many applications over all other forms of capacitor. Their combination of robustness and high tolerance is a rare combination that sets them above all other forms of component. It is only their size and cost that limits their use.

Glass capacitors are used where the ultimate performance is required for RF circuits. The dielectric offers very high levels of performance. Typically a glass capacitor will have a relatively low capacitance values between fractions of a pF up to 2000–3000 pF. As such these capacitors are used mainly in radio frequency circuit design.

On account of their costs, glass dielectric capacitors are reserved only for the most exacting RF requirements, often on low volume products where cost is not as much of an issue as it is in high volume products. The supply of glass capacitors is also limited to a small number of manufacturers and suppliers, and the capacitors may not be available ex-stock.

15.5.1 Glass capacitor construction

The construction of glass dielectric capacitors is relatively straightforward to understand. The capacitor consists of three basic elements: the glass dielectric, aluminium electrodes and the encapsulation. However the assembly of the glass capacitors is undertaken in a manner that ensures the required performance is obtained.

The capacitance between two plates is not always sufficient to provide the required level of performance. Hence most capacitors use a multilayer construction to provide several layers of plates with an interspersed dielectric to give the required capacitance.

Although the glass plates are always flat, and tubular forms of construction are not applicable, glass capacitors are usually available with leads emanating in either a radial or axial form.

15.5.2 Glass capacitor advantages and characteristics

Glass capacitors offer several advantages over types of capacitor. In particular glass capacitors are applicable for very high performance RF applications:

- **Low temperature coefficient:** Figures of just over 100 ppm C^{-1} are often obtained for these capacitors.
- **No hysteresis:** Some forms of capacitor exhibit hysteresis in their temperature characteristic. This is not the case for glass capacitors which follow the same temperature/capacitance when the temperature is rising and falling.
- **Zero ageing rate:** Many electronics components change their value with age as chemical reactions take place within the component. Glass capacitors do not exhibit this effect and retain their original value over long periods of time.
- **No piezoelectric noise:** Some capacitors exhibit the piezoelectric effect to a small degree. This can result in effects such as microphony on oscillators. Use of glass capacitors could help solve the problem.
- **Extremely low loss/high Q:** There is virtually no dielectric loss. This enables very high *Q* circuits to be built using them, provided the other components (e.g. inductors) are not lossy.
- **Large RF current capability:** Some capacitors are not able to withstand large values of current. This is not the case for glass capacitors which are suitable for use in RF high power amplifiers, etc.
- **High operating temperature capability:** Glass dielectric capacitors are able to operate at very high temperatures, even at temperatures up to about 200°C without fear of damage or performance shortfall.

15.5.3 Glass capacitor applications

Glass capacitors can find applications in many areas as a result of their performance characteristics. They do tend to be specialist components and are normally fairly expensive.

- **Circuits exposed to temperature extremes:** With the tolerance to a wide range of temperatures, both high and low, some circuits that may be exposed to very harsh environmental conditions may choose to use glass capacitors. Not only can they withstand high and low temperatures, but they do not change value at these extremes by a great amount. Accordingly remote sensors may choose to use glass capacitors.
- **Applications requiring a high Q circuit:** Many circuits including oscillators and filters may require high *Q* components to give the required performance. Filters will be able to attain their required bandwidth, and for oscillators there are a number of advantages including improvement of phase noise performance, reduction in drift and reduction of spurious oscillations.
- **Low microphony requirements:** Glass capacitors may be used in circuits where microphony may be a problem. RF oscillators including those found in phase locked loops and PLL synthesizers may benefit from their use.
- **High power amplifiers:** The high current capability of glass capacitors may enable their use in RF power amplifiers where other forms of capacitor would not be suitable.
- **High tolerance areas:** In many areas such as filters or free running oscillators the high tolerance and precision accompanied by the low temperature coefficient may be required to maintain the tolerances within a precision circuit.

15.6 VACUUM CAPACITORS

These are used for ratings over 5 kV and up to 50 kV. They are used in equipment such as high-powered broadcast transmitters, amateur radio RF amplifiers and large antenna tuning units and couplers. They also find use in high voltage and high current network matching circuits. One application of vacuum capacitor is in the manufacture of LCD panels – thin-model, big-screen televisions. The LCD technology is used for photovoltaic power generating panel manufacturing. Vacuum capacitors are incorporated in the impedance matching network of RF generators for plasma generation. They are used in chemical ingredients analyzers and magnetic resonance imaging (MRI). RF generators are used in semiconductor manufacturing processes of physical vapour deposition (PVD), chemical vapour deposition (CVD), and etching, with vacuum capacitors used in the impedance matching network. These capacitors play a part in communications equipment, transmission systems for short and medium wave broadcasting, aircraft antenna tuners used in harsh environments, and various mobile communications equipment. They are used in particle accelerators and basic science research facilities.

Fig. 15.4 *High voltage vacuum capacitors.*

15.7 VARIABLE CAPACITORS

A variable capacitor is one whose capacitance may be intentionally and repeatedly changed mechanically or electronically. Variable capacitors are often used in L/C circuits to set the resonance frequency, e.g. to tune a radio (therefore they are sometimes called *tuning capacitors*), or as a variable reactance, e.g. for impedance matching in antenna tuners.

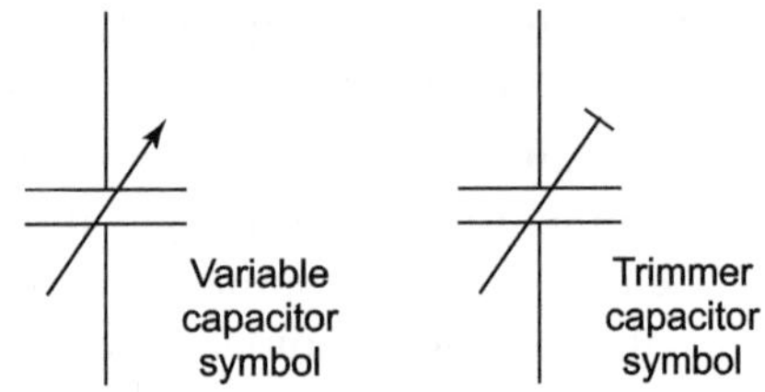

Fig. 15.5 *Variable capacitor and trimmer symbols.*

[A] Mechanically controlled

In mechanically controlled variable capacitors, the distance between the plates, or the amount of plate surface area which overlaps, can be changed. The most common form uses a group of semi-circular metal plates on a rotary axis (rotor) that are positioned in the gaps between a set of stationary plates (stator). The overlap area is changed by rotating the axis. Air or plastic foils can be used as dielectric material. By choosing the shape of the rotary plates, various functions of capacitance vs. angle can be created, e.g. to obtain a linear frequency scale. Reduction gear system is often used to achieve finer control over a larger angle, often several turns. Figure 15.6 shows gang condensers with air dielectric.

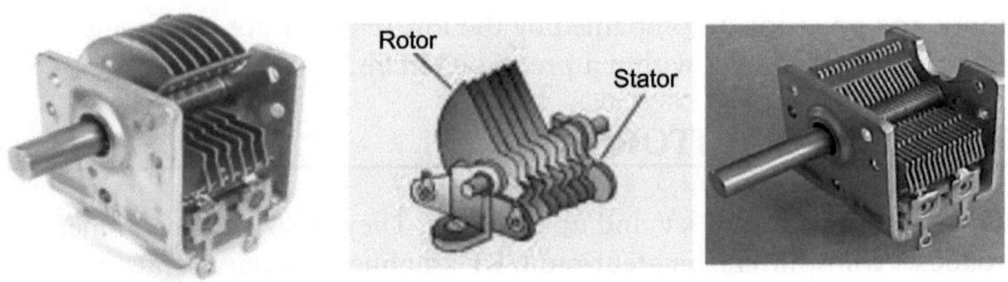

Fig. 15.6 *Mechanically variable capacitors.*

[B] Trimmers

Small variable capacitor operated by screwdriver (for instance, to precisely set a resonant frequency at the factory and then never be adjusted again) is called trimmer capacitors or in old language, a 'padder capacitor'. This capacitor uses a springy material for the plates. The assembly is held together by a small screw. The plates are so springy that if there is no screw to hold, they would fly apart. There is a dielectric material between the plates, usually mica, but polypropylene film or ceramics are also used. The distance between plates is altered by tightening or loosening the screw, thereby changing the capacitance value.

Capacitance variation depends on the number of plates and size of trimmer, and usually ranges from 1.4 to 10.0 pF and from 5.5 to 65 pF.

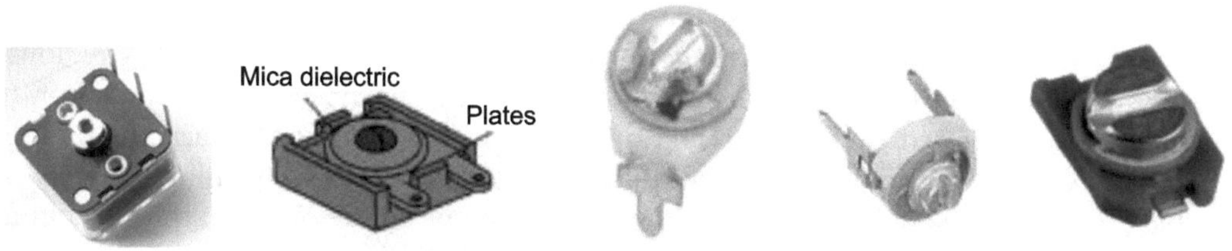

Fig. 15.7 *Trimmer types.*

[C] Vacuum variable capacitor

A vacuum variable capacitor (Fig. 15.8) uses a set of plates made from concentric cylinders that can be slid in or out of another set of cylinders (sleeve and plunger). These plates are then sealed inside a non-conductive envelope such as glass or ceramic under a high vacuum. The operating mechanism for capacitance change is through a vacuum-sealed shaft coming out. The vacuum increases the working voltage and current rating of the capacitor. The most common usage for vacuum variables is in high-powered transmitters such as those used for broadcasting, military and amateur radio as well as high powered RF tuning networks. The elements being under a vacuum, the working voltage can be higher than an air variable of the same size, allowing the size of capacitor to be reduced.

Fig. 15.8 *Vacuum capacitors Left: Glass body 20–1000 pF±5% 100 kV 50 A 30 MHz Right: Ceramic body 350–2500 pF 15 kV to 40 kV current 300 A.*

Vacuum variable capacitors are commonly used in high voltage applications of 5000 V (5 kV) and above. They are used in equipment such as high-powered broadcast transmitters, amateur radio RF amplifiers and large antenna tuners.

[D] Electronically controlled (Varicap)

(i) Varicap diode

The depletion layer thickness of a reverse-biased semiconductor diode varies with the DC voltage applied across the diode. Any diode exhibits this effect (including P/N junctions in transistors), but devices specifically made as variable capacitance diodes (also called varactors or varicaps) are designed with a large junction area and a doping specifically designed to maximize capacitance.

The basis of operation of the varactor is quite simple. It is operated under reverse bias conditions and this gives rise to three regions. At either end of the diode are the *P* and *N* regions where current can be conducted. However around the junction is the depletion region where no current carriers are available. As a result, current can be carried in the *P* and *N* regions, but the depletion region is an insulator. This is exactly the same construction as a capacitor. It has conductive plates separated by an insulating dielectric. In the case of the varactor diode, it is possible to increase and decrease the width of the depletion region by changing the level of the reverse bias. This has the effect of changing the distance between the plates of the capacitor, and the capacitance (Fig. 15.9). They are effectively voltage controlled capacitors, and indeed they are sometimes called varicap diodes, although the term varactor is more widely used these days. The barrier capacitance of a P/N junction is frequency independent, with high *Q* and low noise level.

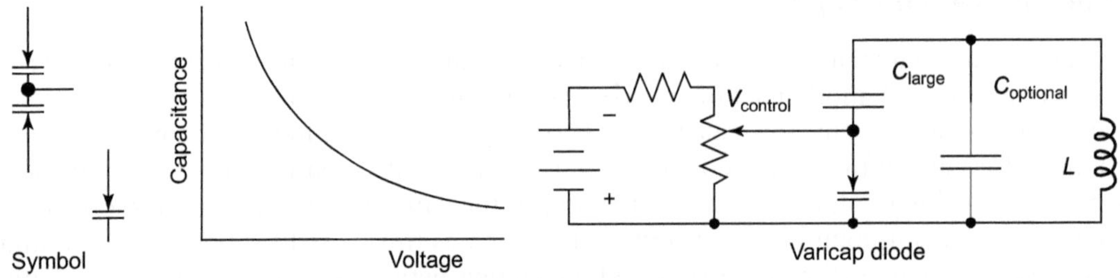

Fig. 15.9 *Varicap diode symbol, characteristic and typical circuit.*

Their use is limited to low signal amplitudes to avoid distortions as the capacitance would be affected by the change of signal voltage, precluding their use in the input stages of high-quality RF communications receivers, where they would add unacceptable levels of intermodulation. At VHF/UHF frequencies, e.g. in FM radio or TV tuners, varicaps are commonly used in the signal path. Varicaps are used as voltage-controlled capacitors, rather than as rectifiers. They are commonly used in parametric amplifiers and voltage-controlled oscillators as part of phase-locked loops and frequency synthesizers. These are not really capacitors in the usual sense, but the capacitor effect of reverse biased diode depletion region is used for these products. Hence these do not form part of usual capacitor technology in the strict sense.

The thickness of the depletion region in the varactor diode is proportional to the square root of the reverse voltage across it. In addition, the capacitance of the varactor is inversely proportional to the depletion region thickness. From this it can be seen that the capacitance of the varactor diode is inversely proportional to the square root of the voltage across it. One of the key parameters for a varactor diode is the capacitance ratio. This is commonly expressed in the form Cx/Cy where x and y are two voltages towards the ends of the range over which the capacitance change can be measured.

Varactors are operated reverse-biased, so no current flows, but since the thickness of the depletion zone varies with the applied bias voltage, the capacitance of the diode can be made to vary. These are low voltage low value capacitors in pF range. For a change between 2 and 20 V an abrupt

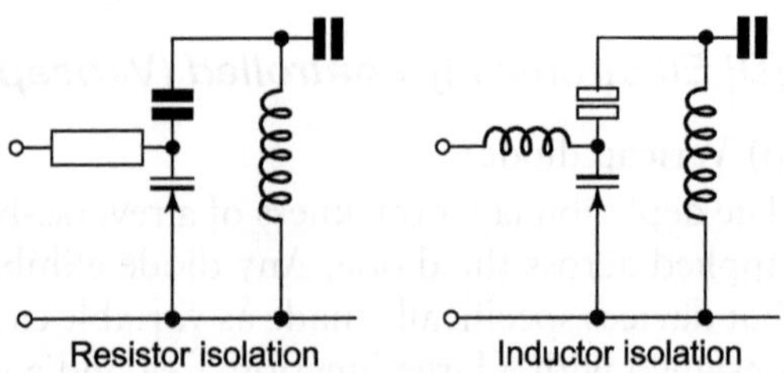

Fig. 15.10 *Applying varactor tuning voltage via resistor and inductor*

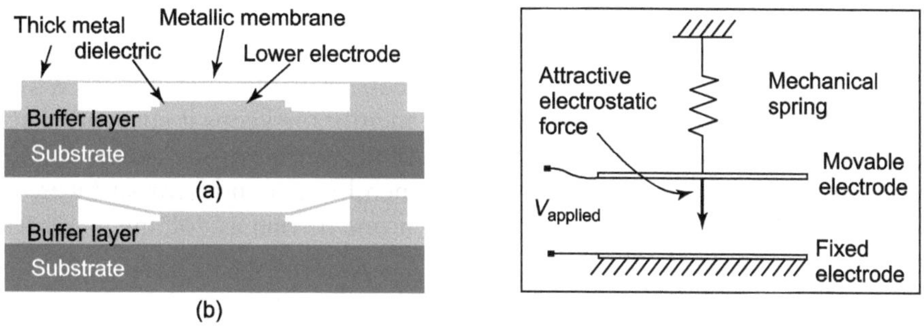

Fig. 15.11 *(a) MEM capacitor switch.* *(b) Mechanical equivalent.*

diode may exhibit a capacitance change ratio of 2.5 to 3, whereas a hyper abrupt diode may be twice this, e.g. 6.

Varicaps are used for frequency modulation of oscillators, and to make high-frequency voltage controlled oscillators (VCOs), the core component in phase-locked loop (PLL) frequency synthesizers that are common in modern communications equipment. Varactor diodes or varicap diodes are semiconductor devices that are widely used in the electronics industry in many applications where a voltage controlled variable capacitance is required. Accordingly they are used in many RF circuits including voltage controlled oscillators, amplification of microwave pulses, tuners, filters.

(ii) RF MEMS (micro-electromechanical system) variable capacitors

The acronym Radio frequency microelectomechanical system (MEMS) refers to electronic components of sum-millimetre size with moving parts which provide functionality. The functionality is achieved in several ways and for various purposes. These are micromachined components made with precision at the nanometer levels. The advent of this technology has enabled new and exciting ways for creating and controlling capacitance with wide tuning range and high Q. This is due to the unique capabilities enabled by micromechanical tuning and the low loss materials used in the construction of these devices.

According to Coulomb's law, the electrostatic force acting between two charges is inversely proportional to the distance between the charges. This means the capacitor plates of a charged capacitor attract each other. For macro-scale objects, this force is normally negligible. However, micro-scale devices may have very small gaps, making electrostatic attraction an important source of mechanical motion. This actuation technology is especially attractive because it uses very little power. On the other hand, large voltages (typically tens to hundreds of volts) are required. Today RF MEMS capacitors are used in large numbers in accelerometers, gyroscopes, microphones, mobile phones, various sensors as well as communication and entertainment industry.

(a) Capacitor switch: The simplest electrostatic actuator consists of a movable plate or beam which is pulled toward a parallel electrode under the application of a voltage difference. This type of actuator is illustrated schematically in Fig. 15.11. The capacitor thus formed on an IC acts as a switch, switching from a capacitor to a shorted connection.

(b) Variable capacitor: In the second type of MEM capacitor (Fig. 15.12(a) the RF MEMs capacitor consists of a lower electrode fabricated on the surface of the IC and a thin aluminium membrane suspended over the electrode. The membrane is connected directly to grounds on either side of the electrode while a thin dielectric layer covers the lower electrode. The air gap between the two conductors determines the MEMs capacitor off-capacitance. With no applied actuation potential, the residual tensile stress of the membrane keeps it suspended above the RF path. The application of a DC electrostatic field to the lower electrode causes the formation of positive and negative charges on the electrode and membrane conductor surfaces. A variant of this is Fig. 15.12(b), where the top beams snap down one by one as voltage increases.

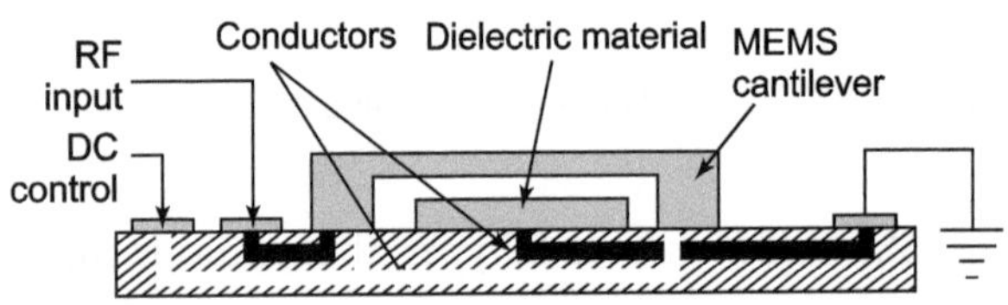

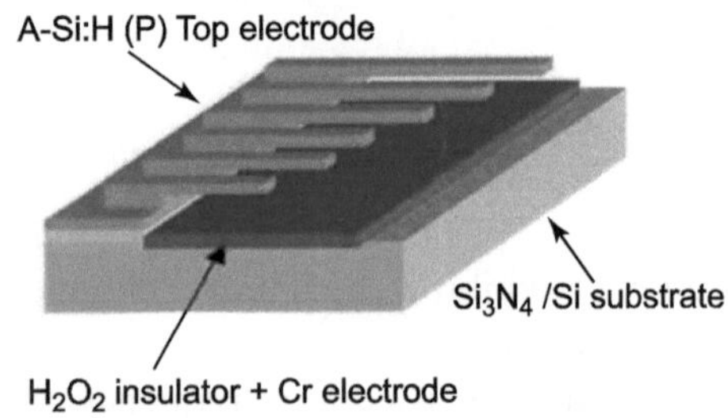

Fig. 15.12 *(a) MEMS varicap construction.* *(b) Multicantilever MEMS capacitor.*

[E] Diaphragm Type (Transducers)

Variable capacitance is sometimes used to convert physical phenomena into electrical signals. Some types of industrial sensors use a capacitor element to convert physical quantities such as pressure, displacement or relative humadity to an electrical signal for measurement purposes.

In a capacitor microphone (commonly known as a condenser microphone), the diaphragm acts as one plate of a capacitor, and vibrations produce changes in the distance between the diaphragm and a fixed plate, changing the voltage maintained across the capacitor plates.

Capacitor manometer vacuum gauge

Deflection of a thin metal diaphragm separating a known pressure from an unknown pressure is a measure of the pressure difference between the two sides. A capacitive diaphragm (Fig. 15.13)

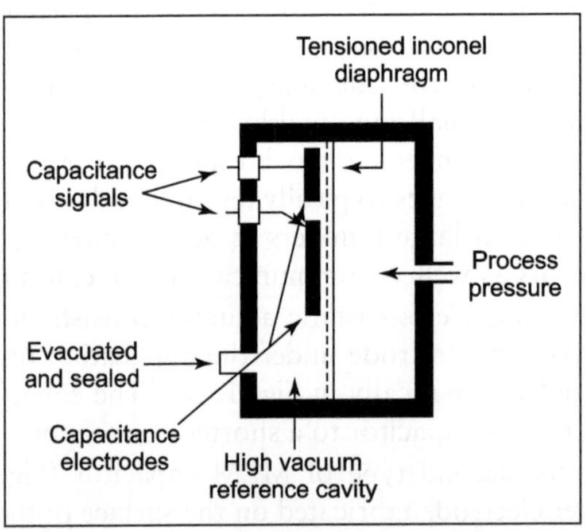

Fig. 15.13 *Capacitance manometer.*

is used to measure vacuum accurately through a change in capacitance brought about on two sides of a diaphragm – one side open to applied vacuum, while the other side is sealed at pre-determined vacuum.

Capacitive sensors can also be used in the place of switches, e.g. in computer keyboards, or 'touch buttons' for elevators that have no movable parts. In elevators, a touch on a fixed plate brings about capacitance change through our body contact, and this sends the signal. Touch screen operation of computer control and mobile phones also make use of diaphragm capacitance.

16

CAPACITORS FOR RFI SUPPRESSION

16.1 NEED FOR RADIO FREQUENCY INTERFERENCE SUPPRESSION (RFI)

Frequencies above 30 MHz tend to radiate directly from the generating circuits, while those below 30 MHz are usually conducted by the AC line and other connections. These are capable of radiating (or receiving) RFI. An AC input filter is usually required to keep these radio frequencies out. The type of filter varies widely depending on the equipment. Where the interference is generated by a single source, it is best located as close to the source as possible. If the interference is widespread but contained in a metallic enclosure (such as in a switched-mode power supply) AC cord entry point is the best location.

Two methods of RF interference suppression by use of capacitors must be considered separately. One system is based on XY capacitors, which are discreet capacitors connected between supply lines, or between supply lines and ground. The second system, based on feedthrough capacitors, takes care of RF interference between an instrument inside and outside of its case via connecting wires, by eliminating the Electro Magnetic Interference (EMI) at the point of entry of wires. We shall consider the XY capacitors first.

Filtering is usually accomplished with capacitors specially designed for AC use, often in combination with chokes or transformers of various designs. RFI is conducted on the AC line either symmetrically or asymmetrically. Symmetric interference can be visualized as a source connected between the main and neutral wires. A line-to-line capacitor (designated as type X) properly applied is effective for this type.

Asymmetric interference comes from a source between either main or neutral, and chassis ground. A line-to-ground capacitor (type Y) provides filtering in such case. While the X capacitor may be of some practical value, the Y capacitors generally must be kept to small values to limit the 50/60 Hz leakage current to ground. A typical Y capacitor value is 4700 pF. Common X capacitor values are 0.1–1.0 μF.

Types of transients found online, as per Unipede (a consortium of power producers), a European agency, are 10, 6, 1.2 and 1 KV, and the duration could be 0.1 – 1000 μs. EN 132400

(IEC384 – 14) defines a total of seven classes of RFI capacitor, three X classes and four Y classes. The required class is determined by the equipment standards for the final product. For typical business equipment and computers covered under EN60950 (IEC950) the applicable classes are X2 and Y2.

Safety agency approvals do not ensure unfailing product performance. Simply stated, equipment may fail after a line transient, provided it fails safely. The Institute of Electrical and Electronics Engineers (IEEE) has devised a performance specification standard IEEE587 which gives reasonable assurance that equipment will survive a severe line transient. IEEE 587 (ANSI/IEEE C62.41 – 1980) (Guide on Surge Voltages in Low Voltage AC Power Circuits) is of particular interest to manufacturers of uninterruptable power supplies, surge protectors, lighting, computers and terminals, business equipment and others who wish to assure their customers of the reliability of their product.

16.2 XY CAPACITORS

XY capacitors are used as radio frequency interference protection as also for effective grounding of electrical and electronic circuits, generators and instruments. These **safety capacitors** are classified by the above standards according to their use in the circuit. European standards EN 132400/IEC 384-14, 2nd edition, identical in content, are used in Europe.

X capacitors are used in line-to-line applications. In this application, failure would not lead to danger of electric shock. These are defined as X1, X2 and X3. Y Capacitors are for use in line-to-ground applications. In this application failure could lead to danger of electric shock. These are defined as Y1, Y2, Y3 and Y4. These are also referred as **safety capacitors** because of their function to save the equipment and the user from impulses and shocks.

These components are rated 1.0 microfarad or less, 85°C or less, and 60 Hz or less. In addition, across-the-line, antenna-coupling and line-bypass components are rated 125 or 250 V AC. Double-protection components are rated 125 V only. In addition, X1 and Y1 components are suitable in circuits rated up to 250 V; Y2 components are suitable in Class II (double insulated) circuits rated up to 125 V and Class I (grounded) circuits rated up to 250 V. The dv/dt rating of these capacitors can range between 1200 – 3000 V µ/s.

[A] Class X capacitors

These are divided into three subclasses: X1, X2 and X3 (Table 16.1) according to the peak voltage of the impulses superimposed on the mains voltage to which they may be subjected in service. Such impulses may arise from lightning strikes on outside lines, from switching in neighbouring equipment, or switching in the equipment in which the capacitor is used.

Table 16.1 *X Capacitor Classification as per EN132400*

Rating	*Rated volts AC RMS*	*Withstand voltage AC RMS*	*Peak impulse voltage in service*	*Peak values of surge voltage V_p (before endurance test)*	*IEC 664-1 category*	*Application*
X1	250	1500	2.5 kV < V_p ≤ 4.0 kV	for $C_R \leq 1.0$ μ, $V_p = 4$ kV: for $C_{R>}1.0$ μF $V_p = 4/\sqrt{Cr}$ kV	III	High pulse application
X2	250	1500	$V_p \leq 2.5$ kV	for $C_R \leq 1.0$ μ, $V_p = 2.5$ kV: for $C_{R>}1.0$ μF $V_p = 2.5/\sqrt{Cr}$ kV	II	General purpose
X3	250	1500	$V_p \leq 1.2$ kV	No test	I	General purpose

[B] Class Y capacitor or RC unit

Suitable for applications where failure of the capacitor could lead to danger of electric shock. Class Y capacitors are further divided into four subclasses: Y1, Y2, Y3 and Y4.

Table 16.2 *Y Capacitor Classification as per EN132400*

Rating	*Voltage range*	*Withstand voltage AC RMS*	*Peak impulse voltage (volts)*	*Type of insulation bridged*
Y1	≤500 V	1500	8000	Double or reinforced
Y2	≥150 V ≤300 V	1500	5000	Basic or supplementary
Y3	≥150 V ≤250 V	1500	None	Basic or supplementary
Y4	<150 V	1500	2500	Basic or supplementary

X2 capacitor Y2/X1 capacitor XY capacitor Y1 Y2 ceramic capacitor

Fig. 16.1 *XY capacitors*

Both X and Y capacitors are often made in one case (XY). Of these, X2Y2 are the most popular. Some configurations of X, Y, and XY capacitors are shown in Fig. 16.1.

16.3 FAILURE MECHANISMS

[A] Active flammability

According to EN 132400 EMI suppression capacitors have to be tested for active flammability. This test is to ensure that the capacitors and surrounding gauze do not ignite at a defined electrical overload.

[B] Failure mechanisms of RFI capacitors

These could be one of the following reasons:

- Weak spots in dielectric
- Poor metallization
- Heated material
- Excessive self-healing

These failures can cause destruction or even fire. Studies on TV fires show that the majority of these occur after the unit has been operated for some time. It is therefore important to know when a capacitor begins to self-heal. In general, an impregnated metallized paper capacitor undergoes self-healing at much higher voltages than a metallized film type. This is because the entire winding is impregnated with epoxy, filling in any weaknesses and voids. A film capacitor cannot be impregnated and therefore may be left with weak spots in the dielectric. Multilayer windings and quality control will reduce the problem, but some weak spots may remain. Therefore for a given value and physical size an impregnated paper capacitor will self-heal at higher voltages.

Impregnated paper leaves the lowest percentage of carbon, followed by polyester. For this reason RFI capacitors are constructed primarily of these two dielectrics. The self-healing properties of metallized capacitors make them preferable over film-foil types in applications where high transients (such as those on the AC line) are found.

Ceramic capacitors do not self-heal. Therefore they must be constructed to withstand surges specified by the various safety agencies. If a transient on the line exceeds the strength of the dielectric it can fail in an unsafe manner (short circuit). Because of their construction ceramic RFI capacitors can be larger than those made with impregnated paper.

The temperature and voltage stability of a capacitor is important, especially in a Y application. All equipment subject to agency approvals has limits on the allowed leakage current to the chassis ground. The permitted leakage generally ranges from 50 μA in medical applications to 0.5 mA in business equipment. Higher leakages are usually not allowed so that in case of a ground interruption operators are not exposed to excessive currents. Because of the normal conduction of 50/60 Hz in a Y capacitor its value is generally limited to 470 pF in medical devices and 4700 pF in other applications.

Safety certified capacitors are application specific high voltage capacitors that are designed to withstand AC voltage conditions and high voltage impulses. These capacitors shunt the energy from an impulse to the ground, providing protection for the circuit as well as the end user. The capacitors are rated 250 V AC, and are governed by IEC 60384-14, IEC 60950 and UL 60950.

16.4 APPLICATION

Safety certified capacitors are used in large quantities in telecom applications utilizing standard phone line connections. Common applications for EMI suppression include:

- DSL and dial up modems, VoIP phones, set-top boxes, point of sales terminals and fax machines.
- Capacitors are used for filtering EMI from tip and ring lines to ground to meet EN 55024 requirements.
- Capacitors bridge the isolation barrier and must withstand high impulse voltages in case of a power surge.
- The capacitors are used on the AC or input side of the circuit for EMI filtering.
- AC/DC power supplies use X1Y2 rated capacitors.

In this kind of application safety certified capacitors are used in two circuits. The first is for filtering EMI from the tip and ting lines to ground. A telephone line carries a lot of noise and must be filtered to meet EN 55024 requirements. The capacitors are used as EMI filters in this case, but since they bridge the isolation barrier between line and ground they must be able to withstand any high impulse voltages that may come down the line. The capacitance values used in this circuit are usually between 100 pF and 1000 pF.

[A] EMI filtering

XY capacitors are often used for suppression or filtering of electromagnetic interference (EMI) in electrical and electronic circuits. The general circuit is as in Fig. 16.2.

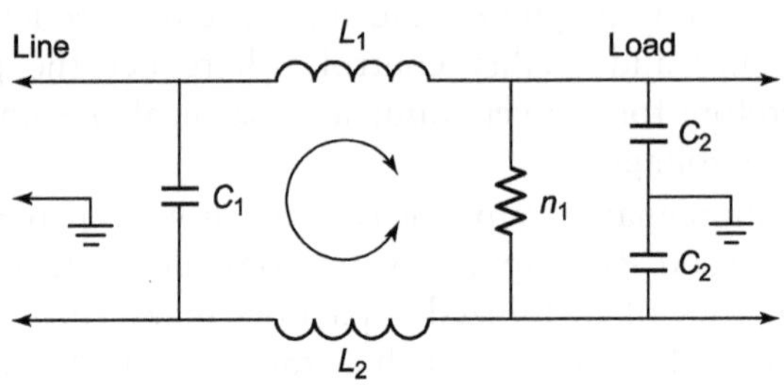

Fig. 16.2 *XY capacitors for EMI filtering.*

[B] Isolation

- The capacitors are used for isolation between telecom network voltage (TNV) and protective earth (SELV) as in Fig. 16.3.
- These applications use either X2Y3 or X1Y2 rated capacitors.

The X1/Y2 capacitors offer higher surge immunity and can reduce the number of failures in the field due to surges on the line. That is why some companies use them as the higher part cost is justified by the reduction in field failures.

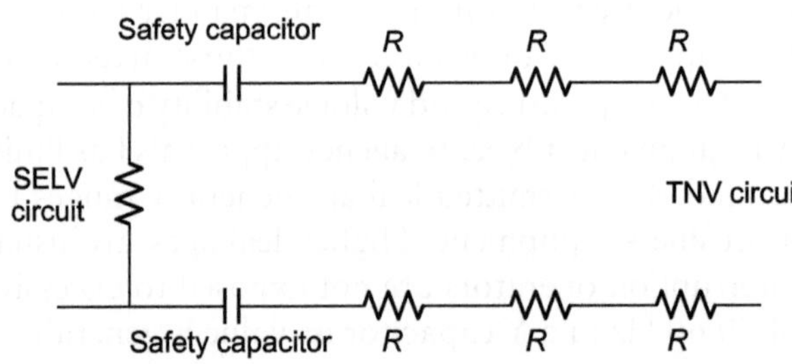

Fig. 16.3 *Isolation by safety capacitor.*

[C] AC/DC application

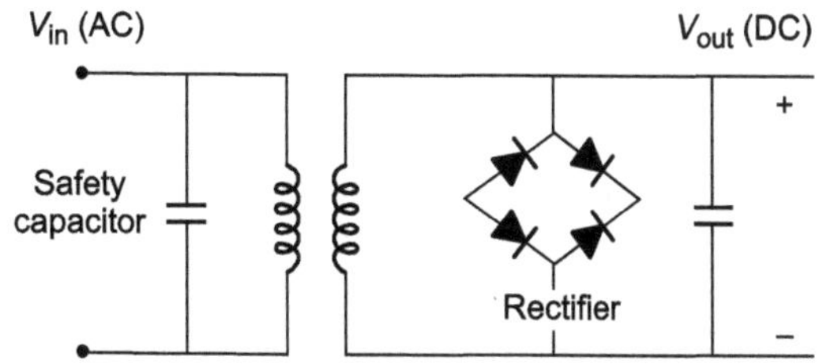

Fig. 16.4 *Safety capacitor in AC/DC circuit.*

Another important application for safety certified capacitors is in AC/DC power supplies. The capacitors are used on the AC or input side of the circuit for EMI filtering. Like phone lines, AC power lines carry a lot of noise as well as surges and transients. The capacitors bridge the isolation barrier, so they must withstand the surges and impulses in case of power surges.

AC/DC power supplies require the use of X1/Y2 rated capacitors. Some power supplies require Y1 rated capacitors, which are not available on the market in a surface mount package. The only surface mount solution available is to use two Y2 capacitors in series as in this case they are considered equivalent to Y1 capacitor.

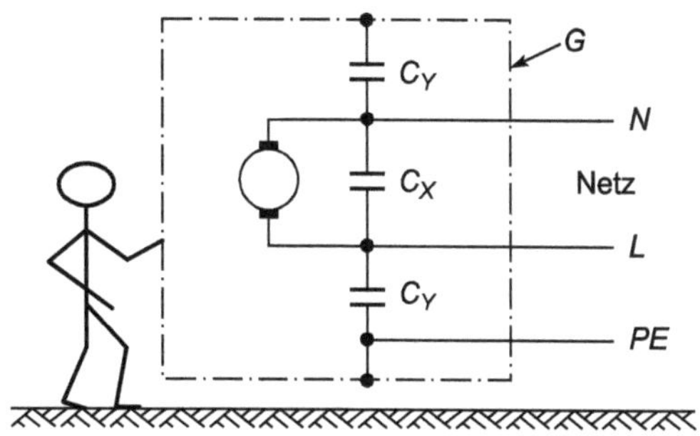

Fig. 16.5 *EMI suppression with X and Y capacitors.*

Depending on the way they are connected, X and Y capacitors are effective against different kinds of electromagnetic interference. X capacitors which are connected between the line phases are effective against symmetrical interference (**differential mode**). Y capacitors which are connected between phase and neutral (zero potential) to ground are effective against asymmetrical interference (**common mode**).

Electrical apparatus is classified roughly into two categories, i.e. (a) household appliances and (b) office appliances including office automation (OA) and others. The standards for noise suppression capacitors to be used in the household appliances are stricter than those in the office appliances and others. In order to avoid any accidents in machine applications which may experience unexpected abnormal surge voltage, or which are subjected to continuous 24-hour use, it is necessary to build in an extra measure of reliability. Here, the strict test conditions conducted by the above-mentioned safety standards organizations can be considered as one of the criteria from a reliability point of view. The product should be selected on the basis of a thorough consideration of such safety standards according to its application.

16.5 DESIGN CONSIDERATIONS

- When protecting switching contacts, always include a resistor in series with the noise suppression capacitors.
- In high-speed circuits, the addition of a noise suppression capacitor may reduce the response time of the circuit. For best response characteristics, do not use a larger capacitor than is absolutely necessary to suppress the noise level.
- Noise suppression capacitors are most effective when located close to the noise source. Excessive capacitor lead length may cause abnormal oscillation and decrease the energy absorption capability.

- When noise suppression capacitors are connected across power lines, it should be checked that the resulting inrush current does not cause opening of the fuse or the circuit breaker. Capacitor value and breaker ratings must be given due consideration.
- Noise suppression capacitors are meant for standard line frequencies and should not be used in circuits where normal operation will exceed 70 Hz.
- Safety capacitors do absorb normal line surges. However they are not intended to absorb high-energy surges such as induced lightning.

16.6 FEEDTHROUGH CAPACITORS

A feedthrough capacitor is one that provides a desired value of capacitance between the feedthrough conductor and the metal chassis or panel through which the conductor is passing; used chiefly for bypass purposes in ultra-high-frequency circuits. Feedthrough capacitors offer EMI suppression into GHz range. These find extensive use in aerospace, defence and security systems, medical equipment like MRI, industrial and commercial establishments, and telecommunication equipment. In most of these applications, RF interference has to be scrupulously kept away.

Fig. 16.6 *Feedthrough capacitor symbol.*

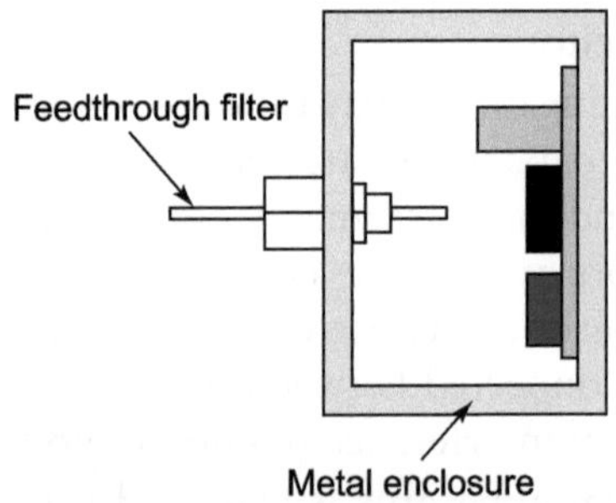

Fig. 16.7 *Feedthrough filter capacitor.*

Feedthrough capacitors were originally designed for DC power lines in RF systems. They allow the DC and low frequency signals to pass through but block the RF energy. Usually the feedthrough capacitors are fitted to the metal case of the RF module on holes where the wires enter the system. They let the signal pass through but will stop the RF from escaping from the device through that hole to the outside wiring. The difference between regular capacitor and feedthrough capacitor is explained by Fig. 16.8.

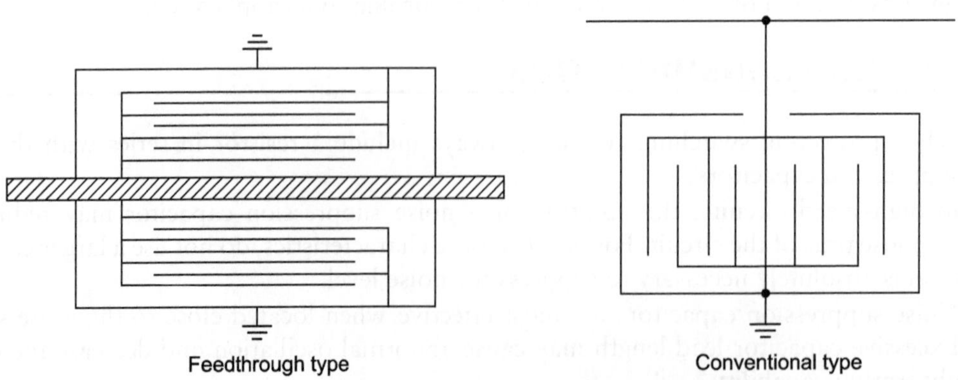

Fig. 16.8 *Feedthrough and conventional RF line filter.*

A feedthrough capacitor is formed between the current carrying wire and the case, while in all others the capacitor is separate from the case.

The dielectric in feedthrough capacitors (Fig. 16.9) may be formed by a dielectric layer of a composite material containing a resin material and dielectric material powder, and arranged between a feedthrough terminal and an outer electrode terminal surrounding this feedthrough terminal, a capacitance being formed between the feedthrough terminal and the outer electrode terminal.

Fig. 16.9 *Feedthrough capacitors.*

A ceramic plate is formed with sleeve-like bush with an internal aperture, the plate is metallized around the bush, and feedthrough conductors are placed through the apertures, the metallized coating around this projection and the feedthrough conductor forming the capacitance; the metallized coating is removed from the zones surrounding the terminal ends of the openings, to provide ceramic insulation material between the feedthrough conductors and the metallized coating. The metallized coating may be removed by grinding. Figure 16.10 shows a feedthrough construction and its mounting on chassis.

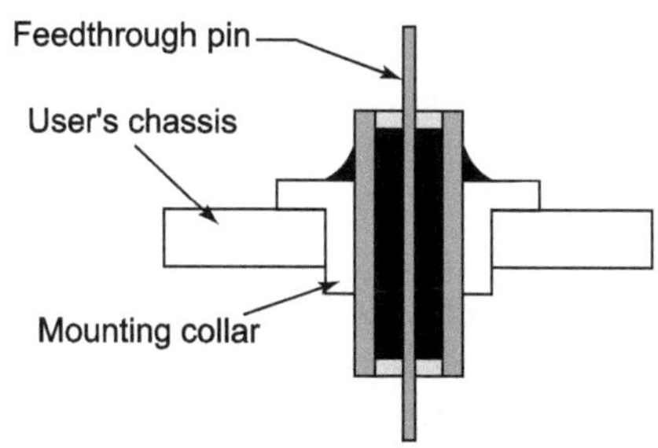

Fig. 16.10 *Feedthrough capacitor construction and fitting.*

[A] Insertion loss

Insertion loss, the basic assessment of EMI suppression filter performance is defined as the ratio of the voltage across the load R_L before addition of the filter to the voltage across R_L after the addition of the filter (expressed in dB).

Four methods of shielding equipment from RF interference from chassis and surroundings are shown in Fig. 16.11.

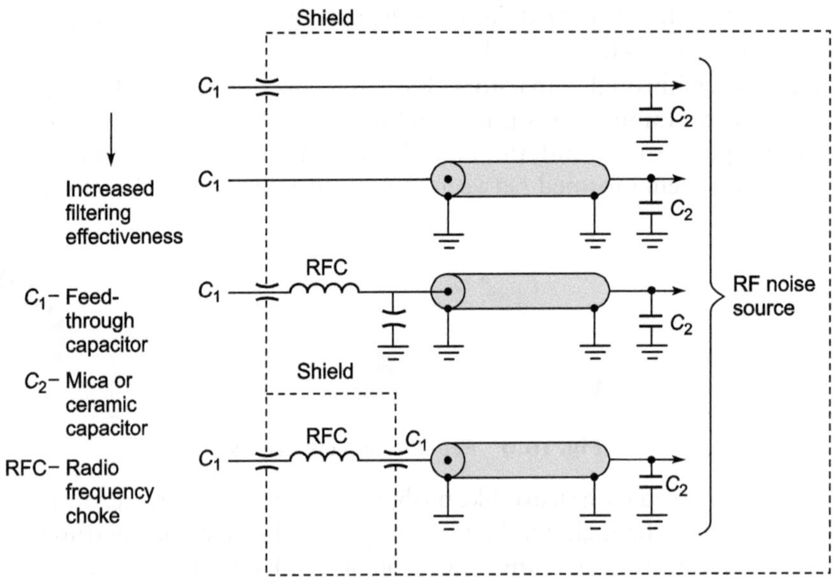

Fig. 16.11 *Methods of shielding RF interference.*

[B] EMI suppression filters as feedthrough

The L-C filter in Fig. 16.12 illustrates the principle used in an EMI suppression filter. Inductors are represented as a series inductor and series resistor, shunted by a capacitor Cs to represent self-capacitance.

Capacitors are represented as an ideal capacitance in series with an inductance and resistance. For feedthrough filters, Le can be taken as being practically zero at power frequency, whilst the residual resistance Rc is typically 5 mΩ in value. The inductance Le dominates the RF performance of leaded capacitors, and is due to their construction and lead inductance. A typical value of 5nH produces a significant resonance in the insertion loss characteristic.

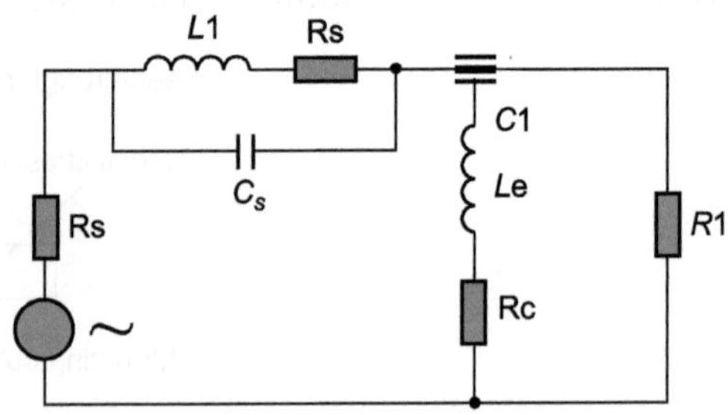

Fig. 16.12 *LC filter.*

[C] Types of filters

RF filters can be of different types.

(1) C filter

A feedthrough capacitor is called a C filter. This is a single element, a capacitor from line to ground, a Y type capacitor, with a through wire connecting the input to output. It maintains this attenuation at higher frequencies. A feedthrough capacitor filter is usually the best choice for filtering lines that exhibit very high impedance. The basic circuit of C capacitors is shown in Fig. 16.13(a).

(2) L **filter**

An L circuit consists of two elements viz. a feedthrough capacitor from line to ground, and an inductor connected in series with it between the input and output terminals. The capacitor can be on either the line or load side of the filter, making it either a capacitive or inductive input. L-circuit filters are usually the preferred choices when the line and load impedances have large differences in impedance (Fig. 16.13(b)). The inductive element is best placed so that it faces the lower impedance. Many times, they are referred as L1 (inductive element is on the end with the threaded mounting neck), or L2 (capacitive element is on the end with the threaded mounting neck).

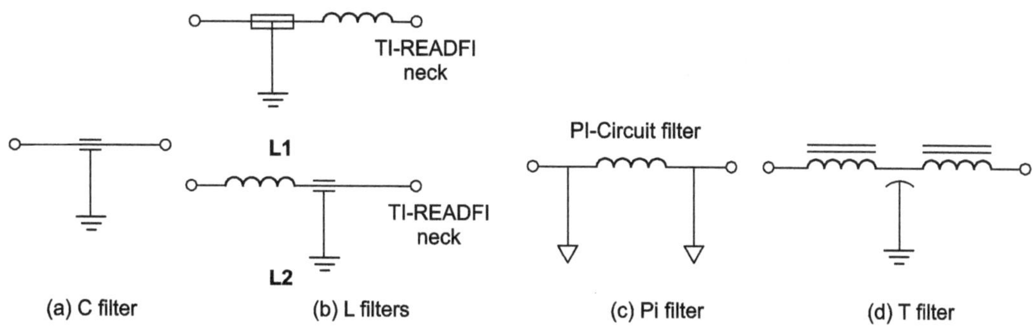

Fig. 16.13 *Filter types*

(3) Pi circuit

Pi circuit is a three-section filter having two feedthrough capacitors, as in Fig. 13.13(c) to ground with a series inductor between them. This is usually symmetrical, but sometimes may be asymmetric, having capacitors of different values. A Pi filter is usually the choice when high attenuation levels are required, and where input and output impedances are similar. It is not used on switching circuits usually.

(4) T circuit

T circuit, as in Fig. 16.13 (d) is a three-section filter with two series-connected inductors between the input and output terminals, and a feedthrough capacitor in between from line to ground. The filter is usually symmetrical (identical inductive elements), but asymmetrical circuit is also used. A T circuit filter is usually a preferred choice where both input and output impedances are low.

(5) LL circuit

LL circuit has a four-section filter with two feedthrough capacitors between line and ground, and two inductors connected in series with them between the input and output terminals. The filter is usually made with identical capacitor and inductor values. Similar to L1 and L2 filters, these also come as LL1 and LL2 varieties, depending upon which element is closest to mounting neck. LL circuit filters are used when extremely high attenuation is required and where input and output impedances differ significantly.

16.7 CONSTRUCTION FEATURES

Most feedthrough capacitors are made from ceramic dielectric, with a high dielectric constant (>3000) to achieve large capacitance for a given volume. Ceramic is a brittle material that can be cracked through mechanical and thermal shocks, which can lead to capacitor failure. So the capacitors need care in handling. Ceramic feedthrough capacitors could be of three geometries, tubular styles, discoidal styles and planar array styles. The latter, used to provide multi-pin filtering in multi-way connectors, will not be specifically dealt with here, although many of the comments regarding use of soldering heat, and so on, are equally valid.

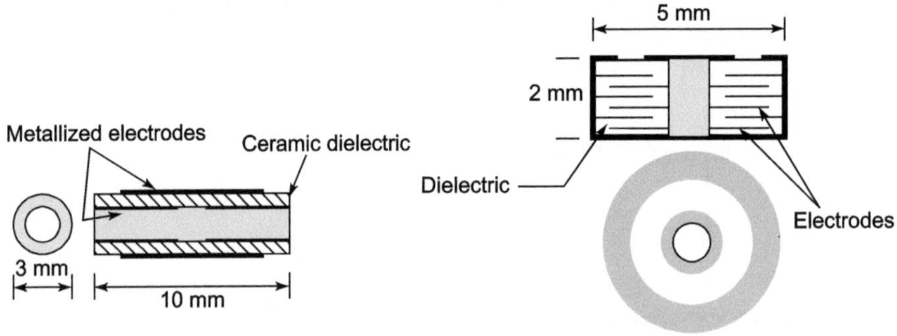

Fig. 16.14 *Tubular and discoidal capacitor.*

Tubular capacitors use the inside and outside surfaces of a ceramic tube, to form a coaxial capacitor: Capacitances up to 5 nF are common, while multilayer ones are also available. Multilayer pi-section capacitances up to 22 nF are in use and C-types can be as high as 1 microfarad. Pi-section split capacitor is achieved by using sections of metallization inside the tube. A metal pin is passed through the centre of the tube, and soldered to the inner electrodes. A ferrite bead placed over the centre pin provides inductance. The outside common ground electrode is soldered into a metal body or collar, for grounding to a bulkhead.

Discoidal capacitors use a multilayer construction to achieve a higher capacitance per unit volume. Overlapping ground and inner electrodes are sintered into one monolithic body, offering capacitances up to around 100 nF in a nominal 5 mm disc, and over one microfarad in a nominal 9 mm disc. Discoidal capacitors are again assembled on mounting collars, or threaded bodies.

16.8 APPLICATION NOTES AND INSTALLATION

The following precautions need to be observed when installing a feedthrough filter capacitor:

- The filter's metal case must be mounted on metal case to effect low resistance contact with the chassis, cabinet, or ground.
- Earthing connections should be short and be of the lowest possible impedance. Wires should not be used for earth.
- Input and output leads of filter should be physically separate to provide clear and best electrical isolation possible.
- Mounting of the filter should be closest possible to the entry of point power line of the device being filtered.
- The most economical filter selection will always be that one with the fewest internal elements.

- Feedthrough capacitor filters are usually the best choice for use with very high impedance lines.
- L circuit filters, or LL filters are usually selected when between line and load impedances differ widely.
- Pi circuit filters are preferred when high levels of attenuation are required and input and output impedances are similar. This filter is not used on switching circuits.
- T circuit filters are chosen when both the input and output impedances are low.
- LL circuit filters should only be used when extremely high attenuation is required and when input and output impedances vary significantly.

17

ENERGY STORAGE AND PULSE CAPACITORS

17.1 ENERGY STORAGE APPLICATIONS

Everyday application of energy storage capacitors is to be found in flash cameras. These compact electrolytic capacitors have the property of very high energy storage to volume ratio. Energy storage capacitors can be used for high current single event discharges if certain precautions in circuit design are implemented. Electrolytic capacitors from 200 V to 450 V, and stored energy levels from 3 J to 150 J are available for these applications. In addition, high voltage capacitors have their own separate application, and are widely used. Both the electrolytic and high voltage capacitors of different constructions have their own separate applications, and are extensively used.

In some applications, a brief but high energy pulse of current is required periodically, rather than a continuous flow of current. Examples are a photoflash unit or an automotive capacitance discharge. This pulse might have a current level of hundreds or even thousands of amperes. A capacitor for this application should demonstrate an extremely low equivalent series resistance (ESR), along with a high current carrying capability.

Energy storage capacitors find extensive use in aerospace, military, defence, household security, and telecommunication, nuclear and medical applications. Often, the choice is between standard power-supply filter capacitors, and pulse-rated type. The pulse-rated will have lower ESR, and larger terminals. However, the standard aluminium electrolytic capacitors have low enough resistance to achieve an under-damped circuit, and are much more commonly available.

Energy storage capacitors (sometimes also called 'reservoir capacitors') are used in power supplies where they smooth the output of a full or half wave rectifier. They can also be used in charge pump circuits as the energy storage element in the generation of higher voltages than the input voltage. These capacitors can be used individually or in banks. The stored energy is given by Joules = $1/2 \times C \times V^2$, where C = capacitance in microfarads and V is charging voltage in kilovolts.

Capacitors are connected in parallel with the power circuits of most electronic devices and larger systems (such as factories) to shunt away and conceal current fluctuations from the primary power source to provide a 'clean' power supply for signal or control circuits. Audio equipment, for example, uses several capacitors in this way, to shunt away power line hum before it gets into the signal circuitry. The capacitors act as a local reserve for the DC power source, and bypass AC currents from the power supply. This is used in car audio applications, when a stiffening capacitor compensates for the inductance and resistance of the leads to the lead-acid car battery. These are covered elsewhere and we will consider specialized applications of energy storage capacitors in the present discussion.

17.2 ENERGY STORAGE CHARACTERISTICS

The capacitor is designed to suit the waveform and duty cycle to which it will be subjected in service and the life it is expected to give. The life expectancy of an energy storage capacitor is a function of the electrical stress, the voltage reversal and the repetition rate of discharges. To obtain maximum peak current, the self-inductance value of the capacitor should be low. Specific designs are available to meet this requirement.

Fig. 17.1 *Some energy storage capacitors.*

Main parameters may include capacitance: 0.01 to 200 MFD, voltage: 1–1000 kV DC, peak ampere: 0.2–500 kA, inductance 17–250 nH and these may be subjected to voltage reversal of up to 80%. Typical life can be from 1 shot to 10^9 shots, depending upon application. Capacitors for pulsed power also include those rated for several KV.

17.2.1 Voltage

The potential energy (PE) depends greatly on voltage squared so it is a great advantage to use higher voltages.

$$PE = \frac{1}{2}CV^2$$

Costs for electrolytics are favourable in the range 100–300 V DC. Alternatives may be compared between capacitors on the basis of cost per joule.

DC capacitors are made up to 150 kV DC, energy up to 7 kJ, and inductance as low as 20 nH, manufactured using biaxially oriented hazy propylene film as dielectric, aluminium foil as electrode and impregnated with non-PCB impregnant. These capacitors are designed for use in energy storage and discharge applications – with insulated casing, open terminal (2 bushings) – with live casing, open terminal (1 bushing). Applications include research, controlled nuclear fusion, lasers, impulse generators, high energy x-rays and metal forming.

17.2.2 Pulse Storage Capacitor Classification

Pulsed storage capacitors are classified into different categories depending on maximum peak current, repetition rate, working temperature, storage temperature or inductance values.

1. **Classification based on max. peak current**
   - Max. peak current up to 25 kA.
   - Max. peak current up to 50 kA.
   - Max. peak current up to 200 kA.
   - Max. peak current up to 500 kA.
2. **Classification based on repetition rate**
   - Up to 5 PPM (Pules per minute)
   - Up to 1 PPS
   - Up to 10 PPS
   - Up to 100 PPS
   - Over 100 PPS.

The most important [multiplying] factor for energy storage for a given capacitance is the voltage that these capacitors have applied to them. Energy storage is measured in joules. For a high value capacitor of 100,000 µF, this can be calculated as follows:

Capacitance µF	*Voltage V*	*Energy J*
100,000	10	5
100,000	75	562.5
100,000	500	12500

17.3 WEAPONS SYSTEMS AND SPACE TECHNOLOGY

In military applications, small rail, reconnection, coil, gauss, quench and reluctance guns can be constructed with appreciable velocity and yet still be handheld with properly designed systems.

Electrolytic energy storage capacitors provide a convenient method of storing electrical energy, which can be released, as required, for a very short duration of time under controlled conditions. Owing to their ability to release energy at very high rates and to produce high peak current and power under discharge conditions, these capacitors are extremely useful in applications like plasma physics and subatomic particle research. Capacitors up to 12,000 µF 400 V ratings are available, which are used for magnetizers, kinetic weapons, power supplies and a number of military and space applications, as well as for on-ship power supply backup circuits.

Oil-filled energy storage metal case capacitors up to 30 µF 4000 V are used in many projects, some of them being excellent for coil guns, wire explosion, can crushing, magnetizing and plasma generation. Large oil-filled energy storage capacitors of up to 13,500 µF, 1300 V, 11,500 J are available. High energy electro-kinetic rail guns, coil guns, high energy discharges, magnetizers, etc. use capacitors rated up to 4000 V with a current rating of over 3000 A on single discharge event. These may store energy up to 250 J. Four of them connected in parallel can give almost 1000 J at over 12,000 A discharge current.

17.3.1 Military Applications

A coil gun is a handheld, battery powered (12 V DC) rifle that is capable of launching a .30 calibre metallic projectile at adjustable velocities.

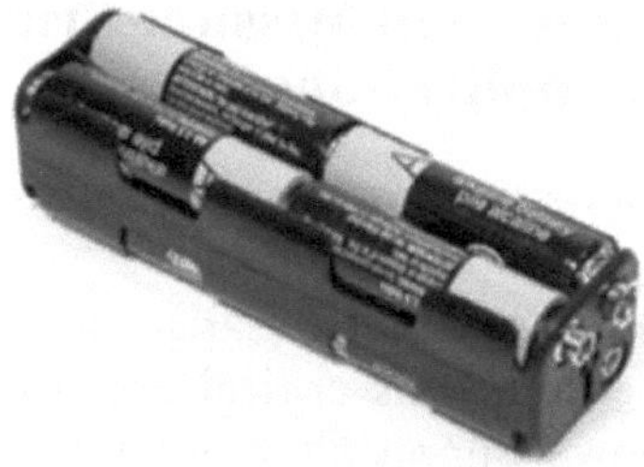

Fig. 17.2 *Capacitor assembly in coil gun.*

The circuit board contains a step-up transformer converter, a voltage multiplier cascade, voltage comparator to set the charge voltage on the capacitor bank, an SCR switching section, control panel and connectors for the 12 V battery pack, capacitor bank, accelerator coil and the fire switch. Other components of the gun are the barrel, breech loading mechanism, battery supply, control panel, display, projectile, pistol grip with trigger assembly and an aluminium stock that contains all of the components.

There are several important factors in choosing an energy storage capacitor for a coil gun.

A new non-lethal gun that may make a crowd disperse, because people think they are being fired, could be in use by 2012 by the US armed forces. An active denial system fires a harmless beam which causes an intense burning sensation. It will be used to disperse large crowds and is set to be deployed soon. Tests showed a person can withstand the beam for up to five seconds before wanting to step away. The nickname already earned by the weapon is the Goodbye Gun.

Energy storage capacitor bank for a gun is comprised of ten 200 V DC, 1500 μF capacitors wired into a configuration of 1000 V DC, 600 μF. The charge time between shots is approximately 16 seconds.

17.3.2 Marine Applications

This is a type of ship component. It stores energy for use by the ship's weapons and gives out pulses of energy. With each shot, energy from the capacitor is transferred to the weapon, draining the capacitor. A capacitor has a maximum design storage energy and also a recharge rate. Powering the capacitor requires energy from the ships' reactor (the 'reactor energy drain'). Large (greater mass) capacitors usually store more energy and can also recharge faster. Capacitor energy and recharge are both affected by the capacitor overload. High-end space application necessitates capacitor overloads.

Groups of specially constructed, large, low inductance, high voltage capacitors (banks of capacitors) are used to supply huge pulses of current for various pulsed power applications. These applications include:

- Electromagnetic forming.
- Radar.
- Fusion research.

- Pulsed lasers (especially TEA lasers).
- Pulse forming networks.
- Particle accelerators.

17.3.3 High-voltage Capacitors Suit Mission-Critical Military/Aerospace Applications

When the military or aerospace companies use the term 'mission-critical', application failure can result in breaches in national security or even fatalities.

The military and aerospace fields are two areas in which Reconstituted mica paper capacitors (RMPCs) are used for numerous critical applications, such as aerospace ignition systems, baggage x-ray systems, power-supply filtering for military and commercial applications, energy storage with high-current discharges for military and commercial detonation systems, voltage multipliers, traveling-wave tubes, radar, particle accelerators, power utilities, exciters for industrial generators, partial-discharge detection and welding equipment. The failure of any of these applications could be devastating financially, militarily, or fatally.

Reconstituted mica paper capacitors (RPMC)

RMPCs are not position-sensitive, nor do they have a polarity, although the outside foil can be identified if necessary. By knowing the outside foil lead, the capacitor connections can be arranged in a way that reduces the potential voltage stresses on the outer layers of the capacitor and its adjacent components or assembly structure.

RMPCs are solid state and contain no liquid to contaminate the electronic device or its surroundings, making them extremely durable to physical shocks and vibration. RMPCs are also more stable than ceramics or most film capacitors and offer a durable platform from which to work when a high-voltage capacitor is required. RMPCs have a –3% maximum drift at –65°C from nominal capacitance reading to 5% maximum drift at 175°C and a temperature coefficient of less than ± 500 ppm°C^{-1} for the –65 to +125°C operating temperature range.

These capacitors can also survive 100,000 Gs of acceleration. Thermal cycling and thermal shock from –65 to +125°C will likewise cause no damage. Capacitance can range from 50 pF to 5 µF with voltage ratings from 1 kV to 75 kV DC and more with special designs. AC voltages up to 20 kV present no problem and can also be found with corona-free (no partial-discharge) ratings. Peak current is usually limited by the inductance of the discharge circuit and not by the rise time effects of the dielectric. The mica used in RMPCs is naturally resistant to radiation. There are satellite and space applications in which some RMPCs have been in continuous use for more than 34 years.

17.4 USE IN EXPLOSIVES

Capacitors with high pulse energy are used in mining, oil exploration and military for use in explosives as detonators. Portable capacitor blasting machines are used for such applications. Civil uses include mining and quarries, where stray electrical currents might detonate normal blasting caps, requiring very precise timing for multiple point commercial blasting. Many of these are one-time but reliable operations, and need high enough pulse energy for the purpose. These must have a very high IR, and be operable over a wide temperature range.

17.4.1 Pulsed Power Military Applications

Groups of large, specially constructed, low-inductance high-voltage capacitors (*capacitor banks*) supply huge pulses of current for many pulsed power applications. These include electromagnetic forming, Marx generators, pulsed lasers (especially TEA lasers), pulse forming networks, radar, fusion research and particle accelerators.

Large reservoir capacitor banks are used as energy sources for the exploding-bridgewire detonators (EBW) or slapper detonators, also known as exploding foil initiators (EFIs), in nuclear weapons and other specialty weapons. Experimental research and work is taking place as a way of using banks of capacitors as power sources for electromagnetic armour and electromagnetic coil-guns and rail-guns.

There are numerous factors to be considered while selecting capacitors for individual use, including the following:

- Desired safe electrical charge storage.
- Holding capacity of charge for desired time without appreciable loss.
- Thermal capabilities to withstand surge currents and resultant heating.
- Operation with partial discharges/corona.

The specialized needs require a detailed study for the following aspects:

- Total energy required
- Capacitance value and charge voltage
- Energy storage time before discharge
- Voltage reversal and peak currents
- Pulse-width and shape, repetition rate
- Inductance limitations
- Lifetime expectancy
- Dimensional constraints

17.5 PULSE CAPACITORS

A pulse capacitor is a capacitor designed primarily for applications with intermittent charges and/or discharges at high values of the charge/discharge current. They are intended for coupling, bypass, filtering, and snubbing or pulse operation in SMPS at low as well as at high AC voltage of high frequency where there is a need for high pulse rise time and high ionization level, e.g. fly-back circuits in TV-sets.

Pulse operation

These capacitors are subjected to fast rise or fall time voltage spikes (high dv/dt), and are exposed to high current pulses ($I = C \times dv/dt$). The current limits for a specific type of capacitor are dependent upon:

- Amplitude and form of the pulse
- Rated voltage of the capacitor
- Capacitance value

Pulsed capacitors also find use in high voltage pulse testing. At repeated pulse operation, self-heating, ambient temperature and cooling set the load limit. Pulse current limits are commonly expressed in the form of maximum permitted dv/dt in volts per microsecond.

17.5.1 Geometrical Configuration of the Winding

Pulsed capacitors could be metallized plastic film capacitors, electrolytics (aluminium or tantalum), ceramics or mica dielectric capacitors, depending upon energy and voltage, together with environment and thermal considerations. The circuit also places certain choice constraints. Reliability is critically important. Size, voltage and capacitance vary widely for different applications. Peak current (dv/dt) is the main concern when metallized film capacitors are used in pulse applications.

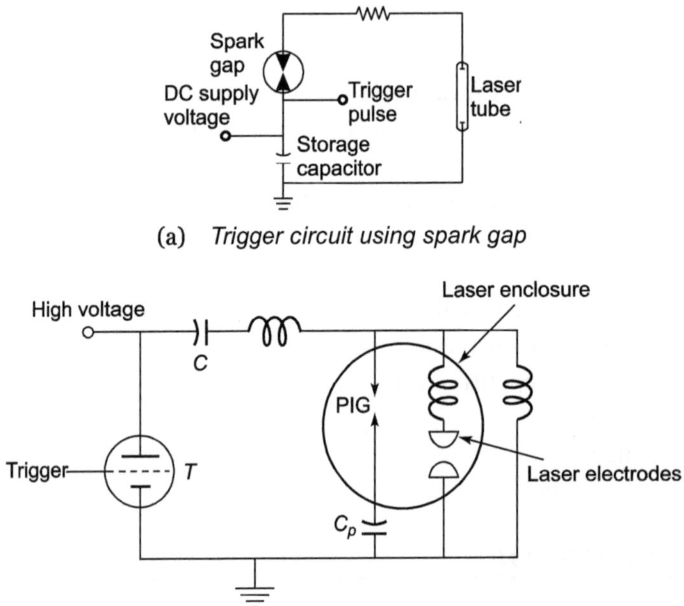

(a) *Trigger circuit using spark gap*

(b) *Basic excimer laser power supply*
T: Thyratron. C: Storage capacitor
C_p: Pulse capacitor. PIG: Pre-ionization gap

Fig. 17.3

17.6 METALLIZED FILM ENERGY STORAGE CAPACITORS

Metallized capacitors in energy storage are found in lasers, beacons, flash and other applications requiring small, lightweight energy storage capacitors. These capacitors are designed to discharge into lamp loads at 1 pps or less at ambient temperatures up to 40°C.

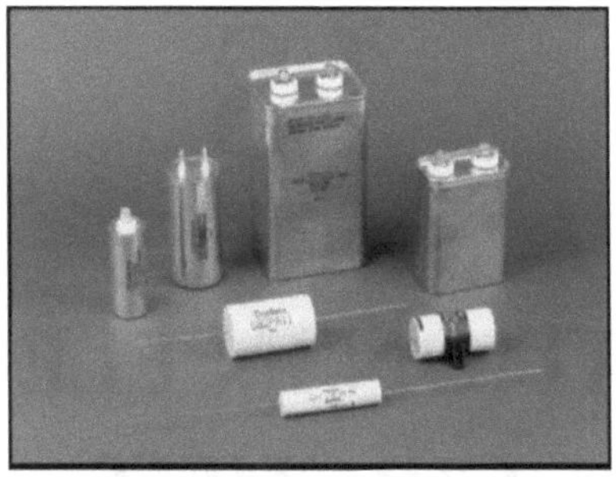

Fig. 17.4 *Metallized energy storage capacitors.*

These capacitors are characterized by:

- Half the size and a third the weight of conventional paper energy storage capacitors
- High energy (up to 1600 J) and high current (up to 3000 A)
- Capacitance range: 10 – 200 μF with tolerance of +20% –10%, ± 10%
- Operating temperature: 0°C to +40°C
- DC voltage range: 2000 – 4000 V DC, dissipation factor: 1.0% maximum
- Discharge rate: 1 discharge per sec. maximum
- Inductance: The typical inductance at resonant frequency is 0.1 μH
- Insulation resistance: 2000 megohm-microfarads at 500 V
- The case could be metal/plastic and in different configurations.

The field of energy storage is ever rising, and finding newer applications, both in civil and military uses.

The advent of ultracapacitors has added a new dimension to such applications. These store so much energy that they can literally be used as stand-alone power supply in place of batteries in extreme temperatures, where batteries cannot be effectively used. They are also used as backup for batteries and extend battery life considerably.

18

APPLICATION IN ELECTRONIC CIRCUITS

18.1 INTRODUCTION

There are many capacitors in an electronic circuit board, scattered throughout the board. There are mica capacitors, ceramic capacitors, tantalum capacitors, aluminium electrolytic capacitors, paper capacitors and so on. All of them have a function in the circuit and one can find widespread applications in the electrical and electronics fields.

Capacitors may be used in three basic ways: (1) as a means of discriminating between higher and lower AC frequencies, (2) a method of storing or releasing energy, and (3) as a method of discriminating between AC and DC. Various capacitor applications are broadly classified below; however, the dividing lines between each application are not always sharply drawn. Main applications in the electronics field include:

- As filter to remove ripple in power supplies
- As inter-stage signal coupling capacitors: A capacitor will not pass DC, but an AC signal will go through
- Tuning resonant circuits, oscillator circuits
- Decoupling in power supplies and amplifiers
- As part of timing circuit, and
- As wave shapers and filters

The table below summarizes various types of capacitors for some common applications:

Table 18.1 *Table of Capacitor Uses and Applications*

Application	*Suitable types and comments*
Power supply smoothing	Aluminium electrolytic: High capacity and high ripple current capability *
Audio frequency coupling	Aluminium electrolytic: High capacitance Tantalum: High capacitance and small size Polyester/polycarbonate: Cheap, but values not as high as those available with electrolytics
RF coupling	Ceramic C0G: Small, cheap and low loss Ceramic X7R: Small and cheaper than C0G, but higher loss than C0G, although high capacitance per volume Polystyrene: Very low loss, but larger in size and more expensive than ceramic

Application	Suitable types and comments
RF decoupling	Ceramic COG: Small, low loss, values limited to max. 1000 pF Ceramic X7R: Small, low loss, higher values available than COG
Tuned circuits	Silver mica: Close tolerance, low loss and stable, but high cost Ceramic COG: Close tolerance, low loss, although not as good as silver mica

Care must be taken to ensure that the ripple current rating of the capacitor meets the requirements of the capacitor application. It is necessary to look at the exact requirements for any capacitor application in a circuit, and choose the capacitor according to the needs and specifications available.

18.2 FILTER/RECTIFIER CAPACITORS

A filter capacitor removes voltage or signal spikes in electronic circuits. Capacitors are used as filter devices due to their ability to absorb and effectively store electrical charges at predetermined values. A filter capacitor is used to absorb or buffer voltage values exceeding set parameters. These capacitors are usually placed across a load or provide a path to ground in circuits. Filter capacitor types include electrolytic, ceramic and tantalum.

18.2.1 RC Filters

One of the oldest, most basic and useful of capacitor applications is filtering or smoothing the output of an AC to DC power supply. In this supply, an AC voltage is rectified by a diode into a varying DC voltage. A capacitor is placed across the output, and serves to 'fill in' the output whenever the output voltage is less than the capacitor. The capacitor subsequently recharges on the next cycle and repeats the process. Thus, the output voltage of the power supply is smoothed out.

It is also sometimes used on extremely high-voltage, low-current power supplies for cathode ray and similar electron tubes, which require very little load current from the supply.

The simple capacitor filter is also used where the power-supply ripple frequency is not critical. Electronic circuit designers have to overcome the voltage and signal ripples.

These may take the form of DC ripple in power supply outputs, audio interference and switch induced arcing. These can create havoc

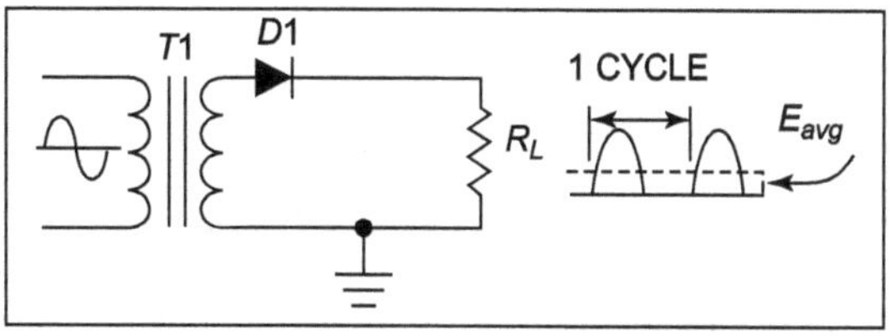

(a) Half wave rectifier without capacitor

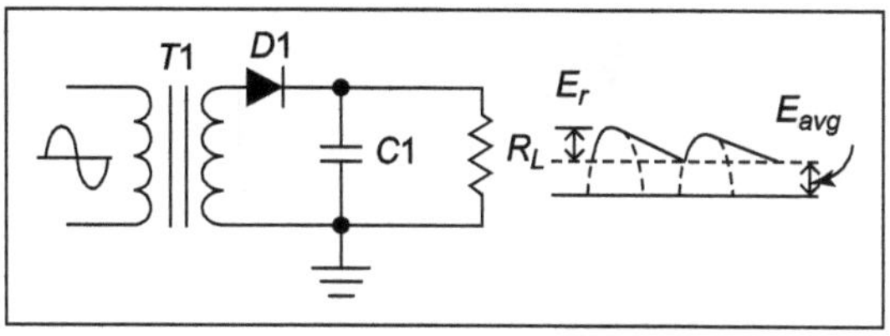

(b) Half wave rectifier with capacitor

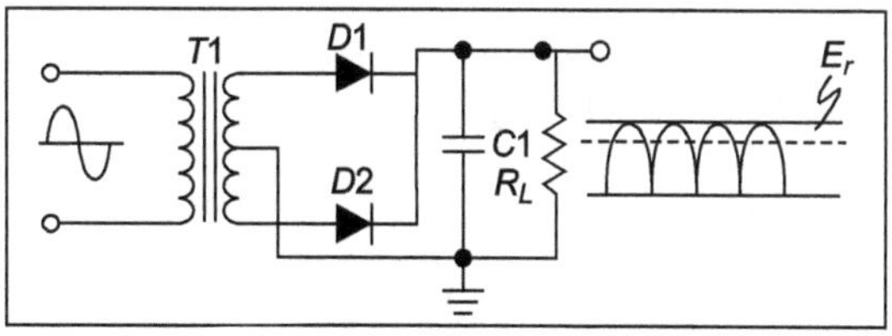

(c) Full wave rectifier with filter capacitor

Fig. 18.1 *Power supply filters.*

in sensitive circuits or high-end audio applications. A capacitor makes an ideal filter component to remove these voltages and signal spikes.

RC filters as shown in Fig. 18.2 are most commonly used with rectified inputs, and the output depends upon the type of filter (full/half wave rectifier) and RC values.

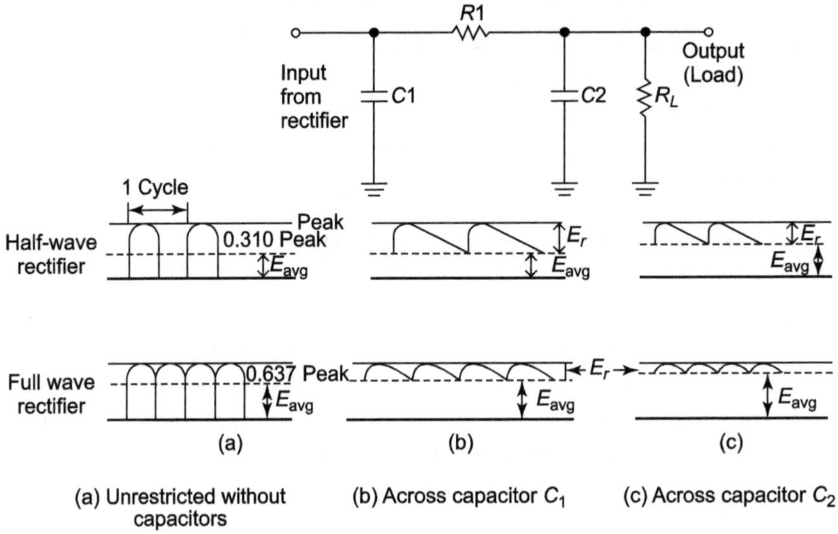

Fig. 18.2 *RC Filter and output voltages.*

Mains filter capacitors are usually encapsulated wound plastic film types, since these deliver high voltage rating at low cost, and may be made self-healing and fusible.

18.2.2 L-section and Pi-filters

These filters get their name from the configuration in the respective circuits as shown in Fig. 18.3 below.

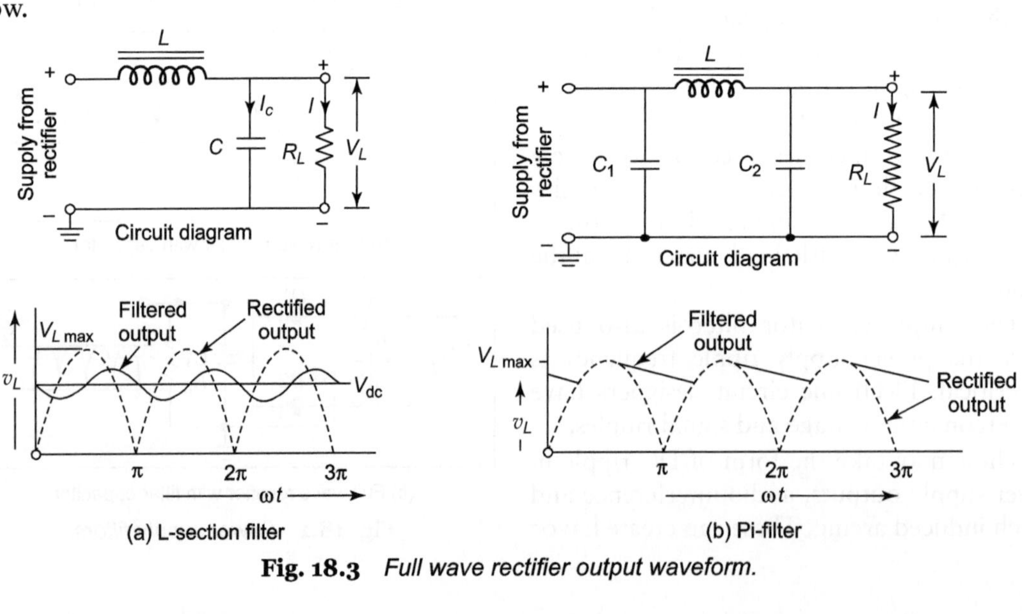

Fig. 18.3 *Full wave rectifier output waveform.*

1. In a pi-filter the DC output voltage is much higher compared to that from an L-section filter having the same input voltage.

2. In a pi-filter the ripples are less. So the values of L are smaller.

3. In a pi-filter, the capacitor is to be charged to the peak value hence the RMS current in supply transformer is larger than in an L-section filter.

4. Voltage regulation of pi-filter being very poor, these are suitable for fixed loads. L-section filters can work satisfactorily with varying loads if a minimum current is maintained.

18.2.3 Line Filter Capacitors

Power supplies receive power from an AC source such as a commercial power line, a motor driven generator or an inverter. Under normal circumstances, the AC power is rectified producing a pulsating DC. The DC is 'smoothed' to eliminate the voltage variations. The simple method is to utilize a single large capacitor or a combination of capacitors and inductors. A polyester capacitor is generally used in a power supply application because of its small size and economic considerations. For switching power supply applications, the polypropylene capacitor is the best selection, because of the low ESR and the high current carrying capability.

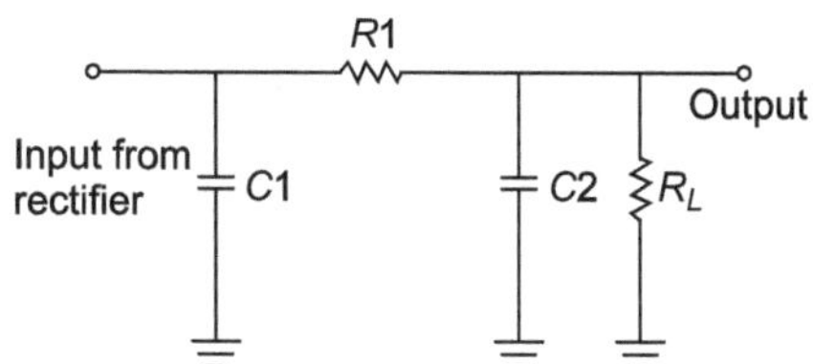

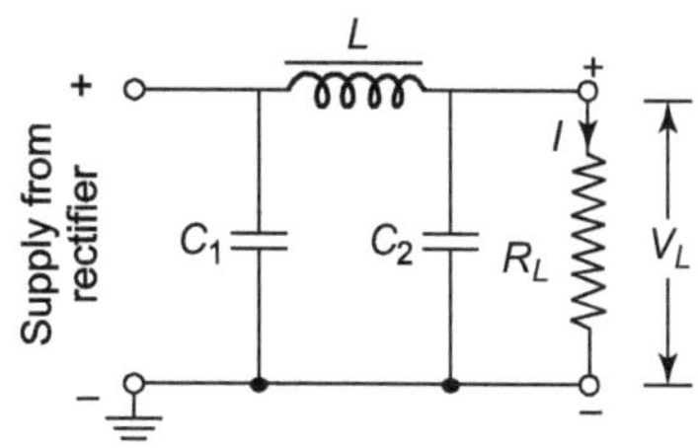

Fig. 18.4 *(a) Output RC filter.* *(b) Output LC filter.*

R in RC filter (Fig. 18.4(a)) may be replaced by low value inductor (Fig. 18.4b) which has a similar effect without the voltage drop. A capacitor alone is sometimes put across a circuit component that uses power to prevent current peaks and stop noise from being propagated through the power supply. The amount of ripple is decided by the supply frequency voltage, output current and the capacitance. An inductor added to this circuit will compensate for voltage sag by inducing a voltage if the current starts to drop.

18.2.4 Band-Pass Filters

Several types of filters are used in electronic circuits, such as low pass filter (LPF) as in Fig. 18.5(a), high pass filter (HPF) as in Fig. 18.5(b), band pass filter (BPF), etc. The reactance of

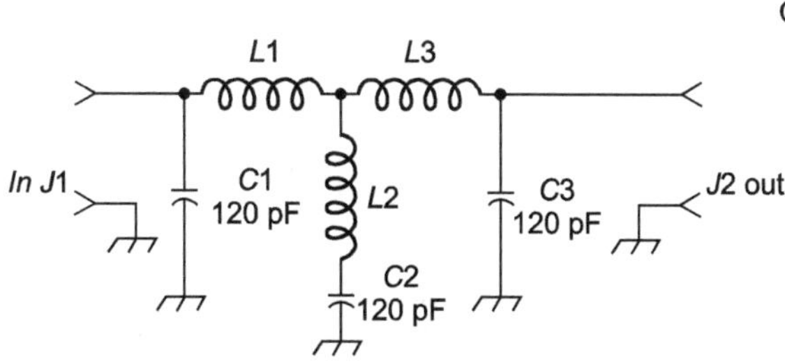

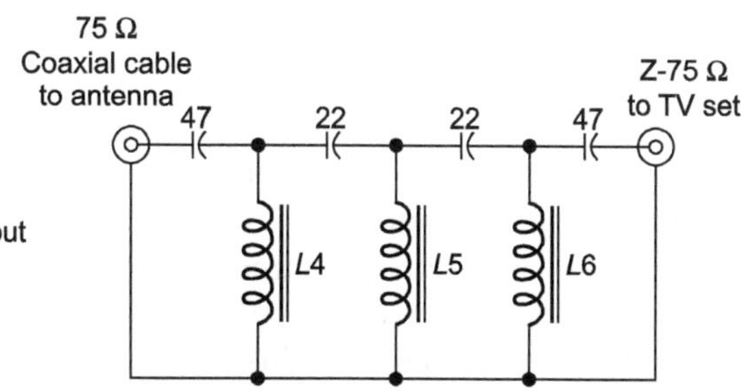

Fig. 18.5 *(a) Low pass filter.* *(b) High pass filter.*

the capacitor being inversely related to the frequency, it can be used to increase or decrease the impedance of the circuit at certain frequencies and therefore does the filtration job. High-pass and low-pass filter designs can be combined to form a band-pass filter. A band-pass filter passes a band of frequencies, typically an entire amateur band, and rejects signals above and below that frequency band.

18.3 TUNING CAPACITORS

These capacitors are used in most circuits for adjusting resonance frequency of a radio or an amplifier circuit (Fig. 18.6). These may be used in series or parallel combination with reactors. Most electronic circuits use them in different circuit configurations to make circuits selective or resonant at desired frequencies. These are particularly useful in audio/video equipment. These are usually variable capacitors with knob for tuning. Sometimes trimmers may be used for pre-setting/fine-tuning the frequency.

These components are so common we are not even aware of them when we tune our radios and TVs, or turn the knobs to adjust our instruments/oscillators. Variable capacitors, also called trimmers, are invaluable in the design of electronic equipment. Variable capacitors are generally employed to provide a range of capacitance and are commonly used in applications where exact capacitance values cannot be obtained using normal design procedures.

The most widely used trimmers are ceramic, glass, air, plastic and mica. Automatic tuning circuits use varicaps where a voltage change on the varicap brings about a change in frequency.

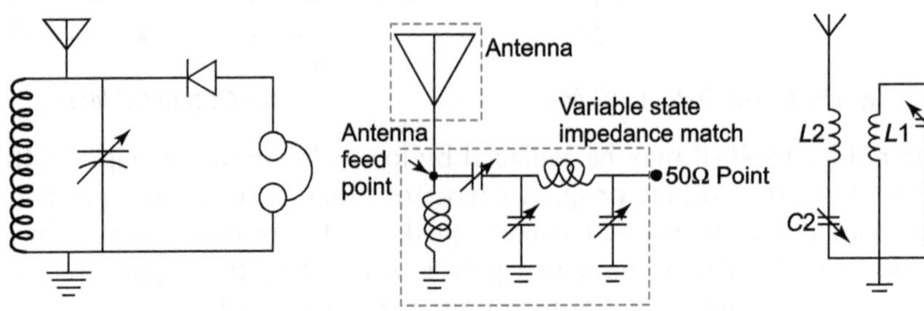

Fig. 18.6 *Series and parallel tunable filters.*

18.4 COUPLING/DECOUPLING (BLOCKING)

The ability of a capacitor to pass an AC signal allows it to couple a section of an electronic circuit to another circuit. They are often used to separate the AC and DC components of a signal. This method is known as AC coupling or 'capacitive coupling'. Here, a large value of capacitance, whose value need not be accurately controlled, but whose **reactance** is small at the signal frequency, is employed.

Sections of electronic circuits may be linked with a capacitor because capacitors pass AC (changing) signals but block DC (steady) signals. This is called capacitor coupling or CR-coupling. It is used between the stages of an audio

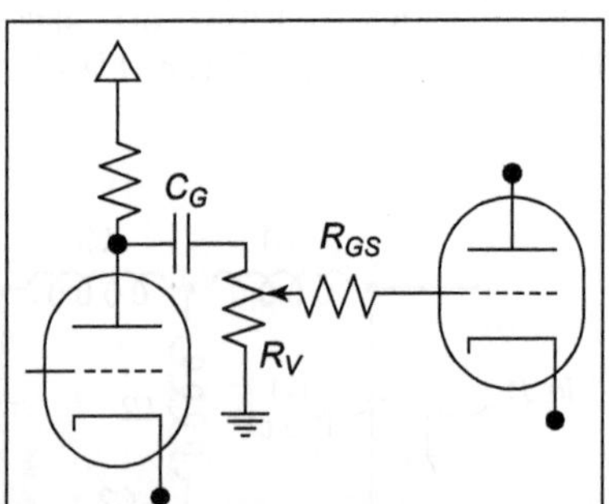

Fig. 18.7 *Coupling circuit – C_0 is the coupling capacitor.*

system to pass on the audio signal (AC) without any steady voltage (DC) which may be present (Fig. 18.7), for example to connect a loudspeaker. It is also used for the 'AC' switch setting on an oscilloscope. Proper selection of coupling capacitors ensures the maximum transfer of RF energy. All capacitors will block DC by definition; however, considerations for satisfying the requirements of a coupling application depend on various frequency-dependent parameters that must be taken into account beforehand.

Coupling capacitors are widely used to 'couple' two circuits together by means of the capacitive reactance common to both circuits. A decoupling capacitor provides a low impedance path to ground to prevent coupling between the stages of a circuit. In Fig. 18.7, the coupling capacitor C_0 acts as short circuit path for all audio frequencies, while blocking DC voltages.

The output voltage for a square wave input to a coupling circuit may vary widely depending upon RC time constant, as shown in Fig 18.8. The polypropylene or polycarbonate dielectric is the common dielectric for this application. The precise behaviour of a capacitor coupling is determined by its time constant (RC). The resistance (R) may be inside the next circuit section rather than a separate resistor. For successful capacitor coupling in an audio system the signals must pass through with little or no distortion. This is achieved if the time constant (RC) is larger than the time period (T) of the lowest frequency audio signals required (typically 20 Hz, T = 50 ms).

A capacitor can be used to block DC voltage since once charged, it is essentially an open circuit to DC while passing AC currents. Effective coupling demands low capacitor reactance over the entire frequency range of interest. Otherwise certain frequencies may be attenuated when compared to the other frequencies.

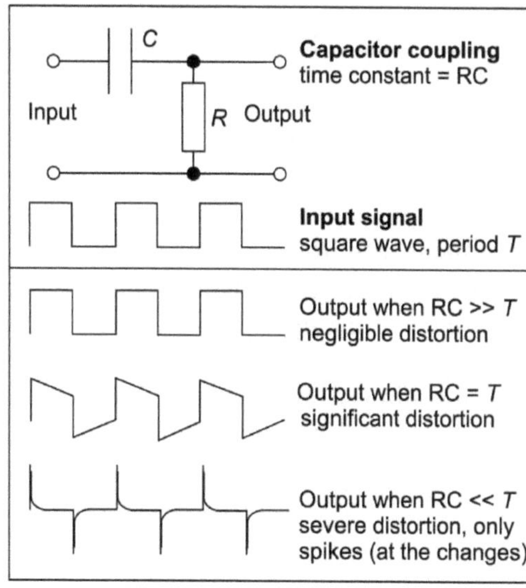

Fig 18.8 *RC coupling output is affected by time constant.*

In Fig. 18.9, input sides contain blocking capacitors, which prevent DC signal, but allow AC to pass through.

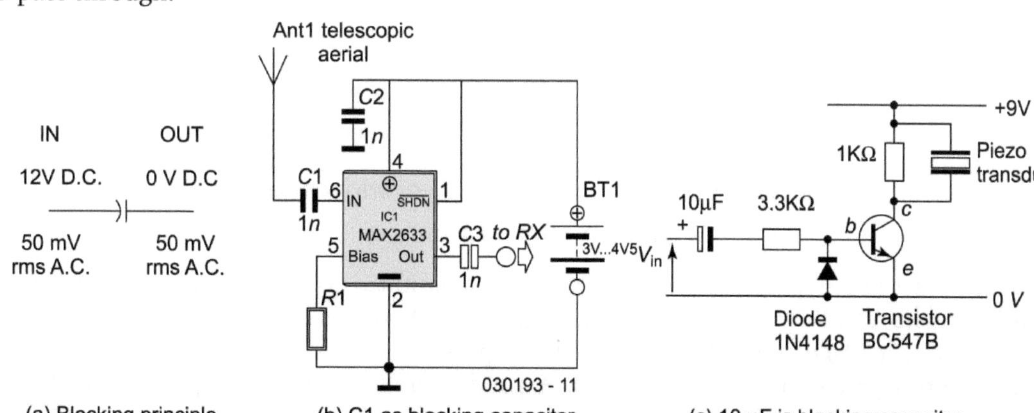

Fig. 18.9 *Blocking capacitors.*

Insertion loss

One of the fundamental considerations for all coupling applications is the capacitor's insertion loss at the operating frequency. By evaluating its magnitude one can readily decide whether or not the subject capacitor is suitable. It is especially important to check the presence of one or more parallel resonances falling within the operating passband. These resonances will generally show up as distinct attenuation notches at their frequencies of occurrence. If a parallel resonance does fall within the operating passband, its acceptability has to be estimated. The magnitude for a given capacitor may be excessive, rendering it unusable for the application. An insertion loss of several tenths of a dB is generally an acceptable criterion for most coupling applications. Losses that exceed several tenths of a dB within the passband could adversely affect the performance of a circuit.

18.5 BYPASS CAPACITOR

The reactance of a capacitor decreases as the frequency increases. Therefore in certain applications it is used in parallel with other components to bypass it at a specified frequency. Capacitors will pass AC currents but not DC. Throughout electronic circuits this very important property is taken advantage of to pass AC or RF signals from one stage to another while blocking any DC component from the previous stage.

By definition, a bypass capacitor is a device employed to conduct an AC current around a component or group of components, and it must offer negligible opposition to the frequencies being bypassed. The capacitor acts like a conduit for the AC signal by passing it to ground. Three of the most important factors to consider when choosing a bypass capacitor are its impedance, its dissipation factor and insulation resistance. When the device is installed the leads must be kept as short as possible to eliminate parasitic inductance. A bypass capacitor stores an electrical charge that is released to the power line whenever a transient voltage spike occurs. It provides a low-impedance supply, thereby minimizing the noise generated by the switching outputs of the device.

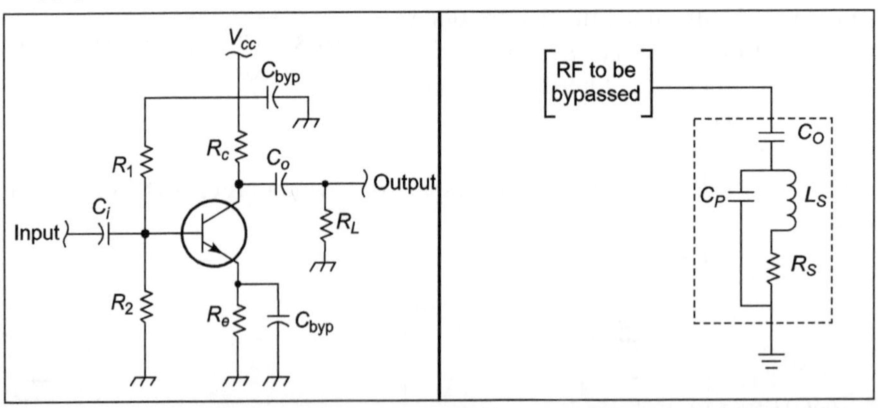

(a) C_{byp} is bypass capacitor (b) Bypass C_o with equivalent C_p, L_s and R_s.

Fig. 18.10 *Bypass circuits.*

In Fig. 18.10 (a), bypass capacitors have been mentioned as C_{byp}. The effect of bypass capacitor on voltage and current is clear from Fig. 18.11. A bypass capacitor is used to dampen the AC component of DC circuits. By installing bypass capacitors, the DC circuit will not be as susceptible to ripple currents and voltages. A good default value for a bypass cap is 0.1 uF.

Higher frequencies require lower values of capacitors. A film type polycarbonate, polyester or polypropylene capacitor should be selected for this application.

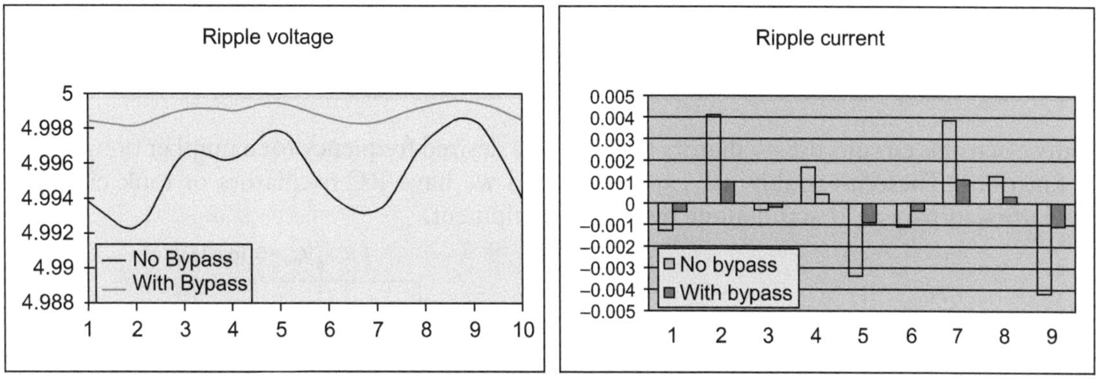

Fig. 18.11 *Effect of bypass capacitor on ripple voltage and ripple current.*

18.6 SAMPLE AND HOLD

In this type of application, the capacitor is used as a temporary storage cell until the time constant is reached, or in a sample and hold circuit, until the next sample is taken. To change the voltage across a capacitor, it is necessary to change the stored charge which takes a finite time. This phenomenon is put to use in timing circuits, such as oscillators, signal generators and latch

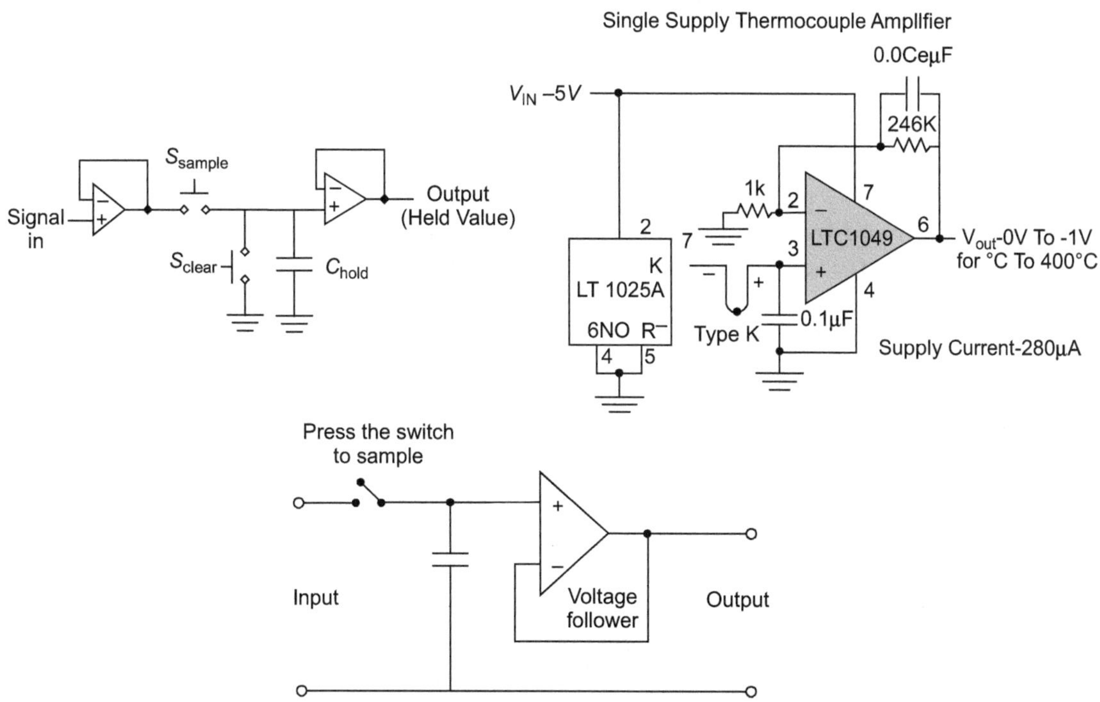

Fig. 18.12. *Sample and hold circuits.*

timers. Capacitors selected for this application must have extreme capacitance stability, high insulation resistance, relatively low ESR and a low dielectric absorption. A polystyrene capacitor would be the proper device to select for this critical application.

18.7 TIMING CIRCUITS

Many electronic circuits use oscillators to generate a desired frequency for a number of operations or functions. These invariably use capacitors, and we have RC oscillators or tank circuit LC oscillators, as part of or stand-alone devices in equipment.

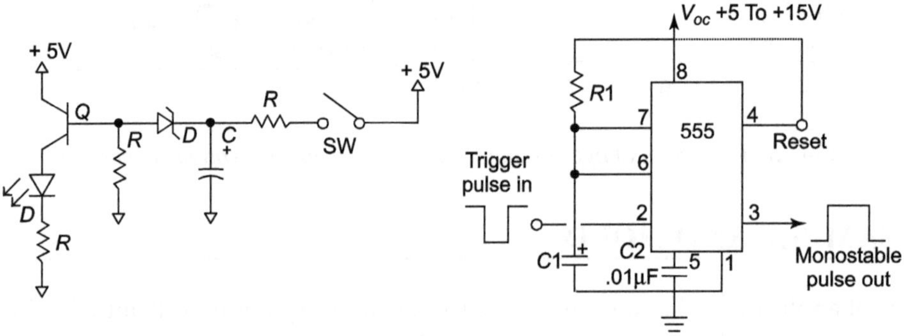

Fig. 18.13 *(a) Basic timing circuit.* *(b) Monostable timer circuit.*

18.8 SIGNAL PROCESSING AND SWITCHED CAPACITORS

Digital signal processing allows inexpensive construction of a wide variety of filters. The signal is sampled and an analogue-to-digital converter turns the signal into a stream of numbers. A computer program running on a CPU or a specialized DSP (or less often running on a hardware implementation of the algorithm) calculates an output number stream. This output can be converted to a signal by passing it through a digital-to-analogue converter. There are problems with noise introduced by the conversions, but these can be controlled and limited for many useful filters. Due to the sampling involved, the input signal must be of limited frequency content or aliasing will occur.

The energy stored in a capacitor can be used to represent information, either in binary form, as in DRAMs, or in analogue form, as in analogue sampled filters and Charge Coupled Devices (CCDs) Capacitors can be used in analogue circuits as components of integrators or more complex filters and in negative feedback loop stabilization. Signal processing circuits also use capacitors to integrate a current signal.

A switched capacitor replaces the resistor, as in Fig. 18.14 (a). The capacitor ratio determines the transfer function at (b), which senses a charge sampled on C_1 and gives output accordingly. Such a circuit requires minimum power, and also provides improved input-output matching.

Contact suppression:

A wound plastic film capacitor plus series resistor is incorporated into a single component envelope for convenience and robustness. This reduces switch arcing and RFI. The most common combination of values is 0.1 µF + 100 Ω.

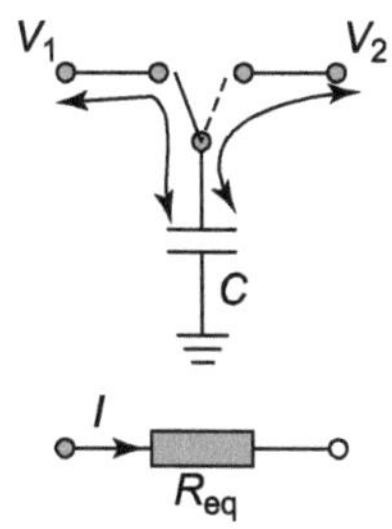

$$q(t) = C.V_1$$
$$q\,(t + \tau) = C.V_2$$
$$\Delta q(t) = C.(V_1 - V_2)$$
$$I = \Delta q(t)/T = C.(V_1 - V_2)/T$$
$$R_{eq} = T/C$$

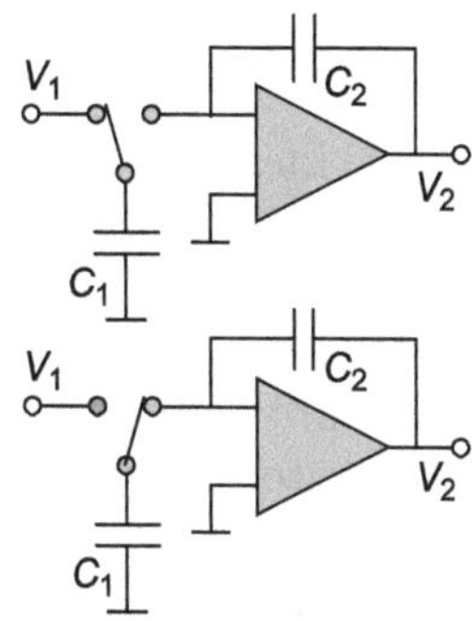

Fig. 18.14 *(a) Switched capacitor as resistor* *(b) sampled voltage on C_1 is sensed by amp. circuit*

18.9 SENSORS

Capacitors are used for a number of metering circuits, based on change in capacitance by varying some property defining the capacitance value. Most capacitors are designed to maintain a fixed physical structure. However, various things can change the structure of the capacitor 1/M the resulting change in capacitance can be used to sense those things.

(i) Change in the dielectric

The effects of varying the physical and/or electrical characteristics of the dielectric can also be of use. Capacitors with an exposed and porous dielectric can be used to measure humidity in air.

(ii) Changing the distance between the plates

Capacitors are used to accurately measure the fuel level in airplanes. Capacitors with a flexible plate can be used to measure strain or pressure. Capacitors are used as the sensor in condenser microphones, where one plate is moved by air pressure, relative to the fixed position of the other plate. Some accelerometers use MEMS capacitors etched on a chip to measure the magnitude and direction of the acceleration vector. They are used to detect changes in acceleration, e.g. as tilt sensors or to detect free fall, as sensors triggering airbag deployment, and in many other applications. They are also used in fingerprint sensors.

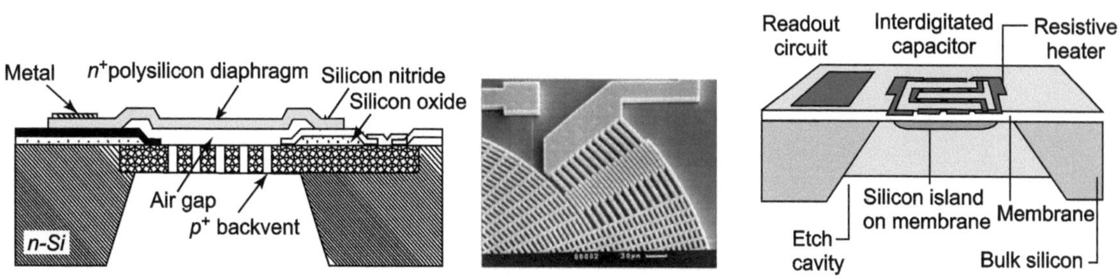

(a) Change in space between plates, (microphone) (b) Change in area (A) (gyroscope) (c) Change in k (chemical sensor)

Fig. 18.15 *Capacitor sensors.*

18.10 RADIO FREQUENCY INTERFERENCE (RFI) CAPACITORS

RFI capacitors are ideal for suppressing unwanted noise from electronic circuits. This minimizes the amount of noise passing from one stage of the circuit to another, thus improving overall circuit performance. EMI interference suppression is also gaining importance worldwide in view of EMI's undesirable effect on nearby equipment and surroundings. Most household and industrial electronic gadgets are required to control EMI to a minimum level, and strict standards have to be followed. Capacitors play a vital part in these circuits.

18.11 NOISE FILTERS AND SNUBBERS

When an inductive circuit is opened, the current through the inductance collapses quickly, creating a large voltage across the open circuit of the switch or relay. If the inductance is large enough, the energy will generate a spark, causing the contact points to oxidize, deteriorate or sometimes weld together, or destroying a solid-state switch. A snubber capacitor across the newly opened circuit creates a path for this impulse to bypass the contact points, thereby preserving their life; these were commonly found in contact breaker ignition systems, for instance. Similarly, in smaller scale circuits, the spark may not be enough to damage the switch but will still radiate undesirable radio frequency interference (RFI), which a filter capacitor absorbs.

Snubber capacitors are usually employed with a low-value resistor in series, to dissipate energy and minimize RFI. Such resistor-capacitor combinations are available in a single package. Capacitors are also used in parallel to interrupt units of a high-voltage circuit breaker in order to equally distribute the voltage between these units. In this case they are called grading capacitors.

19

CAPACITORS FOR POWER ELECTRONICS

19.1 ROLE OF CAPACITORS IN POWER ELECTRONICS

Capacitors are used in power electronics for a wide variety of applications, and often extremely non-sinusoidal voltages and pulsed currents are present. Capacitors are needed for both AC and DC applications, and the choice is governed by voltage characteristics of the circuit. AC capacitors are periodically recharged during operation; DC capacitors are periodically charged and discharged without recharge. Power electronics capacitors face transient currents and repetitive pulses all the time, and the peak currents cannot be measured. It is very difficult to measure fast current pulses directly. It is therefore standard practice to specify dv/dt rating, (peak current can be calculated from the equation $I = Cdv/dt$), and this rating is considered for all applications.

[A] AC capacitor applications

(1) Commutation capacitors

Switched in parallel to a thyristor, these capacitors are designed to quench the conductive state of the thyristor. Since commutating capacitors are periodically and abruptly recharged, the peak current may substantially exceed the RMS value. With IGBTs taking over AC power circuits, communication capacitors are more common today.

(2) Damping (snubber) capacitors

Usually connected in series with a resistor, these capacitors are designed for the damping of undesirable voltage spikes caused by the so-called carrier storage effect during the switching of power semiconductors.

Commutation and snubber capacitors have dv/dt as one of the main characteristics, and maximum current Cdv/dt in each cycle, and hence RMS currents may go very high. For example, a 15 μF 440 V capacitor may have dv/dt rating 30 V/μs, I_{peak} of 450 A (15 μF × 30 V μs^{-1}), and I_{rms} of 16 A. Internal connections have to be heavy, and the design and construction have to take into account the heavy current densities at terminations and tabs, end connections and on the electrodes, as also the size and cooling area for the heat generated in the capacitor. These capacitors may use MPP, MPET or film/foil construction.

These are used in thyristor circuits, and designs are similar to or same as commutation capacitors. In IGBT circuits or high frequency thyristor circuits, capacitors are used to protect thyristors from high voltage switching surges by connecting in parallel with them. These may carry high currents and need to be designed accordingly.

[B] DC capacitor applications

(1) Smoothing capacitors

These are used for the reduction of AC component of fluctuating DC voltage in:

- Power supply
- High-voltage testing equipment, DC controllers
- Measurement and control technology
- Generation of high DC voltage through cascaded circuits

(2) Supporting capacitors, DC-filter or intermediate circuit capacitors

These are used for energy storage in intermediate DC circuits, and are designed to absorb and release very high currents within short periods, the peak value of the current being substantially greater than the RMS value in application like:

- Frequency converters for poly-phase drives
- Transistor and thyristor converters

(3) DC link capacitors

These are mostly used in AC/AC frequency conversion via DC intermediate stage for pulse width modulation (PWM) speed control of motors. The DC link capacitor serves as temporary energy storage between input and output AC stages. Variable frequency drives are becoming commonplace, and these capacitors are being increasingly used.

(4) Surge (pulse) discharge capacitors

These are capable of supplying or absorbing extreme short time current surges and are usually operated at low repetition frequencies. Some applications are:

- Welding applications
- Laser/magnet technology
- Lightening generators

Due to the nature of applications and waveforms encountered, most power electronics capacitors are subjected to voltage and current pulses and high dv/dt, and have to be designed to keep ESR and ESL as low as possible and internal construction and terminations are also given special attention.

19.2 COMMUTATION CAPACITORS

The technological development of thyristors started an ever-increasing use of industrial power electronics. Many applications are based on forced commutation by means of capacitors. Commutation capacitors deliver current pulses necessary to block thyristors. These have to work under severe constraints and complex applied waveforms. The constraints applied to commutation capacitors remain a general type of constraints met in power electronics.

(1) **Dielectric constraints**

- Applied voltage – continuous, RMS and peak value (must remain < V_n)
- Voltage variation rate (dielectric losses increase with high dv/dt)

(2) **Constraints due to ohmic losses and frequency**

- RMS and peak value of currents
- Reactive power (estimation of loss of power using tan δ)

Capacitors for semiconductor-commutation assistance help minimize commutation losses and limit dv/dt. They absorb large pulsed load current at switch opening. They are wound in such a way to provide a minimum series induction (ESL <10 nH) and a low series resistance (ESR), to reduce power dissipation. This is done by using dielectrics which offer very low losses and give high stability of the capacitance versus temperature and time. Terminations, wires and connections to winding need heavy wires/cables and also care in soldering of end connections.

Parasitic inductance of RCD circuit becomes very critical (typically below 100 nH). The capacitors are best made using extended film/foil construction to keep ESL low and have high dv/dt rating. If metallized capacitors are used, winding aspect ratio has to be kept in mind and end spray as well as end connections need careful consideration. The terminations and lead connections also count towards keeping ESL low and dv/dt high.

Classical thyristors are disappearing gradually and being replaced by gate-turn-off thyristors (GTO)/integrated gate-commutated thyristor (IGCT) and IGBT – these active components do not need turn-off commutation capacitors.

19.3 SNUBBER CAPACITORS

A snubber is a device used to suppress ('snub') voltage transients in electrical systems, pressure transients in fluid systems, or excess force or rapid movement in mechanical systems.

Snubbers are any of several simple energy absorbing circuits used to eliminate voltage spikes caused by circuit inductance when a switch—either mechanical or semi-conductor—opens. A snubber eliminates the voltage transient and oscillations that occurs when the switch opens by providing an alternate path for the current flowing through the circuit's intrinsic leakage inductance. Snubbers in switch mode power supplies provide one or more of these three valuable functions:

- Shape the load line of a bipolar switching transistor to keep it in its safe operating area.
- Remove energy from a switching transistor and dissipate the energy in a resistor to reduce junction temperature.
- Reduce ringing to limit the peak voltage on a switching transistor or rectifying diode and to reduce EMI by reducing emissions and lowering their frequency.

Snubbers can be either passive or active networks. Passive snubber network elements are restricted to resistors, capacitors, inductors and diodes. Passive snubbers can control either voltage or current and may be either dissipative or non-dissipative. If energy in the snubber is dissipated in a resistive element it is classified as a dissipative snubber, but if the energy is returned to the input or moved ahead to the output it is classified as non-dissipative. Snubbers are also used to reduce the stress on the switch and improve efficiency in a fly-back topology.

Snubbers are frequently used in electrical systems with an inductive load where the sudden interruption of current flow often leads to a sharp rise in voltage across the device creating the interruption. This sharp rise in voltage, a transient, can damage and lead to failure of the controlling device. A spark is likely to be generated (arcing), which can cause electromagnetic interference in other circuits. The snubber prevents this undesired voltage by conducting transient current around the device.

[A] Simple RC snubber

This is probably the most widely used snubber and is applicable for both rate-of-rise control and damping. In inductively clamped topologies, where there is still some stray inductance, the RC snubber can be used to reduce the peak power dissipation in the switch by controlling the rate-of-rise of drain voltage. The RC snubber, however, will absorb energy during each voltage transition and can reduce efficiency. Also, it will reduce the switching speed of the MOSFET switch. Care must be taken in choosing the value of R and C to optimize the total performance. The main application of a RC snubber is to damp parasitic ringing in the circuit due to unclamped inductance in configurations such as the fly-back converter. In these applications, the value of the resistor must be close to the characteristic impedance of the parasitic resonant circuit it is intended to damp.

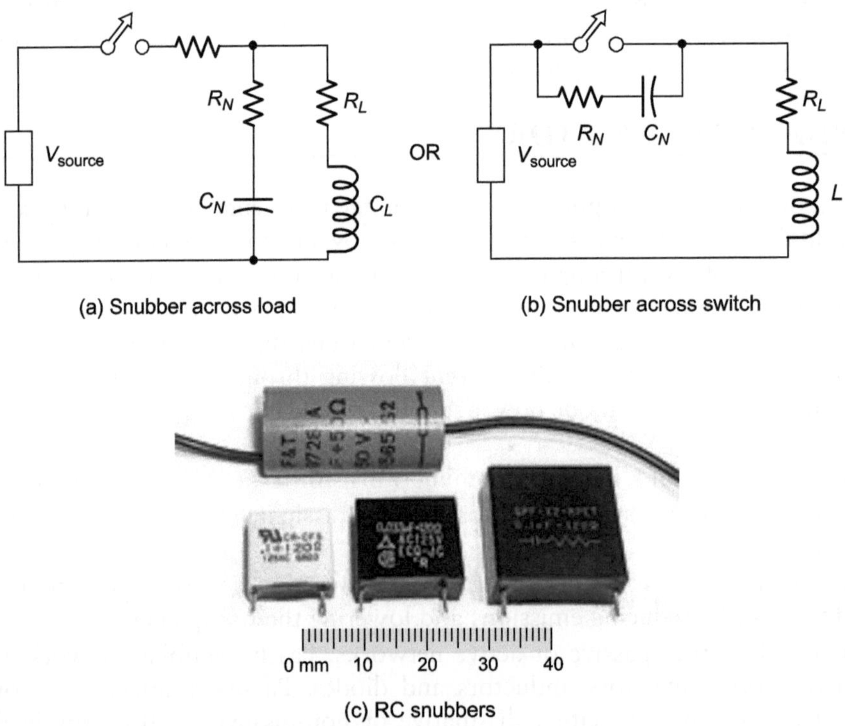

Fig. 19.1 *Simple snubber connections.*

A simple snubber consists of a small resistor (R_N) in series with a small capacitor (C_N) connected across a switching device or a load, as in Fig. 19.1. This combination can be used to suppress the rapid rise in voltage across a thyristor or preventing the erroneous turn-on of the thyristor. It does this by limiting the rate of rise in voltage (dv/dt) across the thyristor to a value which will not trigger it. Snubbers are also often used to prevent arcing across the contacts of relays and switches and the electrical interference and welding/sticking of the contacts. This type of snubber is commonly used with inductive loads such as electric motors. The voltage across a capacitor cannot change instantaneously, so a decreasing transient current will flow through it for a small fraction of a second, allowing the voltage across the switch to increase more slowly when the switch is opened.

While the values can be designed for each application, a 100 Ω non-inductive resistor in series with a 100 nF, or a larger capacitor of appropriate voltage rating is usually effective. The actual voltage rating can be difficult, and can be determined only by measuring the temperature rise of the capacitor. Either of the circuits above may be enough protection for general needs. For extreme situations, both load and contact protection may be required.

[B] RCD clamp

In the clamp mode the purpose of the snubber is to clamp the voltage during turn-off at the drain of the MOSFET (Fig. 19.2). The parallel RC circuit may be returned to ground or to a voltage other than ground (i.e. input voltage if the drain can go above input voltage) since this will reduce the power dissipation in the resistor. The MOSFET switch itself will have to sustain the peak power dissipation during turn-off. The value of the capacitor, C_{clamp}, and resistor, R_{clamp}.

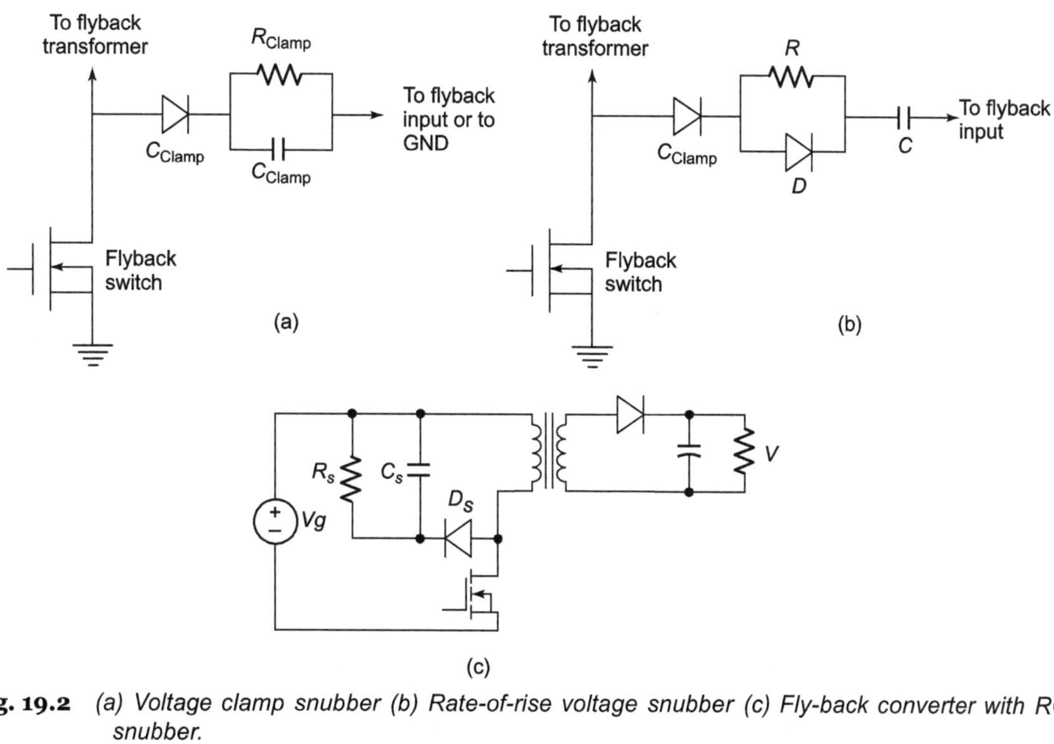

Fig. 19.2 *(a) Voltage clamp snubber (b) Rate-of-rise voltage snubber (c) Fly-back converter with RCD snubber.*

is based on the energy stored in the parasitic inductance, as this energy must be discharged into the RC network during each cycle. The voltage across the capacitor and resistor sets the clamp voltage, V_{CLAMP}.

[C] Rate-of-rise control RCD snubber

When the RCD snubber is used to control the rate of voltage rise at the MOSFET drain, the capacitor must be completely charged and discharged during each cycle to be able to control the rate-of-rise of the drain voltage. The RC time constant of the snubber should, therefore, be much smaller than the switching period (consider the effect of duty cycle on pulse width). Typically, the time constant should be about 1/10th the switching period.

When the switch turns off, the inductor current is diverted through the snubber diode to charge the capacitor. At that time, it is expected that the output rectifier will turn on. When the switch turns on, the snubber capacitor discharges through the snubber resistor and the switch. The capacitance value is obtained from $I_P = C(V_C/t_r)$

Where V_C = voltage across the capacitor and t_r = rise time of the voltage

The resistor is then chosen based on the time constant required. The RC time constant is much smaller than the switching frequency and, therefore, the power dissipated in the resistor is independent of its value. The power dissipation is determined by the value of the capacitance and the switching frequency.

[D] Snubber capacitors for insulated gate bipolar transistor (IGBT)

The primary function of a snubber is to suppress transient voltages in applications where the switching is turned off and a large spike or peak current is generated. When an IGBT switches off, for example, a transient or surge voltage is generated by the parasitic inductance in the power circuit, and that transient voltage needs to be controlled. If this voltage is not controlled, it may exceed the IGBT's rating, causing it to fail. IGBT switches are ten times faster than older Darlington transistors, and have made it possible for inverters in uninterruptible power supply systems (UPS) to operate at 20 kHz or above.

(1) Applications for IGBT

IGBT controls are common in industry today for motor controls (AC motor speed control), UPS systems, inverters/converters, microwave ovens, electric cars, switch mode power supplies, silicon controlled rectifier commutation (SCR), and arc welding equipment. The capacitor's function is mainly **noise suppression and contact protection** in these circuits.

A capacitor alone is not an ideal solution. When the contacts are open, the capacitor gets charged up to the supply voltage. As the contact is made, an inrush of current results, limited only by a residual resistance, and may result in damage. A resistor is therefore added in series with the capacitor to limit inrush current to a safe value. The voltage across the contacts when opened is equal to the load current × resistance (V = IR).

(2) Component selection considerations

A series resistor should be non-inductive, and a good choice is a carbon composition resistor (wire wound resistor is inductive). The capacitor should withstand the extremely high peak currents in snubbers. For capacitance values up to 0.01 µF, the first preference could be dipped mica capacitors. For higher capacitance values, radial-leaded polypropylene, film/foil capacitors may be preferred. Axial-leaded are as good except for the higher inductance intrinsic to axial leaded devices.

For higher voltages and capacitances, polypropylene film/foil capacitors may be preferred. Metallized film capacitors are used to achieve small size, but their use reduces the peak current capability to a third to a fifth of the other high-voltage choices.

The peak current capability is the dv/dt capability times the nominal capacitance. The RMS current capability is usually the current which causes the capacitor to heat up below 15°C. Dipped mica capacitors can withstand dv/dt of more than 100,000 V μs^{-1} for all ratings. PP film capacitors can withstand more than 2000 V μs^{-1}. For high-voltage snubbers, capacitors handling more than 3000 V μs^{-1} are available.

Metallized film types and high-K ceramic types have limited peak-current and transient withstanding capability, usually of the order of 50 – 200 V μs^{-1}. Polyester has 15 times the loss of polypropylene and is fit only for low RMS currents or duty cycles. Voltage and temperature coefficients need to be taken into account. While mica or a DPP type's capacitance is nearly independent of voltage and temperature, by comparison, a high-K ceramic dielectric can lose ¼ of its capacitance from room temperature to 50°C and lose another ¼ from zero volts to 50% rated voltage.

As an example, consider a switch mode converter which needs to snub one of the transistor switches. The switching frequency is 50 kHz and the open-switch voltage is 160 V DC with a maximum switch current of 5 A. The resistor value must be:

$R \leq 160/5 = 32\ \Omega$ and the capacitance value is: $C_s = 1/(160^2)(50 \times 10^3) = 780$ pF

Typically, the calculation of the characteristics of a capacitor for a specific application would be on the following lines. A capacitor with a capacity of 33 µF is needed for a trapezoidal voltage waveform as below: The turn-off waveform at the drain of the MOSFET switch is as shown in Fig. 19.3. Choice of the rated voltage is governed by the following considerations:

The rated voltage of the capacitor must be equal to or larger than the higher one of the two voltages U_1 and U_2, i.e., $U_n > 1000$ V. A capacitor rated 10% above this value is preferred.

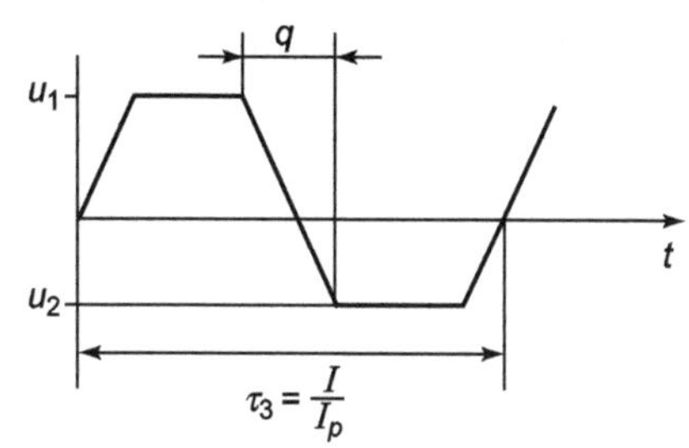

U_1 1000 V
U_2 500 V
F_0 120 Hz
τ 100 µs

Fig. 19.3 *Snubber turn-off voltage waveform.*

Rate of rise of voltage $= dV/dt = (U_1 + U_2)/100\ \mu s = 15\ V\ \mu s^{-1}$

Repetitive peak current I_p $= C dv/dt = 33\ \mu F.\ 15\ V\ \mu s^{-1} = 495$ A

RMS current $= Ip\sqrt{(2t.f_0)} = 76.7$ A

These values as calculated must be within the specification limits of the capacitor.

(3) Requirements of a snubber/commutation capacitor

The snubber and commutation capacitors are manufactured from high quality aluminium foil extended electrodes and hazy PP film and impregnated with dielectric fluid. Internal connections could be with braided cables or tin plated copper strips. This construction provides capacitors that are inherently capable of handling high RMS and peak currents. Metallized construction capacitors are also available, but their dv/dt ratings are much lower on account of current handling limitations, and have to be adequately protected by series resistors.

Snubber and GTO capacitors are characterized by:

- Very low self-inductance
- High pulse reliability
- High RMS current carrying capability
- Excellent self-healing properties (for metallized capacitors)
- High shock and vibration resistance

Capacitors are available with capacitances from 1.0 μF through 100 μF and with nominal voltages from 400 V DC through 1500 V DC.

19.4 DC LINK CAPACITOR

The purpose of DC link capacitor of a PWM converter is to act as energy storage and filter for the DC link voltage. Both output voltage and frequency are controlled in the inverter by pulse width modulation. The DC-link circuit is located between the rectifier and inverter. It is designed to smoothen and filter the voltages and currents supplied by the rectifier. High-rated capacitors or aluminium electrolytic capacitors are required for this purpose.

An AC/AC converter converts an AC waveform such as the mains supply, to another AC waveform, where the output voltage and frequency can be set arbitrarily. An AC/AC converter with approximately sinusoidal input currents and bidirectional power flow can be realized by coupling a PWM rectifier and a PWM inverter to the DC link. The DC link quantity is then impressed by an energy storage element that is common to both stages, which is a capacitor C for the voltage DC link or an inductor L for the current DC link. The PWM rectifier is controlled in a way that a sinusoidal mains current is drawn, which is in phase or anti-phase (for energy feedback) with the corresponding mains phase voltage.

Due to the DC link storage element, the converter stages are to a large extent decoupled for control purposes. Furthermore, a constant, mains independent input quantity exists for the PWM inverter stage, which results in high utilization of the converter's power capability. On the other hand, the DC link energy storage element has a relatively large physical volume, and when electrolytic capacitors are used, in the case of a voltage DC link, there is potentially a reduced system lifetime.

The input-rectifier is normally not controlled, so the DC link capacitor is loaded with the rectified mains voltage, since the rectifier only allows energy flow from mains to the DC link. In case of braking (reverse-energy flow from motor to DC link) a chopper with braking resistor is necessary to protect the capacitor from over-voltage. The chopper loads the DC link with the braking resistor to convert the excess energy into heat. Recent developments with an inverter instead of a simple rectifier allow energy flow in both directions. The output voltage can be adjusted in voltage and in frequency.

(a) DC link capacitors

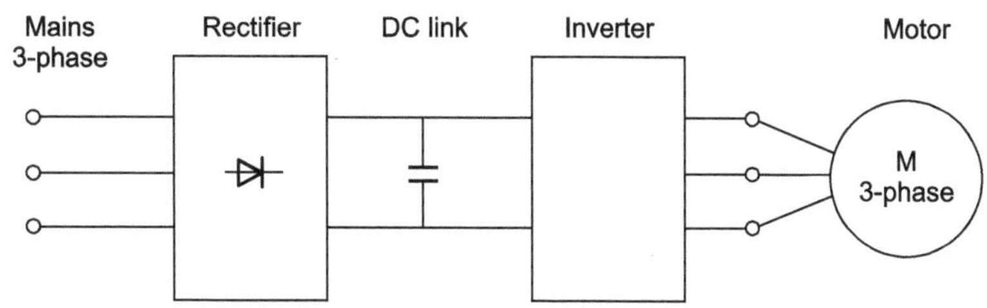

(b) The most common inverter with a DC voltage link circuit

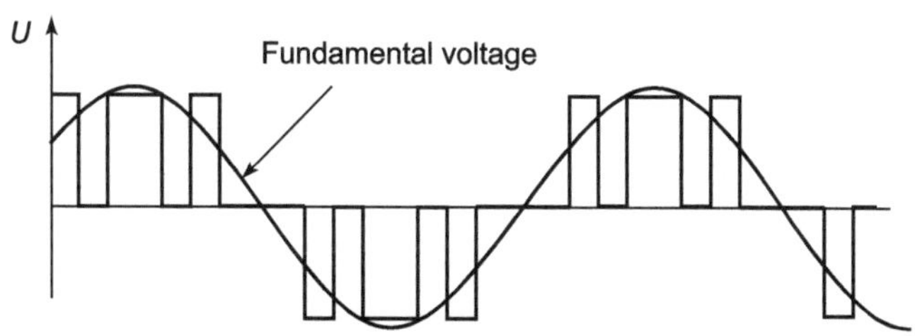

(c) Waveform of DC link capacitor voltage

Fig. 19.4 *DC link capacitors, link circuit and waveform.*

The voltage pulses with fast rise or fall time lead to instantaneous high strong peak current in a capacitor during a very short time voltage transition. This high instantaneous pulse current is a factor for metallized capacitors where pulse current above the permissible limit can destroy the connection to the metallization. For the construction of metallized capacitors, metallization on the film and leads is connected by metal spraying. If excessive instantaneous pulse current flows through the capacitor, the metallization adjacent to the connections may be vaporized due to localized heating at edges. The capacitance value deteriorates or the capacitor becomes an open circuit over a period of time. Thus the pulse current capability, that is, the upper limit of the dv/dt is low in metallized capacitors.

For film/foil capacitors with leads welded directly to the extended foil, the instantaneous pulse current is not limited, and rated voltage and dv/dt are only limited by corona discharge. In pulse applications, power dissipation must be seriously considered, since power dissipation depends on the harmonic content of the waveform. The heat generated by harmonic currents is higher than the sine wave with the fundamental frequency. The AC current or pulse current running through the capacitor will subject the capacitor to self-generated heat.

19.5 CAPACITOR DISCHARGE WELDING

The capacitor discharge welding process is a modification of resistance welding, based on the principle of the capacitor discharge system, with the advantage of very high reproducibility and low operating costs. Two parts are welded through the combined application of mechanical pressure to the parts and discharge of constant amount of electrical energy. Components of the welding process are closely controlled and monitored through electrical control loops. Energy is concentrated and released to the spot where the joining is to take place so that the maximum weld current passes through the minimum of contact area between the two parts.

Capacitor discharge welding is becoming popular because of the precision and accuracy of welding, and tolerances are well maintained without distortion, while adjacent parts are not affected due to localized heating and short weld time. Welding is possible with excellent bonds with many similar or dissimilar materials, as well as metals often considered unweldable. Production speeds are high and no water-cooling is needed. It is characterized by low power requirements and is environmentally clean; there is no smoke or fumes and at the same time the process has low operating costs even at high production rates.

Large capacitor discharge welding machines can have output of 400 kA current, and energy output of 50 kJ. High voltage capacitors between 500 and 3000 V have been commonly used, but recent developments use maximum voltage rating of much lower magnitude.

Typical discharge welder circuit is as per Fig. 19.5. The capacitors are charged by direct current from a rectifier or generator. Welding energy is stored at 50–450 V DC (sometimes even higher voltage) and later discharged through either mechanical or electrical switching methods into the primary of a high current welding transformer.

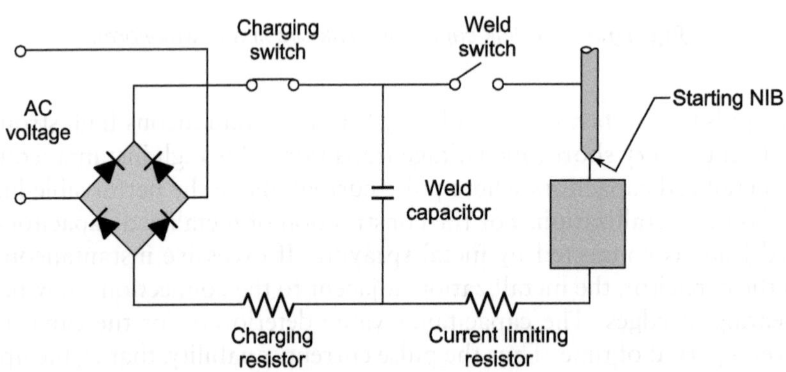

Fig. 19.5 *Circuit for capacitor discharge welder.*

The controls for this weld power supply are:

- ➢ Voltage applied – an adjustment of voltage stored in the capacitors.
- ➢ Amount of capacitance – an adjustment of the number of capacitors.
- ➢ Weld current – an adjustment of the taps on welding transformer to vary the voltage on the secondary winding.

These are a few examples of capacitors for fast charging and sudden discharge, and pulse power applications involving high peak currents, for both AC and DC applications. This is not an exhaustive but a representative survey across various fields of uses of such capacitors. The reader will get an idea of intricacies involved in capacitor design, as well as the selection and usage in various fields.

19.6 CAPACITOR VOLTAGE TRANSFORMER (CVT)

A capacitor voltage transformer (CVT), or capacitance coupled voltage transformer (CCVT) is a transformer used in power systems to step down extra high voltage signals and provide a low voltage signal, for measurement or to operate a protective relay. The device consists of three parts: (a) two capacitors in series, across which the transmission line signal is divided, (b) an inductive element to tune the device to the line frequency and (c) a transformer to isolate and further step down the voltage for the instrumentation or protective relay. The first capacitor is actually a large sequence of smaller capacitors in series to allow even distribution of the signal

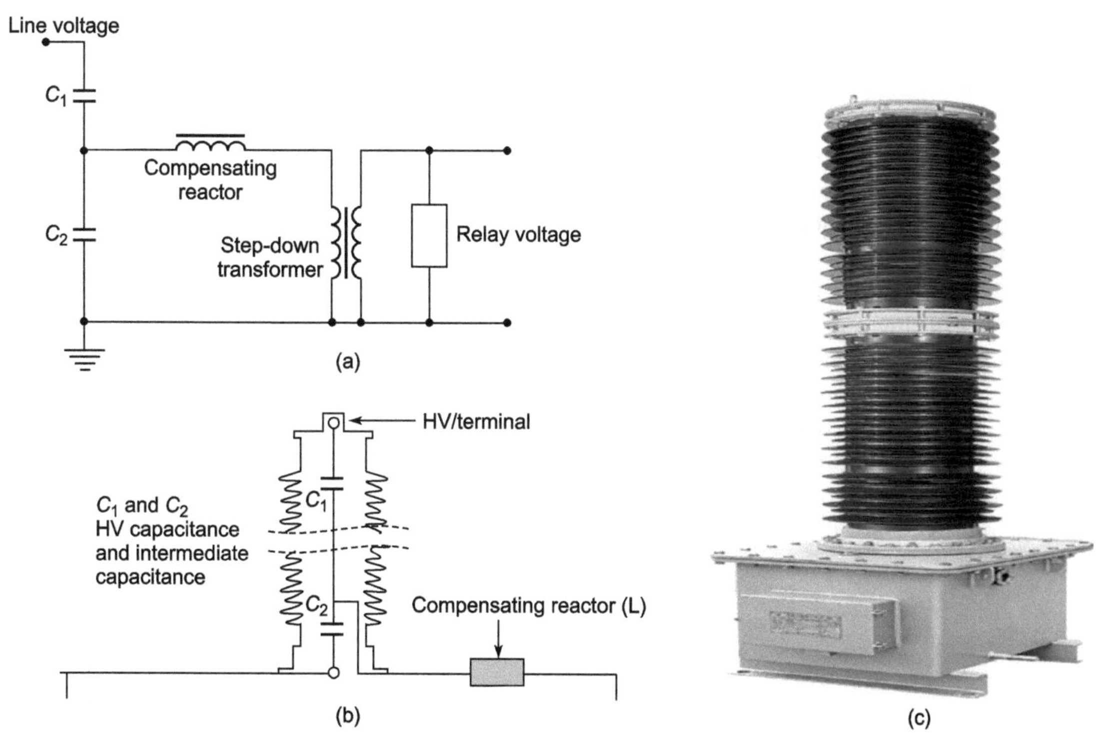

Fig. 19.6 *(a) CVT measurement system. (b) Capacitor arrangement. (c) CVT capacitor.*

across the circuit, as well as for the rapid decrease of the input signal's power. The device has at least four terminals: a terminal for connection to the high voltage signal, a ground terminal, and two secondary terminals which connect to the instrumentation or protective relay.

A potential transformer is the inductive step down transformer used for measurement of voltage and protection whereas a CVT is a capacitive voltage transformer consisting of a stack of series connected capacitors. The voltage across a capacitor is used to measure the voltage. It also serves the purpose of power line carrier communication. The transformer is used for measuring the voltage in a substation or a high voltage transmission line. These same applications require fast, yet secure protection. However, as the requirement for faster protective relays grows, so does the concern over the poor transient response of some CVTs for certain system conditions.

Capacitive voltage transformer components:

A CVT consists of the following components:

- Coupling capacitors (C_1 and C_2)
- Compensating reactor (L)
- Step-down transformer
- Ferroresonance-suppression circuit

When equipped with a communication carrier, the CVT has an additional drain coil, choke coil and carrier switch. CVTs are typically single-phase devices used for measuring voltages in excess of 100 kV where the use of voltage transformers would be uneconomical. In practice, capacitor C_1 is often constructed as a stack of smaller capacitors connected in series. This provides a large voltage drop across C_1 and a relatively small voltage drop across C_2. Capacitor elements consist of aluminium foil insulated with paper and polypropylene film impregnated and filled with mineral or synthetic oil. A pressure release device is provided for each unit of CVT.

The coupling capacitors of the CVT function as a voltage divider to step down the line voltage to an intermediate level voltage, typically 5–15 kV. The compensating reactor cancels the coupling capacitor reactance at the system frequency. This reactance cancellation prevents any phase shift between the primary and secondary voltages at the system frequency. The step-down transformer further reduces the intermediate voltage to the nominal relaying voltage, typically $115/\sqrt{3}$ V.

The CVT is also useful in communication systems. CVTs in combination with wave traps are used for filtering high frequency communication signals from power frequency. This forms a carrier communication network throughout the transmission network.

Capacitor voltage transformers must convert transmission class voltages to standardized low and easily measurable values, which will be used for metering, protection and control of the high voltage system. As such, the need for accurate and reliable voltage transformation is essential. Additionally, capacitor voltage transformers serve as a coupling capacitor for coupling high frequency power line carrier signals to the transmission line.

20

MANUFACTURE OF PAPER/PLASTIC FILM CAPACITORS

Capacitors may be made of wound elements, stacked plates/films, ceramic discs and many other varieties. Electrolytics and others have been covered briefly elsewhere in the relevant chapters. Plastic film capacitors and paper capacitors find extensive use in electrical and electronics industries, and the reader will benefit from knowledge of their manufacturing process.

There are three basic systems available, viz. film/foil, metallized film, and metallized-carrier constructions (MKV). Film/foil construction has the advantage of higher current handling, whereas the metallized structure has self-healing features and smaller size.

20.1 WOUND NON-METALLIZED ELECTROSTATIC CAPACITORS

The initial stage of manufacturing a film capacitor is to start with paper, plastic film, or paper and film, and roll it up with thin aluminium (or sometimes copper foil). As the paper or film may contain microscopic flaws (and the paper always does), multiple layers of dielectric are required rather than one layer of a thicker film. For high voltage, up to seven layers of dielectric may be used at one time. While there have been continuous improvements in the quality and variety of films available, many capacitors are made the same way that they were over 40 years ago.

There are two basic different ways of winding this type of capacitor, known as 'insert tab' and 'extended foil'.

[A] Insert tab construction – inductive winding

Paper/foil capacitors with oil impregnation have been the most common construction of all electrical capacitors. With the 'insert tab' type of winding, small, tinned copper tabs are inserted into the winding at a pre-determined number of turns (Fig. 20.1), one or more on each of the foils; the foils are wound directly over each other, separated by the film. The film is wider than the foil, to permit an 'edge clearance' appropriate to the voltage rating.

For paper/foil capacitors of 250/400 V AC rating, two layers of paper strip are wound on each side of the two foils on a mandrel on the winding machine, and tabs are inserted in contact with foil during winding. These remain in position by winding pressure, and are later soldered

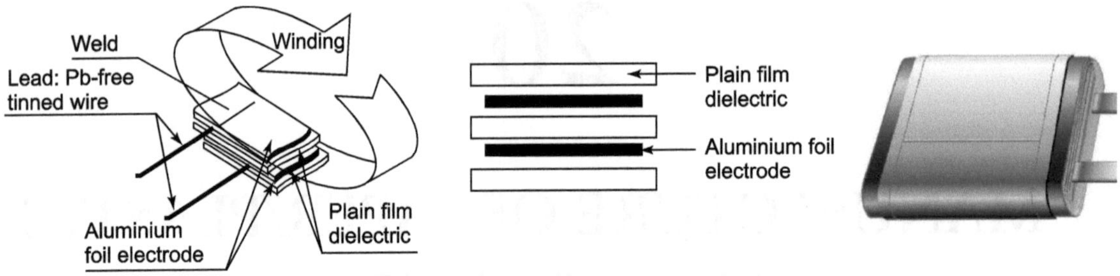

Fig. 20.1 *Tab inserted inductive winding.*

to external contacts during assembly in cans or boxes. The number of layers and tabs could be more for higher voltages and higher currents. The windings are assembled in aluminium cans or MS boxes and hermetically sealed, leaving an inlet for oil. Sealing is achieved by epoxy sealing, crimping of lid, or welding, depending on requirement. They are subjected to high vacuum (approx. 10^{-3} mm Hg or more) for several hours or days at high temperatures of about 120°C to remove all air as well as moisture (either free or loosely bonded with cellulose). This done, processed and degassed oil is introduced in the capacitors by flooding in vacuum or by individual filling (in large sizes or power capacitors). Oil enters the container and impregnates the winding to fill all vacant spaces in the can as well as tiny spaces in the paper, created by removal of air and moisture. After returning excess oil to the oil tank, capacitors are soldered or welded to seal the oil entry point.

Similar process is adopted today for PP/foil/oil or PP/paper/foil capacitors, mainly used for power capacitors. The function of oil is to replace air and moisture removed under vacuum, with a medium of good dielectric strength and a dielectric constant compatible with solid dielectric, so as to have a uniform electric field in the composite system. This draws maximum benefit from the properties of oil and the film/paper.

The windings are wound over a central core, and currents from all over the electrode surface are collected at the inserted terminals. The winding gives rise to an inherent inductance in the capacitor element, though small. This element is called inductive winding for this reason. The current collected at the discreet tab insertion points gives rise to increase in current density at these points, and can be a source of heating for high value capacitors, or during surges. Further, at high frequencies, the inductance may interfere with capacitor working, cause a nuisance in circuit operation, and may also cause resonance effect at certain frequencies.

For small capacitors for electronic circuits, voltages are low and the winding elements are encapsulated in epoxy resin. (In case of most PP/foil elements in small DC capacitors, impregnation is not required.)

[B] Extended foil capacitors

The extended foil capacitor element is wound similarly, but the foils are offset, so that one foil protrudes from one edge of the section, and the other foil from the opposite edge. Ultimately, the ends are soldered over. This provides much lower self-inductance in the capacitor and enables it to discharge huge peak currents – even over 100,000 A. Such currents are common in radar systems and can occur in fault situations in almost anything that can short out an unprotected

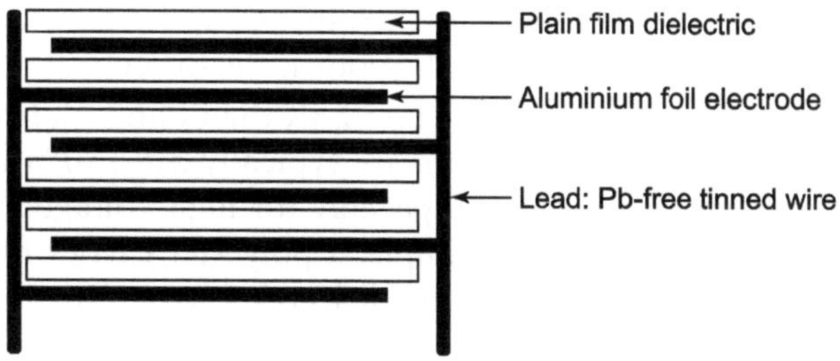

Fig. 20.2 *Extended foil capacitor element.*

capacitor accidentally. It is to be noted that current is also limited by internal wiring. Extended foil capacitors using film/foil are made on similar winding machines and connections are made to extended end surfaces by soldering external contacts. For large windings like commutation or power capacitors carrying high currents and voltages, a similar process is followed. Extended foil capacitors are more expensive than inductive types, are larger in size, but have the benefit of higher operating frequency range compared with inductive winding.

For small capacitors for electronic circuits, the capacitor elements are encapsulated in epoxy resin. Extended foil construction has the benefit of being non-inductive and can carry higher surge currents (higher dv/dt rating). Hence this is a preferred construction for pulsed current applications.

Some electronic wound-film capacitors have a band marked at one end to show that it is attached to the foil on the outside. This lead is connected to ground or some other low-impedance point in certain applications to either prevent noise pickup, or reduce radiated noise.

20.2 METALLIZED FILM CAPACITORS

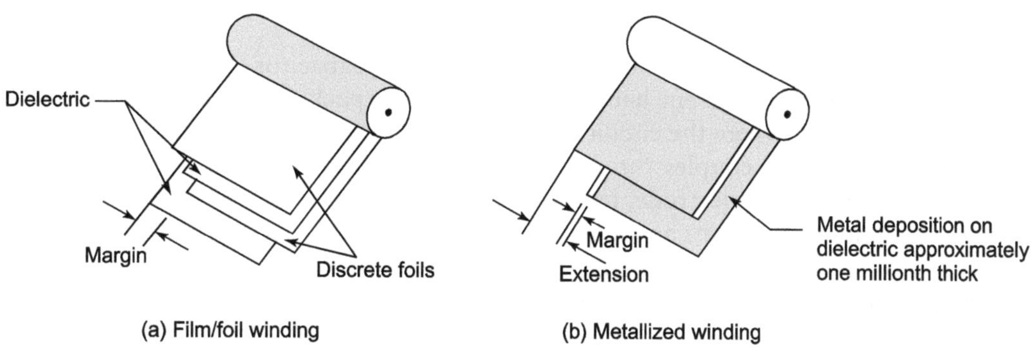

Fig. 20.3 *Basic capacitor winding systems.*

Metallized construction makes for a smaller capacitor, results in self-healing capability, and is the most common construction now used. If a spot on the dielectric breaks down, the arc will usually clear the metal deposit on film from around the hole by vaporizing it (this assumes that enough energy is available to clear the short). Some capacitor films are 'pre-cleared' during manufacturing to blow away pinholes and weak spots in the dielectric.

Metallized film capacitors tend to have a higher ESR than film/foil, however, and somewhat lower insulation resistance. Film/foil capacitors are typically 50–150% larger in volume than their metallized-film equivalent. Some types, especially for high-voltage and pulse applications, have more complicated constructions, for example, a polyester carrier film metallized on both sides and separated by a polypropylene film (the dielectric layer). Other variations include foil and metallized film layers to get reduced ESR, yet allow for self-healing, and even foil and double metallized film both in the same capacitor. In most cases, the idea behind these various constructions is to get the good self-healing of a metallized-film capacitor with the low ESR/high current capacity of film/foil.

20.3 METALLIZED ELECTRODE CARRIER (MKV) CAPACITORS

An alternative to using aluminium foil is to metallize the two sides of the dielectric by vapour deposition. The metallized carrier brings better current handling than a metallized dielectric. The metallized-carrier electrode has half the ohmic losses and twice the thermal conductivity of a single metallized electrode. In an MKV electrode, the short clearing takes place on one side (the defective side), leaving the other side unaffected. Hence capacitance loss per clearing is halved. MKV capacitors have therefore a better stability over time compared to metallized capacitors.

The size and shape of a capacitor has a direct relationship to its ability to dissipate heat and handle current. Its size directly relates to its capacitance value and its voltage rating. Voltage stress on the capacitor dielectric and the reliability considerations of the application dictate the dielectric thickness to be used for a given voltage rating. Advances in metallizing technology have improved the ability to pack more capacitance in a smaller volume for the same voltage. However, this concentrates the heat into a smaller volume and reduces the ability to dissipate the generated heat. For a general capacitor design that has evolved for the needed stress ratings, the size directly relates to the capacitance value.

The electrode system is also a source of heat within the capacitor. The current path length is critical when considering current handling and capacitor package inductance. Current path length includes the point where the circuit current enters the terminal system, flows through the termination and electrode, couples through the dielectric via the electric field, and continues out through the other set of electrode, termination, and terminal system. The user controls the external leads, circuit board paths and items related to packaging. The electrode width (material width) is the primary influence of current path width. Selecting material width requires a blend of packaging, manufacturing and cost considerations.

20.4 WOUND METALLIZED CAPACITOR MANUFACTURING PROCESS

Metallized films are wound on high speed winding machines with metallized margins of films extending out from either end by a small margin. The winding may be done on hollow plastic cores, or on a coreless mandrel. The winding ends of both films are electrically separated from

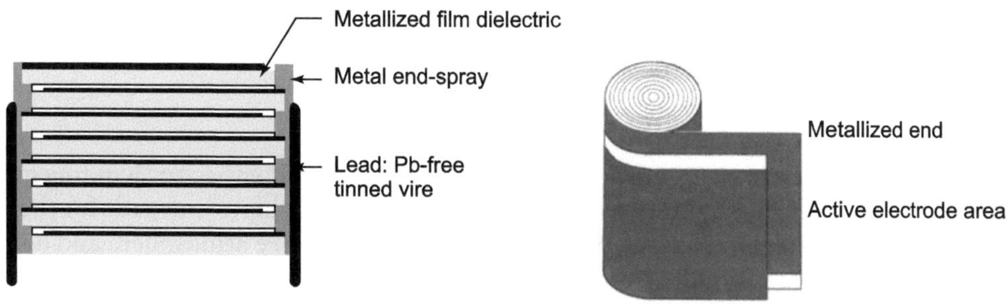

Fig. 20.4 *Metallized capacitor winding.*

each other by a separator film, or by burning out the metallized coating on one or both films. The burning is achieved by passing high enough electric current through the width of the film for the desired length at the start and end of winding. A clear burning leaves the base film fully clear of the metal, and creates an area of high insulation.

The coreless winding is made flat and compressed under pressure and temperature to give a near solid mass. An element with cores wound at high speed and under winding tension fits on the core as a solid mass. These elements are subjected to elevated temperature to solidify the mass further by shrinking. The temperature (between 80 and 130°C) and duration of this oven cure depend on the nature of the film as also its thickness. At the end of this process, the basic capacitor is ready, and further processing is done to seal it hermetically and to ensure proper terminations, casing and mounting.

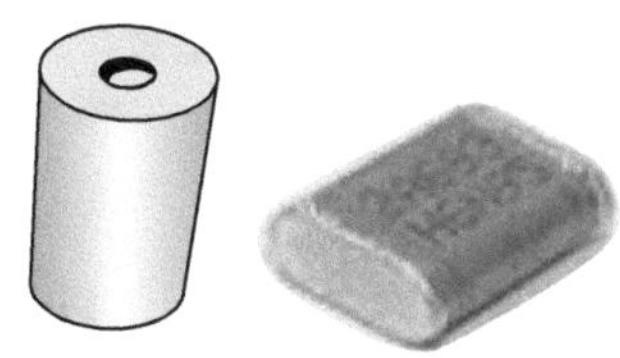

Fig. 20.5 *Round and flattened elements.*

In some cases the elements may be subjected to vacuum and oil impregnation to protect them from effects of corona and sparking and self-healing discharges at the ends by the cooling effect of oil. In high frequency applications or large value capacitors, the oil also helps remove the heat generated in capacitors. These effects have shown to extend the life of metallized capacitors significantly even in capacitors used for fans and motors.

In some higher voltage capacitors, series metallization in sections (Fig. 20.6(a)) is used on one film with metallized ends at both ends, while the second film has metallization over the full

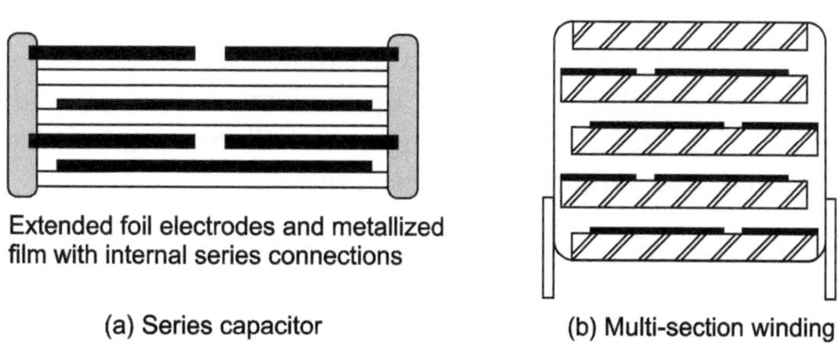

Fig. 20.6 *Series and multi-section capacitors are used for high voltage.*

area, with clear margins on both ends. Multi-section metallized films (Fig. 20.6(b)) are used to increase the voltage rating of capacitors.

[A] Winding

The windings are done on high-speed automatic and semi-automatic capacitor winding machines available from various manufacturers worldwide. The winding machines are made for the specific type of capacitor construction in view. Winding is done in a dust-free atmosphere and controlled temperature and humidity. Winding machines of an Indian manufacturer are shown below for reference.

(a) M/c MPP capacitors (b) Triple concentric winding m/c (c) Film/foil HT capacitor

Fig. 20.7 *Capacitor Winding Machines.*

Three winding machines of different types are shown in Fig. 20.7. The machine (a) winds two metallized elements simultaneously, while (b) is capable of winding up to three MPP elements on one core, one over the other. The one at (c) winds film/foil HT capacitors. The time gap between successive processes of a resin encapsulated capacitor has to be least practicable. This will completely exclude the effect of air and moisture on the capacitor life. Whatever effect the surrounding air has on the element before sealing will certainly affect the long-term performance. Thus the manufacturing process has to be managed with all care and in a controlled atmosphere to ensure long trouble-free performance. Winding being the heart of a capacitor, it needs maximum attention until the element is finally sealed.

[B] MPP capacitor flattening

Coreless elements need to be flattened (unless they are to be used as such) on a hydraulic machine at a good pressure and temperature depending upon element material and size. The winding tension should only be moderate, and not too high. After flattening, the element should not be deformable easily by hand, and should not be over-pressed.

[C] MPP capacitor end spray

The metallized elements from the oven are assembled into jigs and sprayed with zinc metal from a metal spray machine. The process is critical as zinc is sprayed on to a plastic surface. If the spray nozzle is too far away, the atomized zinc spray will not get attached. If the distance is smaller than a critical distance, it may melt the plastic surface and damage it. The distance, the air pressure and speed of zinc wire, are all important parameters to be properly controlled to get optimum results. The process was originally done using a gas torch, which was discarded in favour of arc gun spray. Metal spray was done in a spray booth. The gun was later controlled by X-Y parameter control through air pressure. Excess zinc from the booth is carried away through an exhaust system. Zinc being a noxious material, the operator must be protected from its fine particles by means of gas mask and goggles.

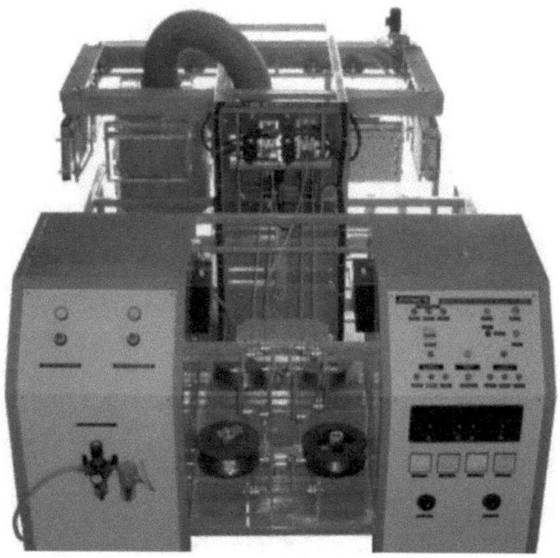

End spray double metallizing machine Sprays both ends simultaneously

Fig. 20.8 *Metal spray machine* (Source: Jognic's).

Nowadays the spray is done using automatic spray system (Fig. 20.8) and a conveyor where element in jigs are loaded outside the spray zone and carried through a continuous spray. The gun moves laterally as elements move under it. Systems are available where both ends of elements are sprayed simultaneously as the element jigs rotate and move on simultaneously through the spray zone via a conveyor.

[D] Short clearing

As a general rule the metallized capacitors have a single layer of dielectric between metallized electrodes. Since the thin dielectric film cannot be ideally perfect over the large area of the capacitor, it follows that there will be weak spots in capacitors which need to be removed to restore full dielectric properties over the entire area of film. This is done by a process called 'short clearing'. The film can have two types of weak links. At some point, there could be pinholes or conducting particles, which may act like a direct short. Others could be points of low breakdown strength due to impurities or weak spots.

The short clearing process involves two steps. First a sudden low voltage AC is applied across the element for a short time, without current limiters. The current during a direct short being heavy, it burns away and vaporizes the metal at the spot and isolates the direct short spots from the surrounding. The capacitor is then subjected to rising DC voltage from a low initial value to maximum value set according to the film and the design voltage. Weak spots at different locations have their own breakdown voltages, and hence get cleared one by one. When full voltage is reached and maintained for a given period, the capacitor element is fully cleared of all defects and is ready for use after testing.

Every event of short clearing isolates a small area around the weak spot, however small, and hence the electrode area is reduced to that extent. This is miniscule, but if the capacitor is subjected to too many surges in life, over a period, the reduction becomes measurable. The tan δ also gets affected over time. This can also happen in case of poorly made capacitors, and can be avoided by proper capacitor design for specific application and by following stringent manufacturing practices.

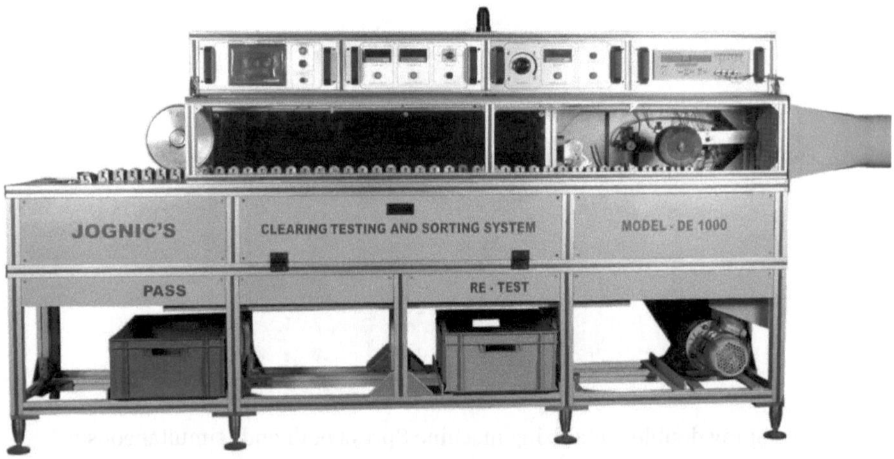

Fig. 20.9 *Automatic short clearing and test equipment for elements.*

The metal layer evaporated during short clearing gets thrown away from the spot. If the energy for vaporization is not enough, the metal may settle closer to the spot and also not vaporize fully. This creates an area of low insulation (IR), through which leakage current can flow. Too many such spots add to the ESR seriously, causing heat generation and thermal runaway, and the capacitor gets damaged irreparably.

Circuit current entering through the termination spreads via the end spray connection to the entire length of the winding. In the broad sense, it does not matter if it is a wound or stacked capacitor; the current path length is the same. In some cases the stacked construction may provide a more intimate contact between the end spray and the electrode, and hence a better dv/dt, but the current path length is still critical. As a general rule, the shorter the current path length, the lower is the inductance of the capacitor and the lower the I^2R losses in the electrode.

[E] End soldering and assembly

The elements after short clearing and testing are soldered at both sprayed ends with lead wires, external or internal, depending on design requirements. In case terminals are to be brought out directly from winding, insulated lead wires of suitable length and size may be used. If external terminations are separate, the inner leads are soldered or joined to these terminals located on the container cover. This generally is the case for electrical capacitors assembled in cans or boxes.

In case of some auto condensers, in most dip type electronic capacitors or in small box type electronic capacitors, the wire leads are welded to sprayed ends and brought out directly for PCB mounting.

Plastic cans or boxes are common for general use, particularly in un-impregnated dry capacitors. These are assembled in cans and encapsulated fully with epoxy or polyurethane (PU) resins. This ensures hermetic sealing and capacitor is insulated from atmospheric moisture or air in the environment in service. In dip-coated capacitors, the elements after thermal pressing are coated under vacuum with a varnish, dipped in epoxy resin and cured in oven to get PCB-mount ready capacitors.

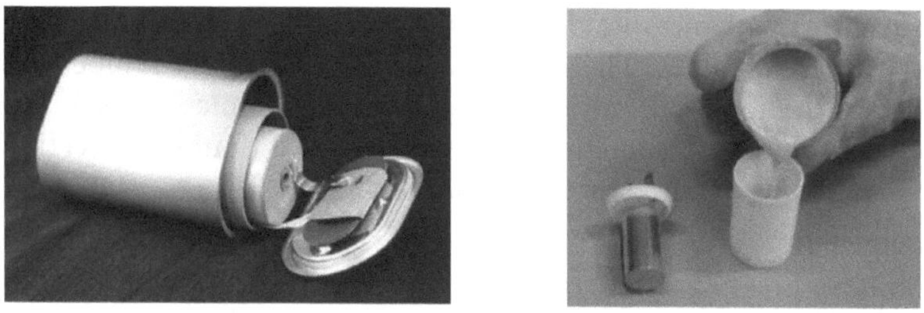

(a) Element assembly in can (b) Resin sealing process

Fig. 20.10 *Soldered assembly and resin sealing.*

The use of filling materials in capacitors is necessary in order to insulate the capacitor electrodes from oxygen, humidity, and other environmental interference. Without such insulation, the metal coating would corrode, an increasing number of partial discharges would occur, the capacitor would lose more and more of its capacitance, and suffer increased dielectric losses and a reduced operating life.

[F] Oil impregnation of MPP capacitors

Large value or power frequency capacitors are often assembled in metal cans or boxes. The elements are insulated from the container from all sides and inner tabs are connected to insulated terminals on cover. Covers are crimped, welded or soldered to the container. An inlet for oil is provided on the cover. The containers are subjected to vacuum and temperature to remove air and moisture, and oil is allowed in them to completely fill the container. The oil inlet is then sealed to complete hermetic sealing.

An elaborate vacuum-drying procedure is initiated immediately after insertion of the capacitor elements into the aluminum case and biologically degradable plant oil or solidifying PU resin is introduced. This provides better protection to the winding from environmental influence and provides an extended life expectancy and stable capacitance.

[G] In-built protection in capacitors

There can be up to four levels of protections provided in an MPP capacitor.

[1] Self-healing properties

This property of metallized film capacitors has been discussed separately in detail. In the event of surges, a miniscule area of metallized electrode evaporates away and the capacitor remains in sound service condition.

[2] Internal fuse

Non-metallized capacitors using separate electrodes use fuses in leads of elements. Particularly in large capacitors or power capacitors, internal fuses are provided in series with each winding element, and in the event of a fault being developed in any element, the series fuse blows off, keeping other elements in working condition. The size and rating of fuses are carefully selected to avoid faulty operation or momentary surge currents.

National standards establish upper limits of temperature, voltage and current which a power capacitor may bear. If these limits are exceeded in a permanent way, the self-healing mechanism is not able to act effectively and the overloaded capacitor can fail.

Unlike the capacitors with metallic electrodes, self-healing capacitors present fault impedance that can have very different values. These impedance values produce fault currents that can vary between the order of the rated current and several hundred times of itself. To avoid this type of failure, power capacitors are sometimes provided with a double protection system in each element – internal fuse plus overpressure system.

[3] Pressure disconnector

In the event of failure, pressure inside a capacitor goes up. A pressure disconnector snaps the internal connection wires to elements when the pressure inside a capacitor exceeds a certain value. This may be brought about by can deformation under pressure, or by the action of pressure sensor fitted on the can.

[4] Use of segmented metallized films

The film is formed of separate segmented sections connected to each other with a narrow path, which acts as a fuse in the event of failure of a section.

The operating temperature of the capacitor is to be observed, which means the maximum temperature at the core of the winding has to remain within permissible limits. This is done by adjusting the temperature by considering core-to-skin and skin-to-ambient differentials. Reliability essentially doubles for every 10°C drop – so even small improvements in cooling help. High-power operation requires planning for reliability.

Capacitor size reductions decrease the surface area for heat removal. In turn, this increases the skin temperature for a given set of operating conditions. Although size reductions are desirable, they should be balanced against reliability. Forced air-cooling helps. Even small amounts can increase the heat removal from the capacitor surface.

20.5 TESTING OF CAPACITORS

Capacitors have to be tested during and after manufacture, and by the user on sampling basis, to ascertain that no deficiencies are present in them, which may hamper their long-term performance in service conditions for which they have been targeted. Different types of capacitors have varying levels and types of tests, and most have been well defined in the national and international standards. Some tests are empirical, based on manufacturers' experience and set norms, which are not mentioned in any standards, but are practically relevant.

Testing cannot transform a capacitor designed beyond safety limits, and a well-designed capacitor inadequately tested is equally unreliable. These two operations should be viewed as separate but interwined concepts with absolute availability as the goal for both. The present discussion will limit itself to capacitors in power frequency applications used in the electrical industry.

Routine tests on capacitors are done on every piece during various stages of manufacture, to ensure there are no major defects left during manufacturing process and that they generally comply with relevant specifications. Type tests are done on a sample basis, with a dual purpose. First, any new design (or a design change) is validated by a manufacturer by type tests to confirm the soundness of the design to meet or exceed the minimum standard requirements, and to ensure long-term performance in the designated environmental and load conditions. Second, these are done on production samples from time to time to ensure their continued quality and conformity with design data and to ensure uniformity in production.

[A] Various test equipment must be specifically designed for capacitors on account of special requirements put forth by capacitive load conditions on supply. For example, a capacitor of 25 KVAR 440 V, when tested at 770 V AC between two terminals (which works out to a load of 215 μF), draws a current of over 91 A from supply at 440 V mains, which is practically difficult. Similarly, a 72 μF motor capacitor will need mains current of 31 A at 770 V or over 35 A at 880 V AC. These problems arise in both routine as well as type tests, for tests between two terminals or in three phase tests. Endurance tests demand 1.25 to 1.35 times rated voltage to be applied for several days, even up to 1500 hours or more, and this puts a heavy load on the supply system.

This load needs to be compensated at the test location itself, as the load cannot be passed on to the supply cables within the manufacturing unit. This is apart from the fact that the manufacturer will end up paying for low power factor for the capacitive load being fed into the supply utility mains. Reactors are used to compensate the capacitive currents to keep the currents within manageable limits wherever capacitors of higher values involving considerable currents are tested. Values of reactors need to be calculated separately for each rating of capacitors under test.

Figure 20.11 shows automatic testing machines for MPP capacitors at final stages of capacitor production. These test capacitors for value, tan δ, HT between terminals for specified time, and sort them in different bins.

Fig. 20.11 *Automatic routine testing machine for finished capacitors.*

[B] Often stray capacitance, resistance and inductance affect the capacitance measurements, particularly at higher frequencies. Most power frequency capacitors specify capacitance measurements at power frequencies not exceeding 120 Hz for this purpose. Since capacitor values are often affected by test voltage magnitude, they are specified to be tested at rated voltage and frequency. Again, this also requires special equipment design, since most of the standard RLC instruments do measurements at low voltage under 1 V and at 1 kHz. However, measurements commonly in practice are not much affected by voltage, and capacitance measurements may be done with standard accurate measuring instruments at power frequency. Further, these instruments are specially designed for capacitors, as most common RLC instruments do not measure tan δ with required accuracy for electrostatic capacitors.

A list of Indian and IEC Standards is given in separate annexure for reference. Table 20.1 below produces some tests for capacitors of common electrical grade capacitors for ready reference.

Table 20.1 *Tests for Electrical Capacitors as per Indian Standard Specifications*

Application	*Fans (440 V)*	*Motors (440 V)*	*Lighting (250 V)*	*KVAR (440 V) Self-healing type*	*KVAR Non-self-healing type*
IS No.	1709 – 1984	2993 – 1998	1569 – 1976	13340 – 1998 13341 – 1998	13585 – 1994
AC HV test between terminals	1.5 × Vn for 10 s	2 × Vn for 10 s	1.5 × Vn for 10 s	1.75 Vn for 10 s	2.15 Vn AC or 4.3 Vn DC for 10 s
Test voltage	660 V	880 V	375 V	770 V	950 V AC/1900 V DC
AC HV test term. and case	2000 V for 1 min.	2000 V for 1 min.	2000 V for 1 min.	3 kV for 10 s	3 kV for 10 s
Capacitance at rated voltage	± 5%	± 5%	± 10%	–5% to + 10%	–5% to + 10%
Tan δ max. at rated voltage	0.002	0.002	Not specified	0.0025	0.0010
Ins. res. at 500 V (1) Term-term (2) Term-body	 3000 s 100 MΩ	 N.A.	 N.A.	 N.A.	 N.A.
Leakage test for oil-filled capacitors	5° above Tc	10° above Tc	10° above Tc	10° above Tc	10° above Tc
Endurance test	1.25 Vn × 500 h	1.25 Vn × 300 to 3000 h as per on category	1.25 Vn × 500 h	1.25 Vn × 1500 h, with 1000 charge-discharge cycles after 750 h	1.25 Vn × 1500 h, with 1000 charge-discharge cycles after 750 h
Self-healing test for metallized capacitors	1.4 Vn 1h 1.6 Vn 1 h 1.8 Vn 1 h 2.0 Vn 1 h Safe mode failure criterion.	1.75 Vn for 10 s, then increase to max 3.5 Vn until 5 self-healing discharges	1.75 Vn for 10 s, then increase to max 3.5 Vn until 5 self-healing discharges	1.75 Vn for 10 s, then increase to max 3.5 Vn until 5 self-healing discharges	N.A.
Destruction test	10 Vn DC, then 1.3 × Vn cycles. Failure should be safe mode	10 Vn DC, then 1.3 × Vn cycles. Failure should be safe mode	N.A.	10 Vn DC, then 1.3 × Vn cycles. Failure should be safe mode	10 Vn DC, then 1.3 × Vn cycles. Failure should be safe mode
Thermal stability at Tc + 5°C	N.A.	~1.2 Vn for 48 – 72 h	1.25 Vn 6 h	~1.2 Vn for 48 – 72 h	~1.2 Vn for 48 – 72 h

Tests on terminals, solderability etc. are not included for brevity. Full test details are also excluded for the same reason. The relevant standards may be referred to for these. The tests are minimum standards set under relevant standards.

[C] Manufacturers will try to exceed the standards above by devising their own in-house empirical tests. Tests like direct charge-discharge tests are conducted by many, involving cycles of charging capacitors to DC voltage of 100 V μm^{-1} of film thickness and discharging through very low impedance, and checking the capacitance drop and tan δ variation after a specified number of cycles. Higher voltage KVAR capacitors have not been considered, being beyond the scope of discussions. Electronic grade capacitors requiring special additional tests are also not included.

Usually capacitance values are within specified tolerance limits as mentioned above, but certain equipment designs may require much closer tolerances. Capacitors used for comparison will have to be within very close tolerance, of even up to 0.1–0.2%. These are then used with component comparators. Such capacitors are either measured on well-calibrated in-house equipment, or in a calibration laboratory.

If the capacitors pass all the tests mentioned as above in relevant standard on regular basis, and if the manufacturer keeps conducting the tests at regular intervals, it can be safely assumed that all capacitors from his unit will have a uniformly acceptable quality. For this reason, a manufacturer seeking an ISI mark or any relevant international standard compliance mark is required to carry out such tests at minimum specified intervals, and maintain a record.

[D] For tests as per IS: 2993, current calculations of AC high voltage testing of capacitors between terminals for 440 V rated capacitors work out as per Table 20.2 below:

Table 20.2 *AC High Voltage Test Parameters*

C μF	*A at 880 V*	*1-phase supply current A at 220 V AC*	*Load at 400 V primary*
10	2.8	11.2	6.2
20	5.6	22.4	12.3
30	8.4	34.0	18.5
36	10.1	40.5	22.5
72	20.2	81	45

The supply mains currents are of quite high order, as can be seen from above. Standard test equipment for endurance test and high voltage tests are not readily available, and may have to be designed in-house, or procured from suppliers specialized in such equipment.

Endurance tests and thermal stability on KVAR capacitors also involves measuring container hot spot temperature inside the test oven, for which a suitable arrangement must be devised.

Certain capacitors like fan regulator capacitors are not covered by such standards, and separate tests have to be devised to confirm their performance under rated as well as severest working conditions. Otherwise most applications will have their own standards established by national bodies or best acceptable industry practice.

[E] Destruction test: This test is done on a number of capacitors, the purpose being to verify that the capacitors fail only in safe mode. Conditions are created to make the capacitor fail, by applying very high DC voltages and then applying rated AC voltage. If the capacitor does not explode, burn, or cause fire producing heat, it passes the test. A DC voltage source feeds a voltage equal to 10 times the AC rated voltage (with current limiting resistor to safeguard the source). An AC rated voltage is then applied for specified time.The capacitor cannot withstand this voltage, and usually opens up by self-healing, by its in-built pressure interrupter, or by a thermal fuse. An AC rated voltage application thereafter will cause the interruption to happen, or if capacitor is not capable of doing so, a muslin cloth around it will char, indicate high body temperatures, indicating failure.

[F] Self-healing test: This test is performed to check the soundness of self-healing property, and that the capacitance loss due to self-healing on account of surges or pulse conditions in service does not appreciably alter the capacitance value. This test was earlier done with DC voltage, but the revised standards specify AC voltages for the test to simulate actual working conditions. This however puts a demand on current and instruments and supplies have to be designed for this. The test is generally done as follows:

- Check the initial capacitance and tan δ values.
- Connect the capacitor to a variable supply source, whose voltage can go to minimum 3.5 times the rated capacitor voltage.
- Connect self-healing discharge detection device in the circuit. The detection circuit may detect the self-healing discharges by the acoustic method or by an oscilloscope.
- Apply an AC voltage double the rated voltage for 10 s across capacitor terminals. If less than 5 discharges are detected, increase the voltage gradually, to a maximum of $3.5 \times V_n$, until 5 discharges are detected. In case 5 discharges are not detected, the test will be repeated with another capacitor.
- Reduce the voltage to 0.8 times the maximum voltage reached, and maintain for 10 s. There should be no more discharges during this period.
- Measure the capacitance and tan δ values. The variations observed should be within specified limits.

[G] Solderability of leaded capacitors: The solderability of the terminal leads is tested in accordance with IEC 60068-2-20, test Ta, method 1. Before the solderability test is carried out, the terminals are subjected to an accelerated ageing procedure (in accordance with IEC 60068-2-2: 4-h exposure to dry heat at 155°C). Since the ageing temperature is far higher than the upper category temperature of the capacitors, the terminal wires should be cut off from the capacitor before the ageing procedure in order to prevent the solderability being impaired by the products of any capacitor decomposition that might occur.
Solder bath temperature: (235 ± 5)°C
Immersion time: (2.0 ± 0.5) s

Immersion depth: Distance from standoff surface or capacitor body: (2.0 + 0/–0.5) mm

Evaluation criterion: Wetting of wire surface by new solder ≥ 90%, with free-flowing solder.

The discussion above gives an idea of test procedure for electrical grade capacitors for applications mentioned. Tests for all types of capacitors have been specified in relevant standards, and the manufacturer carries out these tests accordingly. The user has to be aware of the test standards so as to ensure an acceptable quality of the product.

21

SELECTION GUIDE FOR CAPACITORS

21.1 INTRODUCTION

Capacitors are used in wide temperature extremes and both undersea as well as atop high mountains at very low pressures. Many of the places are difficult for maintenance of equipment. Some demand very high reliability and very good mean time between failures (MTBF). Some demand stability over a large frequency range, while others need good capacitance stability with low loss. Working voltages may vary from 2.5 V DC to 11 kV DC and above, and from 16 V AC to 11 kV AC. frequency of application may be from a few cycles to megahertz range. Pulsed power and IGBT/IGCT circuits place capacitors in the most vulnerable conditions, and these are becoming commonplace.

There are many types and grades of capacitors with a variety of dielectrics, electrodes, lead configurations, mountings, sizes etc. The choice of capacitor for a particular application or use is of paramount importance. Even if a correct value is chosen for a particular application or use, the selection of the correct type is of equal importance. It is critical that the capacitor is matched to the application.

21.2 CONSIDERATIONS FOR SELECTING A CAPACITOR

When a circuit is being designed, it is necessary to consider the specific application requirements and working conditions as also the expectations from the equipment before arriving at a suitable choice of capacitor – its type, dielectric, terminations, size constraints etc., while giving due weightage to cost. It will be pertinent to go through some of the factors governing this choice.

[A] Electrical parameters

- **Capacitance value and loss angle:** All dielectrics have different loss angles, and often a certain value of capacitance may not be available in a dielectric.
- **Voltage rating (DC/AC):** Voltages from as low as 2.5 V to as high as a few thousand volts are possible, and each voltage range may demand different material. Some are not suitable for DC, while others may not go with AC.

- **Current** (**DC pulse/AC**): Pulsed power and IGBT applications require high surge current capability and a high dv/dt rating.
- **Insulation resistance:** IR value may affect the performance in a number of circuits requiring high input impedance, or those used in sensitive environments.
- **Equivalent series resistance** (ESR): ESR should be as low as possible. Sometimes a compromise may be made, if it does not affect the performance.
- **Dissipation factor** (**DF**): Heat dissipation may become a significant consideration. This may also add to damping effect, or play spoilsport by introducing unwanted AC resistance.
- **Capacitance change with temperature:** This is particularly important where capacitance value change can affect the circuit performance.

[B] Mechanical parameters

- **Size:** Available space in the equipment or the PCB.
- **Terminal configuration:** Axial leads add to the flexibility of usage, but radial leads mean minimum lead length and a low ESR and ESL, in addition to ease of PCB mounting. Large capacitors may be provided with solder terminals, wire leads, nuts and bolts, spring clamp (Amp) type terminals. Many electrolytic capacitors have the can as a negative terminal.
- **Type of mounting:** Large capacitors have to be fixed or mounted on the PCB or in equipment with brackets, or with screws, nuts or bolts. PCB mount may not generally need additional mounting arrangement.

[C] Environmental parameters

- Operating temperature range
- Indoor or outdoor usage
- Moisture resistance
- Shock and vibration
- Chemical resistance

[D] Applications

The following points must be considered:

- **Duty:** continuous, short time, intermittent, cyclic, occasional
- **Reliability:** Military, medical, telecommunication, aerospace applications demand extreme reliability. Cost becomes secondary.
- **Safety:** Failure of capacitors may cause serious accidents. A capacitor failure at the end of its useful life may be required not to harm the surrounding.
- **Transients:** Occurrence of surges, lightning or other transient voltage or currents.

Each capacitor type, however, has certain unique characteristics and, even within a particular type, the appropriateness for a given application will depend upon the specific dielectric. For example, tantalum capacitors have no wear-out mechanism and are particularly suited for applications requiring long life and stability. The expected lifetime of aluminium capacitors can be doubled for each 10°C of temperature reduction, but it is important to keep these devices away from cleaning solvents.

It is not always easy to determine which dielectric type will best suit a given application. The application may require a minimum capacitance value or very low ESR. The capacitor cost and size must also be considered, as well as its packaging type. End-of-life reliability issues may also be important. Each capacitor type has its own set of characteristics that will make it the preferred choice for a given application.

Capacitors are widely used in electronic circuits to perform a variety of tasks, such as smoothing, filtering, bypassing etc.

21.3 SPECIAL CAPACITORS

[A] Feed-through capacitors

These capacitors are used in cases where conventional capacitors are not effective for filtering at high radio frequencies. Feed-through capacitors are three terminal devices that do not exhibit the series-resonant characteristic of the conventional capacitor. This enables them to suppress radio-frequency interference over a wide range of frequencies and they are especially valuable in filtering power-supply and control-circuit wiring in shielded high-frequency equipment.

[B] High-energy storage capacitors

These capacitors are constructed with oil-impregnated paper and/or film dielectrics. Their primary use is for pulse forming networks which employ voltages greater than 1000 V. For slightly lower voltages special electrolytic capacitors can be used.

[C] Commutation capacitors

These are constructed from oil-impregnated paper and film dielectrics. They are mainly used in triggering circuits since they are characterized by fast rise times (the time it takes capacitor to rise from 10% to 90% of its maximum voltage) and high current transients and peak voltages associated with switching.

21.4 PACKAGING

Capacitors come in a wide variety of packaging styles. The most common styles are moulded, glass-encased, chip, potted, coated, and dual-in-line packaging (DIP). Moulded capacitors are rectangular-chip capacitors that can be moulded into radial or axial-lead rectangular packages or axial-lead cylindrical packages. Glass-encased capacitors can be single or multi-layered chips with axial leads attached sealed into a glass tube. These look a lot like moulded capacitors. Chip capacitors are thin, flat rectangular capacitors without leads or body encasement so that they may be put into microelectronic circuits. Potted capacitors, in many ways, are synonymous with moulded capacitors. The only difference is that potted capacitors are oven cured. Coated capacitors, more commonly known as dipped capacitors, come in rectangular and disk styles with radial leads and are dipped in liquid resin. Coated capacitors find great usage where exact dimensions can be compromised. DIP capacitors are single or multi-layered capacitors processed into integrated-circuit type packages. Mica chips come in button styles. This package is composed of a stack of silvered-mica disks connected in parallel.

21.5 CAPACITOR RATINGS

Capacitor ratings as regards voltage, current and frequency, along with working environments need to be carefully considered while choosing a capacitor for an application, as often these may be critical and adversely affect the performance.

21.5.1 Voltage Ratings

There are two types of voltage ratings to consider when evaluating capacitor performance: DC and surge voltage and AC voltage. In the case of DC and surge voltage ratings, the thickness of the dielectric determines the maximum surge and DC voltages that may be applied. AC voltage ratings are usually specified for ceramic capacitors. This rating corresponds to the AC voltage required to make the sum of the given DC voltage and AC voltage less than the rated DC voltage.

In addition to these ratings there are certain types of electrolytic capacitors in which the applied voltage is of primary concern. Electrolytic capacitors are sensitive to the effects of voltage because they are highly polarized devices. Even if applied voltage is less than the maximum voltage specified, the voltage drop across the ESR of the capacitor will shorten the capacitor's life expectancy through an accelerated effect of internal heating.

21.5.2 Current Ratings

Current ratings to consider are the leakage and ripple currents. Leakage current is the stray DC current of relatively small value which flows through the capacitor when voltage is applied across the terminals. Ripple current is the AC component of a unidirectional current. For electrolytic capacitors, there is also a maximum allowable charge and discharge current rating.

21.5.3 Frequency

Since there is an internal inductance in a capacitor there will be a resonant frequency. Depending on capacitor type, this frequency may or may not fall in a range that is a problem for the designer. This problem would arise because the designer would want the capacitor to block or minimize DC current, and at resonance the internal impedance is a minimum which causes maximum DC current.

21.5.4 Dissipation Factor

This can be important in many cases, as apart from heat dissipation issues, circuit performance itself may be affected because of equivalent resistance of capacitor.

21.5.5 Dynamic Environments

Dynamic environments can mechanically damage or destroy a capacitor. The main dynamic environments are in the form of shock, vibration and acceleration. The movement of a capacitor assembly inside a case can cause capacitance fluctuations, electrode attachment failures, and dielectric and insulation failures. A capacitor's susceptibility to dynamic environments is dependent on its physical construction; the larger the complex elements in the capacitor, the lower the frequency of response of the elements.

21.5.6 Barometric Pressure

The pressure dictates the altitude at which a hermetically sealed capacitor can safely operate. This altitude is dependent on the design of the end-seal case-wall, the voltage at which the capacitor will be operated, and the type of impregnant used in the dielectric material. As the altitude increases, the dielectric strength across the end-seal will decrease. If the altitude is increased with barometric pressure reduced, then the pressure inside the capacitor will increase the mechanical stress on the case and seal until failure occurs.

21.5.7 Radiation

Radiation particles can degrade the electrical performance of capacitors. The leading cause of radiation-induced capacitor defects is dimensional changes in the inter-electrode spacing. This change is due to gas evolution and swelling. Changes due to radiation are more pronounced in organic-dielectric capacitors. Capacitors using organic materials like polystyrene, polyethylene terephthalate and polyethylene are less satisfactory in a radiation environment by nearly a factor of ten than those capacitors employing inorganic dielectrics. The electrolytic capacitors (aluminium and tantalum) are capable of extended radiation exposure with tantalum being more radiation resistant. Another defect from radiation occurs when the dielectric in the capacitor experiences a noticeable increase in its conductivity in an ionizing-radiation environment. This results in very dangerous discharging of a charged capacitor.

21.5.8 Frequency Response

Impedance of a capacitor varies with frequency as shown in Fig. 21.1.

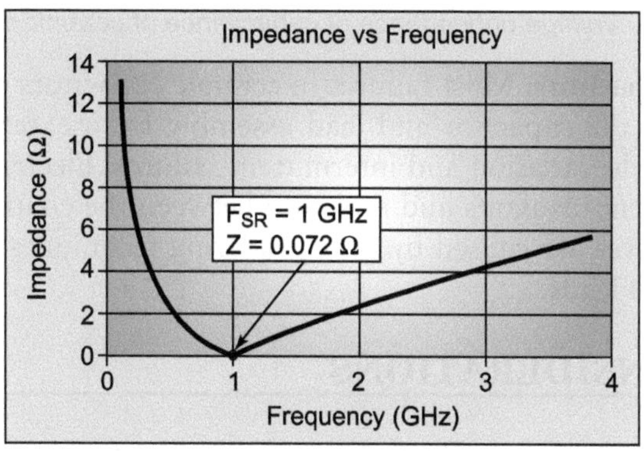

Fig. 21.1 *Frequency dependence of impedance*

Dissipation factor or ESR could become a major consideration in some cases. This can depend on frequency as well as ripple currents. Table 21.1 depicts data on different types of dielectrics.

Table 21.1 *Dissipation Factor @ 25°C Ripple Current for Different Types of Dielectrics*

	Value	*DC V*	*(1 kHz)*	*(100 kHz)*	*(100 Hz)*
Aluminium	1000 µf	6	10.0%	15.70%	0.025
Tantalum	680 µf	6	6.0%	9.80%	0.023
Ceramic-X7R	10 µf	100	2.5%	0.94%	0.015
MLP-PET	10 µf	100	1.0%	0.69%	0.011

21.5.9 Voltage Coefficient of Capacitance

For ceramic capacitors, the voltage applied to the capacitor also affects the capacitance value (Fig. 21.2). The electric field strength across the dielectric changes the effective dielectric constant K of the material. This is not an issue for stable dielectrics such as NP0 or when the percentage of applied voltage is low when compared to the rated voltage. This characteristic is termed as voltage coefficient of capacitance (VCC).

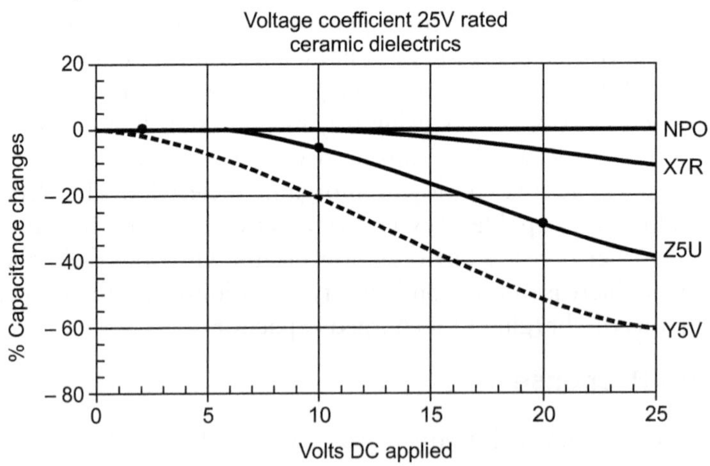

Fig. 21.2 *Voltage dependence of capacitance of ceramic capacitors.*

Note for Ceramic Capacitors: Most failures in ceramic capacitors are caused by encasement materials used to protect the capacitor and lead assembly from external environments. Other failures include electrical degradation and intermittent failures. Electrical degradation is caused by thermal expansion of encapsulants and moisture between the coating and capacitor section. Intermittent or open failures are caused by poor soldering techniques and terminal design that result in loose or detached leads.

21.6 DESIGN CONSIDERATIONS

The reliability of a capacitor is dependent upon the degree of success achieved in housing the capacitor element in a mechanically and environmentally secure enclosure. Capacitors with internal lead construction must be mechanically and electrically sound before the encasement is applied. Encapsulated, dipped or moulded capacitors cannot withstand dynamic environments such as high levels of shock and vibration. For mechanical integrity, metallurgical bonds and reinforcing materials should be used.

When considering which capacitor best performs a specific circuit task, options depend on the cost of the capacitor and the capacitor's physical and electrical properties vis-à-vis the task it has to perform. If precision is a must, then mica, glass, ceramic and film (polystyrene) capacitors may be employed, which possess exceptional capacitance stability with respect to temperature, voltage, frequency and life. Semi-precision circuits can use paper/plastic film capacitors (with foil or metallized dielectric), and these presently constitute a large portion of applications. If precision is of no importance whatsoever, then the least expensive general purpose capacitors are selected, and these may deliver satisfactory performance.

Where suppression of radio-frequency interference is required, RFI and feed-through capacitors are the best. For heavy currents (60 – 40 Hz power supplies), paper or film dielectric capacitors should be used for suppression, and ceramic and button-mica high-frequency capacitors will work for low currents. Ceramic chip capacitors are electrically and physically the best suited for microelectronic circuits. If a capacitor needs to be used as a transmitter, then gas, vacuum or ceramic capacitors may be advised. These capacitors possess the necessary high radio frequency (RF) power-handling capability, high RF current and voltage rating, low loss, low internal inductance and very low ESR.

On the other end of the spectrum, cold temperatures can present problems as well. Electrolytic capacitors change their capacitance significantly on exposure to temperatures below 25°C. Aluminium electrolytics lose capacitance at –55°C. Equipment at low temperatures should be given time for the capacitance to rise once the equipment has been powered up.

Capacitors used in bypass applications are used as shunt elements and carry RF energy from a specific point in the circuit to ground. Proper selection of a bypass capacitor will provide a very low impedance path to ground. A real capacitor will exhibit some impedance to ground due to its reactance and inherent parasitic elements. Satisfying capacitive bypass application requirements requires careful analysis of various frequency dependent capacitor parameters such as series resonant frequency (FSR), equivalent series resistance (ESR), and the magnitude of the impedance. The ESR and impedance should always be evaluated at the operating frequency.

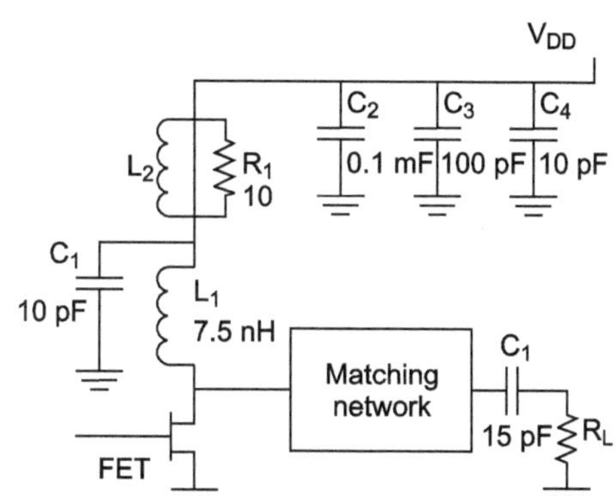

Fig. 21.3 *Bypass capacitors in a 1.9 GHz FET broadband bias network*

The capacitors in Fig. 21.3 serve to suppress RF energy from getting onto the VDD supply line while providing high impedance at the drain in order to maintain optimum in-band RF gain. They also function to keep noise generated by the power supply from appearing on the drain of the FET. High-speed switching environments created by switch mode power supplies (SMPS) will generate noise on VDD supply lines. Instantaneous current generated by fast rising and falling switch pulse edges can easily cause the VDD supply line to ring. The resultant noise can include frequencies of up to several hundred megahertz. RF noise generated by SMPS switching is continuous and will generally occur up to frequencies equal to 0.35/PE, where PE = pulse rise or fall time (s). For example a switched pulse with a rise and fall time of 1.5 ns will yield spurious spectral components up to 233 MHz.

Drain Bias Network

As illustrated in Fig. 21.3, the FET's drain bias network consists of series inductive elements having an impedance of $1/\omega L$ and shunt capacitive elements with an impedance of $1/\omega C$. Proper selection of bypass capacitors in the bias network is essential as they will serve to decouple RF energy from the VDD supply line to ground over a wide range of frequencies.

Since capacitors exhibit a small parasitic inductance there is an associated series (self) resonant frequency F_{SR} where the magnitude of the inductive and capacitive reactances is equal and hence the net impedance is equal to a small ESR value. Accordingly one will ideally select a capacitor with a low F_{SR} at or close to the desired bypass frequency. This preference is based on establishing a low impedance path with minimal or zero net reactance thereby making it ideal for bypassing applications.

Parallel Resonance Frequency F_{PR} usually occurs at more than twice F_{SR} for most multilayer ceramic capacitors. At the capacitor's FPR, the impedance is likely to be high and inductive (R+jL) and may not provide an adequate RF path to ground. To alleviate this, several capacitors are selected such that their self-resonant frequencies are staggered in order to cover a wide range of frequencies with reasonably low loss. The number of required capacitive elements depends on the loss and impedance characteristics of each element over the intended frequency band segments.

21.7 KNOW YOUR LOAD

Charge a bulk capacitance to an initial high voltage, and allow it to discharge to a pre-defined level as current is being delivered to the load during the temporary overload condition. At the end of this overload (pulse) the capacitor will be depleted to V_f and will need to be recharged to V_i before the next discharge cycle. The amount of power the bulk storage capacitance must support is equal to the power delivered to the load, minus that supplied by the input source during discharge. The efficiencies of switching converters should be considered so as to properly estimate the storage capacitor required.

When deciding to use a capacitor bank, one must decide how much discharging is permissible. One way to decide this is to directly connect the load to the capacitor bank. The voltage swing on the capacitors must be within the operating limits. Typical semiconductor loads can only tolerate a 3–5% range around their set voltage. This severely reduces the permissible voltage drop and needs the storage capacitance to be quite large. The benefit is that no additional post regulation of the capacitors voltage is required.

21.8 SPECIAL DATA FOR KVAR CAPACITORS

In addition to the above, for KVAR capacitors, it is necessary to note the following:

- Daily/yearly fluctuations of temperature and maximum service temperature
- Rain/monsoon conditions
- Most capacitors are for indoor mounting, except those designed for outdoors. The location of capacitors and adequate ventilation available should be checked.
- Incidents of lightning/surge conditions
- Accessibility for maintenance
- Presence of harmonics
- Voltage fluctuations encountered

A proper evaluation of capacitor requirements and their selection is possible only after a careful examination of these factors.

22

CAPACITOR FAILURES AND THEIR MITIGATION

Capacitors, like most other components, are likely to fail in service at some time or the other. It is essential to design them for a reasonably long life. The user must also take certain steps to avoid failures due to handling or usage. It is therefore necessary to understand the causes and mechanism of failure.

The lifetime of a capacitor is statistically predictable and is dependent on voltage, currents and temperature stresses to which it is subjected. The manufacturer carries out endurance tests and type tests to ensure a reasonably long life. The user must see that working conditions remain well within the capacitor specifications. Further, the user himself can ascertain the overall performance by testing his product for any anomalies. Capacitor technology has advanced to a point where most failures can be avoided by proper system design. It is therefore necessary to understand the mechanism of failures in capacitors.

22.1 FAILURES DUE TO MANUFACTURING DEFICIENCIES

Paper/foil or film/foil capacitors mostly fail in short circuit mode. There is no self-healing mechanism, and extreme stress at a weak spot causes an irreversible breakdown of insulation. The capacitor then gives a short circuit path to the current at this point, and the resulting heavy current, in an explosion, sometimes leading to accidents or fires.

[1] The manufacturer must take all care to avoid any chance of impurities being introduced during capacitor manufacture. The paper/film, oil (or the impregnant used) must be free of any traces of contaminants, moisture or dust particles. The aluminium foil must also be pure and free of any dust particles etc. The manufacturing process is monitored to ensure the purity of materials throughout until the capacitor is finally sealed.

[2] The electric field has to be uniform over the entire surface and thickness of the dielectric. Any sharp edges or burr will have high stress concentration, accelerating the failure. An electric field tends to have a discontinuity in uniformity near electrode edges, and free margins at either end (distance between two electrode edges) play an important part. In this area the dielectric is not in play, and failure or spark may occur over the air/oil medium. The terminal extensions also come out through this area. The contact points of

any terminations inserted in the winding are also points of stress concentration, and care must be taken to cause minimal disturbance in the field at these points. Extended foil construction scores on this point.

[3] Inadequate control on winding parameters. Uneven or non-uniform winding tension causes wrinkles/creases in winding, as also displacements or slippages. These create weak spots, where failures occur due to uneven voltage stresses or flashovers. High winding tensions due to improper machine setting or freezes of rollers, especially dancing rollers on tension controller sensors, cause excessive localized pressures. Winding stability critically depends on winding parameters, aspect ratio (diameter/width), and thermal treatment efficiency.

Temperatures up to 6000 K and pressures up to 300 bars for a few nano seconds are encountered at the self-healing site, resulting in deposition of carbon. The carbon deposition and damage to the adjoining region of dielectric layers results in areas of low insulation resistance. Currents flowing at the site of self-healing under such conditions cause premature failure of the dielectric.

[4] Zinc metal is sprayed on the ends of the element to make continuous good electrical contact with electrode metallization on the dielectric. Uniform registration of film is required for proper adhesion of sprayed metal. Too hot a metal causes shrinkage or even melting of film. Too cold a metal spray results in weak mechanical contact. Changes in air pressure during spraying process cause variation in particle size and temperature of sprayed metal, both of which affect the contact quality directly. The rate of deposition of metal affects the temperature of windings.

[5] Quality of each raw material has to be consistently good, as purity of materials is vital for long-term performance of capacitor. Minor variations in dielectric properties can have an adverse effect on life. This issue is important, as a variety of materials are available commercially. Further, due to competition, there is always a pressure to optimize dielectric consumption. Selection of dielectric material has to be done after a thorough analysis and tests. All raw materials have to be compatible with each other (e.g. oil, soldering flux, resins).

[6] Soldering or lead welding process at the sprayed ends of winding is also a critical process. The process has to be quick enough to keep film temperature at ends within limits, while ensuring a proper solder/weld quality. The pressure of welding, under-welding/over-welding (or too long a solder time and soldering iron pressure and temperature) are often causes of failures in vibration tests and increase in tan δ in field testing. The current handling (dv/dt) ability of end connections is becoming increasingly an important and critical parameter in power electronics.

[7] Effect of humidity cannot be overemphasized. Any moisture entrapped in the capacitor element, or ingress of moisture in service, will affect the metallization as well as the end connections adversely, increasing the tan δ, and reduce the IR and breakdown strength. The winding should not be exposed to humidity or air at any stage, either during manufacture or in service. The sealing material has to be chemically inert, compatible with the dielectric, and should not cause stress due to be undue shrinkage. Improper handling of materials during manufacture and non-compliance with set manufacturing norms can affect the capacitor adversely.

[8] Electrical testing and measuring techniques: Quality of testing jigs and fixtures plays an important role in the measurement of capacitance, tan δ and IR. Temperature, humidity and dust are vital environmental factors which must be controlled during measurements, particularly on unfinished capacitors during process.

22.2 FIELD FAILURES/PRECAUTIONS IN SERVICE

The first and foremost consideration is selection of a suitable capacitor for a given application. The installation, operation and maintenance are next in order of importance. Service conditions and severity need careful study for proper choice of a capacitor and its use.

[1] The environment may act detrimentally to capacitor performance. For example, the aluminium case of capacitors used in extremely humid atmosphere or near the sea in a salty atmosphere gets chemically affected and develops holes through which capacitor liquid may ooze out, causing capacitor failure. Metallic dust or carbon deposited at terminals may also damage the capacitor. Paint on KVAR capacitor containers may erode or fade, exposing the base metal to atmospheric corrosion. Chemicals in the surrounding atmosphere may sometimes play spoilsport for capacitor paint and containers.

[2] Sometimes the capacitors are located on the hot bodies of machines or motors. Power capacitors located indoors with series reactors in a substation with poor ventilation can cause temperatures to go beyond the rated temperatures of capacitors, causing them to fail. The choice of capacitors in a locale must also take account of the maximum prevailing temperatures in a year. A wrong choice of capacitor for the temperature category can lead to problems. As a thumb rule applicable above 50°C, a rise or fall of 10°C in case temperature leads to a reduction or increase in life by a factor of two.

[3] Excessive heat: A capacitor mounted too close to a transformer in a constant voltage transformer may fail due to the heat from the transformer. Insufficient ventilation or cooling in an enclosed place or in a hot and sultry enclosure will increase capacitor temperature beyond its bearing limit. In such cases, the core of capacitor elements gets extremely hot due to lack of heat dissipation, leading to failure.

[4] Effect of moisture/water: It has been observed in epoxy sealed capacitors (particularly those used in desert coolers) that water droplets or moisture settles on epoxy surface, and a tracking leakage path develops between terminals over the epoxy surface. The leakage current gradually increases and ultimately the track burns out completely, forming a short circuit. The capacitor element inside is undamaged and sound, and the capacitor may be found to be working after clearing the carbon and soot in the burnt area.

[5] In extremely cold weathers, temperatures may go well below the minimum capacitor rating, which it may not withstand. The minimum working temperature of a capacitor has to be well within its rating, and choice of the dielectric and construction should be compatible with actual working environment.

[6] Physical damage to the capacitor may allow air or moisture ingress in the sealed contents, causing the dielectric to deteriorate.

[7] High frequency harmonics in thyristor controlled systems are often a source of overheating and failures. A high harmonic level means higher core temperatures in the capacitor beyond design ratings.

[8] Surges beyond the bearing limit of a dielectric can become a reason for capacitor failure. These could be, for example, due to switching it on abruptly, or transients in circuit. A good capacitor design has to take this into account, and the user also has to put enough protection in place to avoid such conditions or minimize their effect. Current limiting resistors or inductors (reactors) are generally employed for this purpose.

When this event occurs in MPP capacitors repeatedly, a number of low resistance paths are developed over time due to repeated self-healings, leading to thermal runaway, and the element may burn, often causing an explosion, if the construction is not burst-proof.

[9] In power capacitor banks, the capacitor-reactor system must have adequate air circulation, or forced air cooling may be necessary, particularly for indoor installation. Many a capacitor has failed due to non-observance of this simple precaution.

[10] Where capacitors are switched on/off frequently to keep the power factor within acceptable limits, the unit switched in offers a momentary short circuit to the mains, causing high current inrush. Further, any capacitors already on at that instant get suddenly discharged through this new unit momentarily before attaining stability again. Both these factors cause heavy current density on metallized ends of MPP capacitors, causing short-clearing to take place, and losing some capacitance in the process. Frequent occurrence of this phenomenon can cause the capacitor to degrade and fail.

This is a common occurrence in APFC panels, and hence it is preferable to use non-metallized capacitors in such places. Rolling mills, sugar factories, cable industries etc., where frequent surges in load currents are common, are typical examples where film/foil or mixed dielectric/foil capacitors are preferred.

[11] Too high a ripple content on an electrolytic capacitor generates heat, causing thermal runaway leading to failure. Electrolytic capacitors also do not have very high over-voltage capacity like electrostatic ones.

[12] Voltages much beyond capacitor ratings may appear due to voltage fluctuations or unbalanced system voltages. Capacitor voltage has to be selected with enough safety margins. As a general rule, every 10% drop in voltage doubles the life of capacitor. Similarly, every 10°C rise decreases the life of capacitor by half, or doubles it for every 10°C fall in temperature.

[13] Loose contact due to improper cable terminations, cable size and lugs can lead to capacitor failure.

[14] Electrical properties of the insulating system change with age and continuous electrical stress. The principal contributor to the unexpected breakdown of the high voltage equipment is the insulation failure. Unlike magnetic, conducting and insulating materials in electrical equipment, the insulating material in capacitor is more prone to service stresses like thermal, electrical, mechanical or environmental stress.

[15] High voltage capacitor banks are often installed outdoors, and their surfaces, bushings and terminals are subjected to the vagaries of nature. These need to be cleaned, their discharge resistors checked, and outer surface repainted from time to time. They also need to be checked periodically for insulation to earth, capacitance and tan δ values to detect any electrical degradation.

[16] Peak voltages across a capacitor generally present themselves as high frequency spikes as a result of utility events or state changes in a UPS. These spikes damage the capacitor film

and eventually lead to premature capacitor failure. A peak AC voltage above 1.8–2.1 times the design AC voltage is the DC voltage equivalent that can lead a capacitor to fail short.

22.3 PREVENTION OF FAILURES

Maximum life can be obtained from a capacitor by observing the basic criteria as mentioned above, as also choosing the right capacitor for a given application. KVAR capacitors need little regular maintenance other than cleaning for accumulated dust and moisture periodically. They also should be checked for any deterioration of outer paint, cracks in bushings etc., and any physical damage. It is possible to avoid unexpected operational breakdown by measuring electrical properties such as capacitance and tan δ periodically. Dissipation factor (tan δ) is one of the most powerful off-line non-destructive diagnostic tools to monitor the condition of solid insulation of capacitors.

Capacitance and tan δ values obtained on new insulation are treated as benchmark readings. Then by measuring and comparing the periodical readings of the capacitance and tan δ of the insulating material with these benchmarks, one can know the rate of deterioration of the health of the insulation. Knowing the rate of deterioration helps predict the future unexpected breakdown of the insulation of high voltage equipment and plan the maintenance schedule, and repair the insulation before actual flashover, saving high cost of replacement of material. After repair, the quality of insulation can be checked before returning the equipment to service.

When to check capacitance and tan δ

[1] During the manufacturing process preferably at each stage.

[2] In service, where the frequency will depend on rate of change of tan δ, voltage vs. time (month/year).

[3] Frequency of testing depends on history of past failures on the machine, and environmental conditions. Higher humidity, temperature and pollution would require more frequent measurement of tan δ.

22.4 PROTECTION AGAINST OVERLOAD AND FAILURE AT THE END OF USEFUL SERVICE LIFE

[A] Explosion-proof capacitors

A.C. capacitors consist mainly of polypropylene (up to 90%), and their energy content is relatively high. They may rupture and ignite as a result of internal faults or external overload (e.g. temperature, over-voltage, harmonic distortion). It must therefore be ensured, by appropriate measures, that they do not pose any hazard to their environment in the event of failure or malfunction of the safety mechanism.

(a) One method is the use of a pressure interrupter mechanism in the construction. The mechanism physically disconnects the terminal internal connections in the event of thermal build up (Fig. 16.1). The can is made in a way that it or the cover bulges under pressure, and the resultant pull force on one of the element leads, which is made weak by notching at a place, breaks off the wire to disconnect the circuit. Figure 22.1(a) on the left shows the normal connection and operation of a pressure sensitive interrupter. Note the shape of the terminal cover. Figure 22.1(b) shows capacitors which have failed and the safety mechanism has disconnected the capacitor inside by bulging of the top cover.

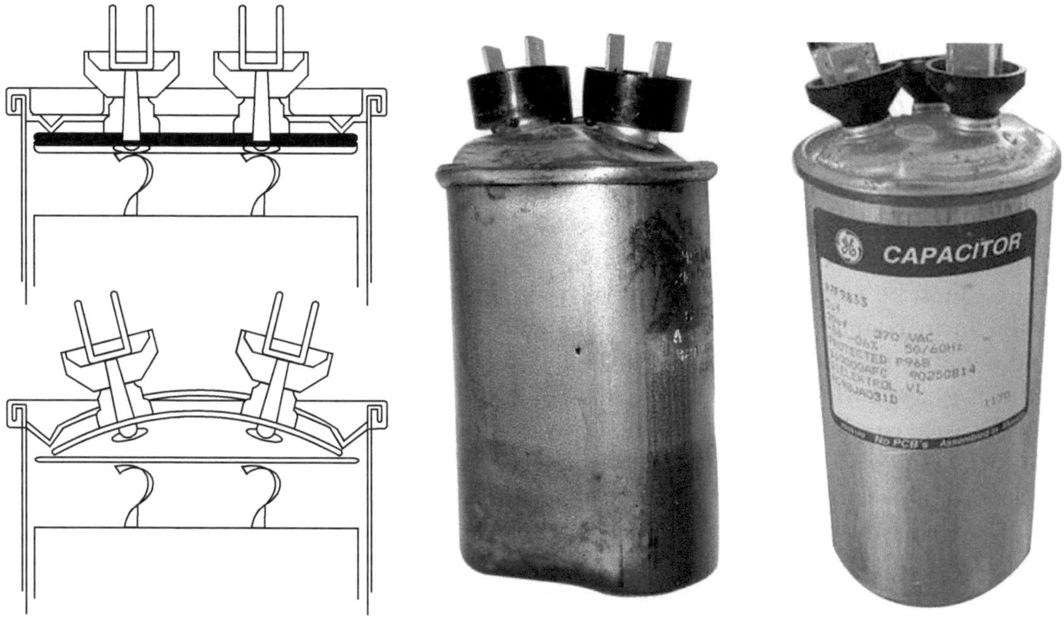

(a) Burst-proof construction (b) Capacitors failed safely with bulged tops

Fig. 22.1 *Burst proof construction and nature of failure.*

(b) In many designs, the can is provided with a double groove, and an area provided between terminal cover and a deep groove at a distance, which can expand in the event of pressure

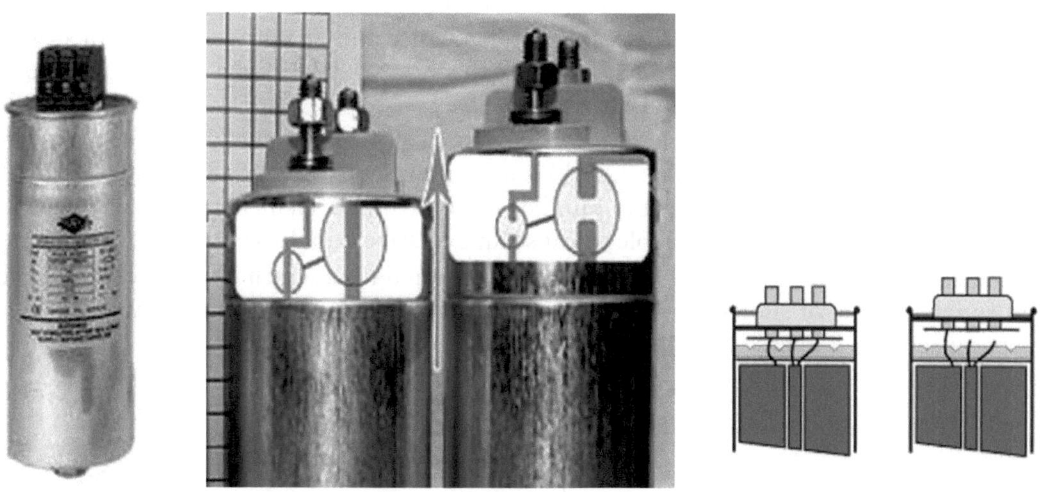

(a) Capacitor (b) Single-phase construction (c) 3-phase construction

Fig. 22.2 *Expansion groove can with 1-phase and 3-phase disconnector*

build up, and cause the internal notched lead wire to snap. This is clarified in Fig. 22.2 for both single and three phase capacitors. In Figs. 22.2(b) and (c), the sketches on the left show a normal configuration, and the right ones show the condition after failure, clearly showing the safety mechanism.

[A] Mounting considerations for the pressure sensitive interrupter

The proper operation of the pressure sensitive interrupter requires that the cover be allowed to bulge without restraint. If this expansion is restricted, it may interfere with the pressure sensitive interrupter mechanism. The following should be considered when mounting capacitors containing the pressure sensitive interrupter.

There must be a clearance of at least 12–15 mm between the tops of the terminals and/or the assembled wire connector and a plane perpendicular to the capacitor terminals. In addition, care should be taken to ensure that there is still adequate electrical clearance between the terminals and the overhead surface after pressure sensitive interrupter operation.

In the event of over-voltage or thermal overload or ageing at the end of the capacitor's useful service life, an increasing number of self-healing breakdowns may cause rising pressure inside the capacitor. To prevent it from bursting, the capacitors are fitted with a 'break action mechanism'. This safety mechanism is based on an attenuated spot at one of the connecting wires inside the capacitor. With rising pressure the case begins to expand, mainly by opening the folded crimp and pushing the lid upwards. As a result, the prepared connecting wire is separated at the attenuated spot, and the current path is interrupted irreversibly. It has to be noted that this safety system can act properly only within the permitted limits of loads and overload.

[B] Protection with internal fuses

Due to their advantages, whenever possible, internal fuses are commonly used to protect 'All-film' high voltage capacitors. In this technology, each elementary capacitance forming the capacitor is protected by its own internal fuse, as in Fig. 22.3. In the event of an elementary capacitance fault, the internal fuse eliminates the corresponding capacitance and guarantees continuous capacitor operation. Figure 22.3(a) shows a high voltage capacitor, where 1 is the discharge resistor, 2 is the internal fuse and 3 the elementary capacitance.

Given the high number of elementary capacitances forming the device, the power loss resulting from the first fault as in Fig. 22.3(b) is negligible (less than 2%). The external unbalance protection is only activated when the number of 'blown-out' elementary capacitances in the same capacitor is significant and liable to cause an excessive unbalance.

Left capacitor [Fig. 22.3(c)] is protected by an external fuse, which disconnects the capacitor from supply in case of shorts. The right capacitor has fuses in each element, and the defective element is separated in case of fault.

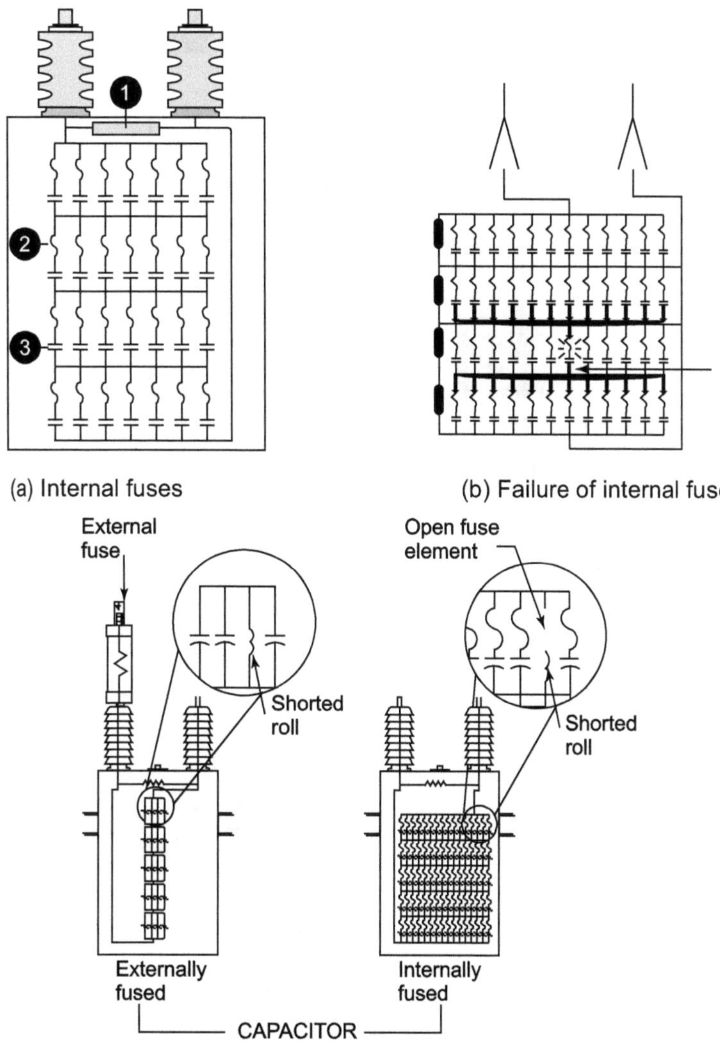

(c) Externally fused capacitor

Fig. 22.3 *Internally and externally fused capacitors.*

[C] Protection with pressure monitoring device

Protection with a pressure monitoring device is of interest whenever the capacitor cannot be protected correctly (due to electrical characteristics or cost problems) with internal fuses or by unbalance monitoring.

This protection is individual for each capacitor. It is formed of a pressure switch (Fig. 22.4) sealed on the capacitor case. This pressure switch is composed of a 'membrane' sensitive to pressure rises generated in the case due to an elementary capacitance blow-outs and a pressure switch used to trigger the bank control device.

Fig. 22.4 *Pressure switch on capacitor body* (Source: ABB)

Figure 22.5 shows actual failures of capacitors in photographs, where the capacitors have burnt, cracked or exploded.

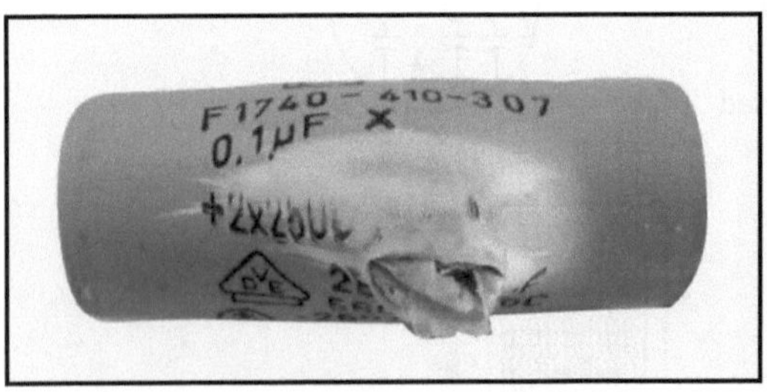

(a) Capacitor burnt and burst

(b) Electrolytic capacitor cracked

(c) MPP Capacitors exploded in service

Fig. 22.5 *Destructive failures of capacitors.*

Often the failures are a result of misapplication or unusual service conditions, as summarized below.

[D] Misapplications of capacitors leading to failure

The more common types of misapplications that result in failures are:

- Ripple current or voltage above specification
- Application voltage beyond surge voltage specified
- Temperature exposure beyond specified limits
- Unusual mechanical shocks and vibration
- Corrosive and abrasive particles in cooling air, conducting dust in cooling air, oil or water vapour
- Presence of corrosive substances
- Explosive gas or dust radioactivity
- Unusual storage or transport
- Unusual humidity (tropical or subtropical region)
- Excessive and rapid change of ambient temperature or humidity
- Superimposed radio frequency voltages

One must take care in selection of a capacitor for an application, operate the capacitor within its rated conditions, and observe due care in maintenance. The capacitors will then continue trouble-free service for its entire lifetime. Capacitor life is significantly increased by judicially observing some simple precautions.

APPENDIX A

Some Common Dielectrics and Their Behaviour

Dielectric material	*Dielectric constant*	*Dielectric strength (kV/mm) DC*	*Loss tangent*
Air	1.0059	0.4–3	0
Vacuum	1.000	3	0
Pure cellulose or paper	5.9 – 6.0	70	0.0053
Polypropylene	2.25 – 2.3	400	<0.0002
Polyester	3.2	200	0.001
Polystyrene	4 – 2.7	20	<0.0005
Polycarbonate	2.9 – 3.0	280	0.0014-
Teflon	2.1	60	<0.0005
Polyphenylene Sulphide (PPS)	3.3	400	0.0002
Polyethylene naphthalate (PEN)	2.9	330	0.0005
HDPE	2.65	500	<0.0002
PVC	3.2	725	0.0115
LDPE	2.3	300	<0.0002
Nylon	3.88	400	0.014
Aluminium oxide	8.5	.1000	0.10
Tantalum Pentoxide Oxide (Ta_2O_5)	27	300	0.04
Glass (silicon)	4.8 – 8	9.8 – 13.8	0.0005
Mica	5.9 – 9.0	64	0.005
Ceramic	Up to 50,000	40	Variable
Castor oil	3.7	12	0.0007
Rapeseed oil	3.28	25	0.0003
Transformer oil	2.2	12	0.005
Dioctyl phthalate (DOP)	5.6	50	0.005
Polychlorinated biphenyl (PCB)	5.9	50	0.0025
Phenyl xylyl ethane (PXE)	2.5	60	0.0012
M/DBT	2.66	70	0.003
Paraffin	2.2	10	0.0002
Distilled water (for comparison)	78.5		0.005
Barium titanate	1200 +	2	0.01
Benzene	2.284	–	–
Glycerine	42.5	–	–

APPENDIX B

Tests for Capacitors as per Indian Standard Specifications

Application	*Fans (440 V)*	*Motors (440 V)*	*Lighting (250 V)*	*kVAR (440) self–healing type*
IS No.	1700-1984	2993 – 1998	1569 – 1976	13340 – 1998 13341-1998
AV HV test between terminals	$1.5 \times V_n$ for 10 s	$2 \times V_n$ for 10 s	$1.5 \times V_n$ for 10 s	$1.75 \times V_n$ for 10 s
Test voltage AC RMS	660 V	880 V	375 V	770 V
AC HV test term. and case	2000 V for 1 min.	2000 V for 1 min.	2000 V for 1 min.	2500 V for 1 min.
Capacitance	$\pm$ 5%	$\pm$ 5%	$\pm$ 10%	$\pm$ 5% to +10%
Tan δ max.	0.002	0.002	Not specified	0.0025
Endurance test	$1.25\ V_n \times 500$ h	$1.25\ V_n \times 300$ to 3000 h as per category	$1.25\ V_n \times 500$ h	$1.25\ V_n \times 1500$ h, with 1000 charge – discharge cycle after 750 h

Note: For tests as per IS: 2993, current calculations for 440 V rated capacitors work out as under:

CµF	*Amp. at 880 V*	*Single phase supply current amp at 220 V AC*	*Load at 400 V primary*
10	2.8	11.2	6.2
20	5.6	22.4	12.3
30	8.4	34.0	18.5
36	10.1	40.5	22.5

It is seen that current requirements for capacitor tests between terminals is quite high.

APPENDIX C

Lighting Capacitors

[1] Selection Charts for Lighting Capacitors

Lamp power	*Lamp power with ballast*	*Parallel capacitor 250 V*	*Series capacitor 450 V*
W	W	µF	µF
4	10	2	
6	12	2	
8	14	2	

Lamp power	*Lamp power with ballast*	*Parallel capacitor 250 V*	*Series capacitor 450 V*
10	14	2	
13	19	2	
15	25	4.5	
15	19.5 (1)	4.5	
16	21	2.5	
18	27	4.5	2.9
18		4.5	
20	30	4.5	2.9
20	25 (1)	4.5	
22	27	5	3.2
30	39	4.5	3
32	42	5	3.6
36	45	4.5	3.6
38	48	4.5	3.6
40	49	4.5	3.6
40	54	4.5	4.4
58	69	7	5.7
65	76	7	5.7
65	80	9	6.8
115	135	18	12.2
140	160	18	12.7

(1) 2 lamps in series on 220 V

[2] High-Pressure Mercury Vapor Lamps

Lamp power W	Lamp power with ballast W	Parallel 250 V μF	Capacitor 380 V μF
50	59	7	
80	89	8	
125	137	10	
150	170	20	
250	266	18	
250	275	32	
400	425	25	
400	385	35	
700	735	40	
1000	1045	60	
2000	2070		37

[3] Low-Pressure Sodium Vapor Lamps

Lamp power W	Lamp power with ballast W	Parallel capacitor 230 V μF
18	25	5
50	62	8
70	83	12
80	89	8
100	115	12
125	137	10
150	170	20
250	266	18
250	275	32
400	425	25
400	450	50
1000	1090	100

[4] High-Pressure Sodium Lamps

Lamp power W	*Parallel capacitor 230 V μF*	*Lamp power W*	*Parallel capacitor 230 V μF*
50	10	400	50*
70	12	400	45*
150	20	1000	100*
250	36	1000	120*

*Depending on lamp type

[5] Halogen–Metal Vapor Lamps

Lamp power W	*Lamp power with ballast W*	*Parallel capacitor 230 V μF*
35	48	6
70	88	12
150	170	20
250	275	32
400	385	35
400	440	45
1000	1050	85
2000	2070	37*
2000	2080	60*
3500	3650	100*

APPENDIX D

[1] Guideline Values for Capacitor Selection for Motors

Motor power rating (kW)	*Capacitor output selection (kVAR)*
up to 3.9	Approximately 55% of nominal motor power
4.0 – 4.9	2
5.0 – 5.9	2.5
6.0 – 7.9	3
8.0 – 10.9	4
11.0 – 13.9	5
14.0 – 17.9	6
18.0 – 21.9	7.5
22.0 – 29.9	10
30.0 and above	Approximately 35% of nominal motor power

[2] Typical Power Factor of Partially Loaded Motors

S. No.	*% loading of motor*	*Power factor*		
		1 HP	10 HP	100 HP
1	25		0.62	0.72
2	50	0.50	0.76	0.83
3	60	0.62	0.80	0.84
4	70	0.68	0.83	0.85
5	80	0.72	0.85	0.87
6	90	0.76	0.87	0.90
7	100	0.80	0.88	0.91

(*Source:* PHD Chamber of Commerce and Industry)

[3] Capacitor Selection Table for PFC of Electrical Motor

Max. Capacitor kVAR for use with 3-phase 50-cycle Induction Motors

	3000 RPM		1500 RPM		1000 RPM		750 RPM		500 RPM	
MOTOR RATING HP	kVAR	Line current using	kVAR	Line current using	kVAR	Line current using	kVAR	Line current using	kVAR	Line current using
10	2.5	9	4	11	5	17	5	23	7.5	28
15	2.5	9	5	11	7.5	16	7.5	21	10	26
20	5	9	5	10	7.5	15	10	20	12.5	24
25	5	9	7.5	9	10	14	10	19	15	22
30	7.5	9	10	9	10	13	12.5	18	15	21
40	10	9	10	9	12.5	12	15	16	17.5	19
50	12.5	9	12.5	8	15	12	20	15	22.5	17
60	15	9	15	8	17.5	11	22.5	14	25	16
75	17.5	9	17.5	8	20	11	27.5	13	30	15
100	22.5	9	22.5	8	25	10	35	12	37.5	14
125	25	9	27.5	8	30	9	40	11	13	13
150	32.5	9	32.5	8	37.5	9	4.5	11	55	13
200	42.5	9	42.5	8	45	9	60	10	67.5	12

APPENDIX E

[1] Guidelines : Capacitor Selection Table for Power Factor Correction of Transformers

Transformer rating kVA	*Capacitor output in kVAR at transformer primary voltages*		
	5–10 kV	*15–20 kV*	*25–30 kV*
50	4.0	5.0	6.0
75	5.0	6.0	7.5
100	6.0	7.5	10.0
160	10.0	12.5	15.0
250	15.0	16.7	20.0
315	16.7	20.0	25.0
400	20.0	25.0	30.0
630	30.0	33.3	40.0
1000	45.0	50.0	55.0
1250	50.0	55.0	60.0

[2] Guidelines: Capacitor selection Table for Power Factor Correction of Welding Transformers

Welding transformer rating (kVA)	*Capacitor rating (kVAR)*
1	1
2	2
3	2
4	3
5	4
6	4
7	5
8	6
9	7.5
10	7.5
11	8
12	9
13	10
14	10
15	11
16	12
17	13
18	13
19	14
20	15
Above 20 up to 22	16
Above 22 up to 24	17.5
Above 24 up to 26	18
Above 26 up to 28	20
Above 28 up to 30	21
Above 30 up to 35	24
Above 35 up to 40	27.5
Above 40 up to 45	32.5
Above 45 up to 50	35

(*Source:* KSEB Officers' Association)

APPENDIX F

[1] Capacitor Selection for Improvement of Power Factor

Actual power factor	*Target power factor*									
	0.70	**0.75**	**0.80**	**0.85**	**0.90**	**0.92**	**0.94**	**0.96**	**0.98**	**1.00**
0.40	1.27	1.41	1.54	1.67	1.81	1.87	1.93	2.00	2.09	2.29
0.45	0.96	1.10	1.23	1.36	1.50	1.56	1.62	1.69	1.78	1.98
0.50	0.71	0.85	0.98	**1.11**	**1.25**	1.31	1.37	1.44	1.53	1.73
0.55	0.50	0.64	0.77	0.90	1.03	1.09	1.16	1.23	1.32	1.52
0.60	0.31	0.45	0.58	0.71	0.85	0.91	0.97	1.04	1.13	1.33
0.65	0.15	0.29	0.42	0.55	0.68	0.74	0.81	0.88	0.97	1.17
0.70	0.00	0.14	0.27	0.40	0.54	0.59	0.66	0.73	0.82	1.02
0.75		0.00	0.13	0.26	0.40	0.46	0.52	0.59	0.68	0.88
0.80			0.00	0.13	0.27	0.32	0.39	0.46	0.55	0.75
0.85				0.00	0.14	0.19	0.26	0.33	0.42	0.62
0.90					0.00	0.06	0.12	0.19	0.28	0.48

The table above shows the values for typical power factors according to the formula:

Q_c = P * {tan [acos (pf1)] – tan [acos (pf2)]}

Q_c = required capacitor output (kVAR)

pf1 = actual power factor

pf2 = target power factor

P = real power (kW)

The required capacitor output may be calculated as follows:

Select the factor k by matching point of actual and target power factor.

Calculate the required capacitor rating with the formula:

$Q_c = k \times P$

For example: Actual power factor = 0.70, target power factor = 0.96, real power = P = 500 kW,

Q_c = k * P = 0.73 × 500 kW = 365 kVAR

[2] Nomogram for Power Factor Correction

A capacitive kVAR required for improvement of power factor at a given load can be made from the nomogram in the figure below. However, in most cases the capacitor bank rating has to be carefully selected after due consideration of rated voltage of the system, system over-voltages, harmonics in the system, rating of series reactor if any, etc. Besides the correct sizing of the capacitor bank, the reliability also depends on the right selection of associated equipment such as the circuit breaker, series reactor, protective relays, etc. Above all, it is essential to check that the capacitor installation will not cause dangerous stress to the user's system due to resonance.

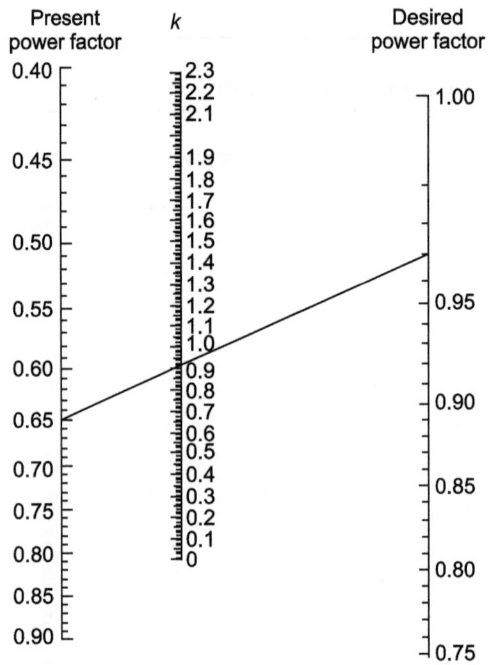

NOMOGRAM

A nomogram is used for calculating the necessary capacitor power rating Q (kVAR), required for improving the power factor of a load P (kW). For example, to calculate the capacitor power rating required to increase the power factor of an installation from 0.65 to 0.97 if the load P is 1500 kW, from the nomogram K = 0.92 and hence Q = 0.92 × 1500 = 1350 kVAR.

APPENDIX G

Cross Section of Connecting Cable Between Main Supply and Capacitor Bank, Fuse Rating

	Rated voltage 230 V, 50 Hz			*Rated voltage 400 V, 50 Hz*			*Voltage 415 V, 50 Hz*		
Output kVAR	*Rated current A*	*Fuse A*	*Cable/ mm²*	*Rated current A*	*Fuse A*	*Cable/ mm²*	*Rated current A*	*Fuse A*	*Cable/ mm²*
2.5	6.3	16	2.5	3.6	10	1.5	3.5	10	1.5
5.0	12.6	25	4	7.2	20	2.5	7.0	20	2.5
6.67	16.7	35	6	9.6	20	2.5	9.3	20	2.5
7.5	19	35	6	10.80	20	2.5	10.4	20	2.5
8.33	21	35	6	12	20	2.5	11.6	20	2.5
10.0	25	50	10	14.4	25	4	13.9	25	4
12.5	31	63	16	18	35	6	17.4	35	6

Output kVAR	Rated voltage 230 V, 50 Hz			Rated voltage 400 V, 50 Hz			Voltage 415 V, 50 Hz		
	Rated current A	Fuse A	Cable/ mm²	Rated current A	Fuse A	Cable/ mm²	Rated current A	Fuse A	Cable/ mm²
15.0	38	63	16	21.7	35	6	20.9	35	6
16.7	42	80	25	24.1	50	10	23.2	50	10
20.0	50	100	35	28.9	50	10	27.8	50	10
25.0	63	125	50	36.1	63	16	34.8	63	16
30.0	75	125	50	43.3	80	25	41.7	80	25
33.3	84	160	70	48.1	80	25	46.3	80	25
40.0	100	160	95	57.7	100	35	55.6	100	35
50.0	125	250	120	72.2	125	50	69.6	125	50
60.0	-	-	-	86.6	160	70	83.5	160	70
66.7	-	-	-	96.3	160	70	92.8	160	70
70.0	-	-	-	101	160	70	97	160	70
75.0	-	-	-	108	160	70	104	160	70
83.3	-	-	-	120	200	95	116	200	95
100.0	-	-	-	144	250	120	139	250	120

The cross section for connecting cable and fuse rating has to be selected in accordance with the standard practice. Guideline values for operation under normal conditions and at an ambient temperature of 40°C are given above. Higher values may have to be selected if conditions differ from normal (e.g. higher ambient temperature or high harmonic distortion).

APPENDIX H

Calculation of the Required Rated Capaciotor output in Detuned Filter Circuts

Factors to be multiplied with the required output per step.

Supply voltage : 400 [V]	*Detuning factor in [%]*						
Rated voltage *of capacitor [V]	5	5.5	6	7	12.5	13	14
440	1.150	1.143	1.137	1.125	-	-	-
525	1.637	1.628	1.619	1.602	1.507	1.499	1.481
Supply voltage: 415 [V]	Detuning factor in [%]						
Rated voltage *of capacitor [V]	5	5.5	6	7	12.5	13	14
440	1.068	1.062	1.057	-	-	-	-
525	1.520	1.512	1.504	1.488	1.400	1.392	1.376

Supply voltage : 400 [V]	*Detuning factor in [%]*						
Rated voltage *of capacitor [V]	5	5.5	6	7	12.5	13	14
525	1.352	1.345	1.338	1.324	1.246	1.239	1.224
Supply voltage: 480 [V]	Detuning factor in [%]						
Rated voltage *of capacitor [V]	5	5.5	6	7	12.5	13	14
525	1.136	1.130	1.125	1.113	-	-	-
660	1.796	1.787	1.777	1.758	1.654	1.645	1.626

Example

Required output per step at supply voltage	50 kVAR
Supply voltage	400 V
Detuning factor	7%
Rated voltage of the capacitor	440 V
Factor of the table	1.125

Required rated output of the capacitors: 50 kVAR × 1.125 = 56.25 kVAR

* For filter circuits the capacitor rated voltage has to be chosen always higher than the supply voltage. i.e.: Fundamental voltage increased by the reactor and harmonics.

APPENDIX I

Financial Benefits of Capacitor Installation

Consider a load of 100 kVA at 70% power factor. If a capacitor of 40 kVAR is added to the system, the load can be increased to 80 kW without adding extra capacity to generating equipment. When lagging load is fully compensated to get unity power factor, all the active power is supplied by the generating system and all kVARs are supplied by capacitors.

	Load	100 kW	Assumed cost of capacitors				Rs/kVAR – 225.00		
Present P.F.		*Target P.F. 0.9*			*Target P.F. 0.94*			*Target P.F. 0.98*	
	kVAR reqd.	*Cap. Cost Rs.*	*% Returns p.a.*	*kVAR reqd.*	*Cap. cost Rs*	*% Returns p.a.*	*kVAR reqd.*	*Cap. cost Rs*	*% Returns p.a.*
0.4	180.46	40603.50	61.57	192.31	43269.75	59.74	208.46	46903.50	56.78
0.45	150.15	33783.75	59.20	161.54	36346.50	57.37	178.46	40153.50	53.87
0.5	124.6	28035	57.07	136.9	30802.50	54.71	153.1	34447.50	51.19
0.55	103.6	23310	54.60	115.38	25960.50	52.30	131.08	29493	48.69
0.6	84.6	19035	52.53	96.92	218070	49.76	113.08	25443	45.72
0.64	71.54	16096.50	50.48	83.85	18866.25	47.58	100	22500	43.37
0.68	60	13500	47.93	71.54	16096.50	45.49	87.69	19730.25	41.07
0.7	53.6	12060	47.38	65.69	14780.25	44.42	81.15	18258.75	40.24
0.74	42.54	9571.50	45.18	54.62	12289.50	42.11	70	15750	37.82
0.78	30	6750	45.58	49.2	11070	35.48	60	13500	34.89

Load 100 kW Assumed cost of capacitors Rs/kVAR – 225.00

Present P.F.	*Target P.F. 0.9*			*Target P.F. 0.94*			*Target P.F. 0.98*		
	kVAR reqd.	*Cap. Cost Rs.*	*% Returns p.a.*	*kVAR reqd.*	*Cap. cost Rs*	*% Returns p.a.*	*kVAR reqd.*	*Cap. cost Rs*	*% Returns p.a.*
0.8	26.6	5985	41.77	38.84	8739	38.35	54.23	12201.75	33.87
0.84				28.46	6403.50	35.60	43.85	9866.25	31.03
0.88				17.69	3980.25	32.80	33.85	7616.25	27.40
0.9				12.07	2715.75	31.34	28.07	6315.75	25.85
0.94							16	3600	21.71

(1) Returns have been calculated based on savings per year in maximum demand for the same load shown as percentage of investment. Billing rate for maximum demand has been assumed @ Rs. 15/KVAR.

(2) Capacities of transformers, cables and switchgears are released for additional loads. Savings made on this account are considerable. These are not considered in the above calculations.

(3) The calculation does not take into account penalties imposed by Electricity Boards, e.g. if monthly average power factor falls below 85%, say, to 80%, there shall be a surcharge of 2% for each 1% by which it falls below 85%, i.e. 10% for 80% power factor.

(4) The boards also give incentives for power factor above 0.98 by way of reduced tariffs.

(5) The power factor however should never be leading, as this is damaging to the system. It may damage motors due to over-excitation. Short circuit impedance of system goes down, resulting in extremely high short-circuit fault currents.

(*Source:* PHD Chamber of Commerce and Industry)

APPENDIX J

CODE MARKINGS ON CAPACITORS

[1] Standard Configurations and Marking Styles

5-band ceramic capacitors Radial/axial lead ceramic capacitors (6 band / dot) system)

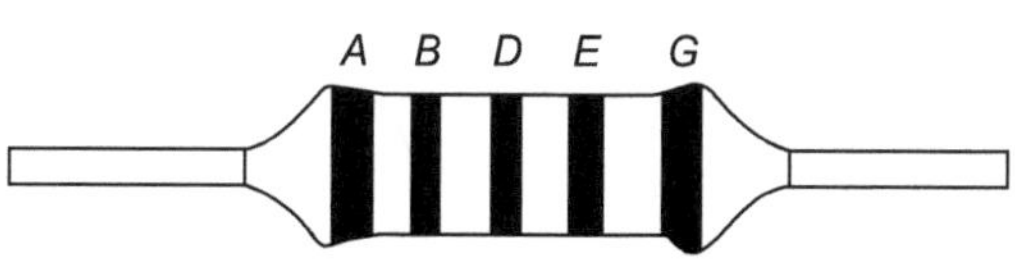

5-band ceramic capacitors

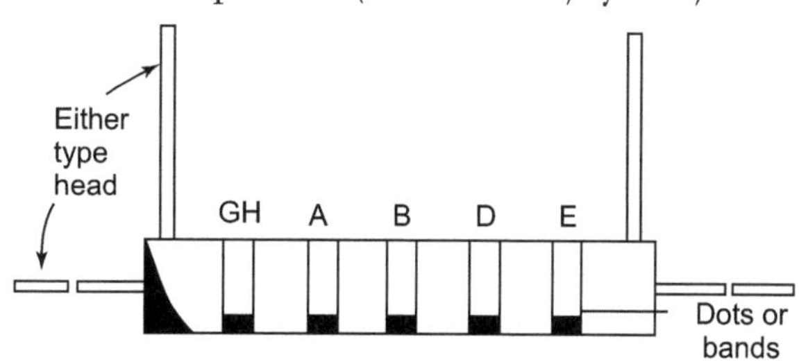

Radial/axial lead ceramic capacitors (6 band/dot) system)

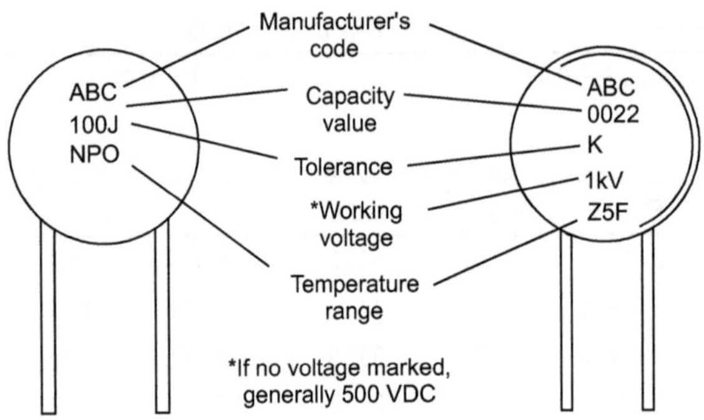

Ceramic disc capacitors

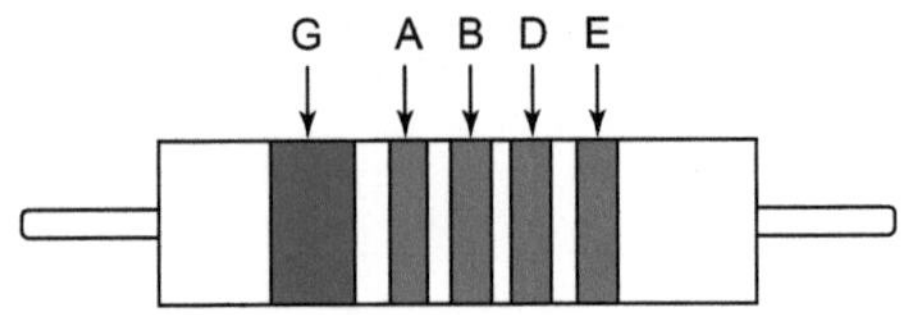

5-dot/band ceramic capacitors (one wide band)

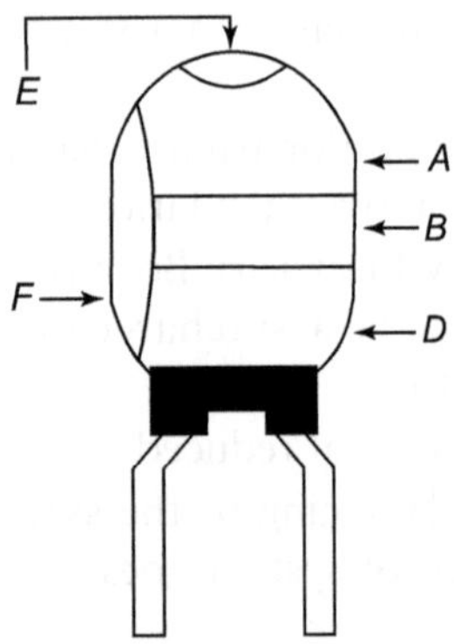

Dipped tantalum capacitor

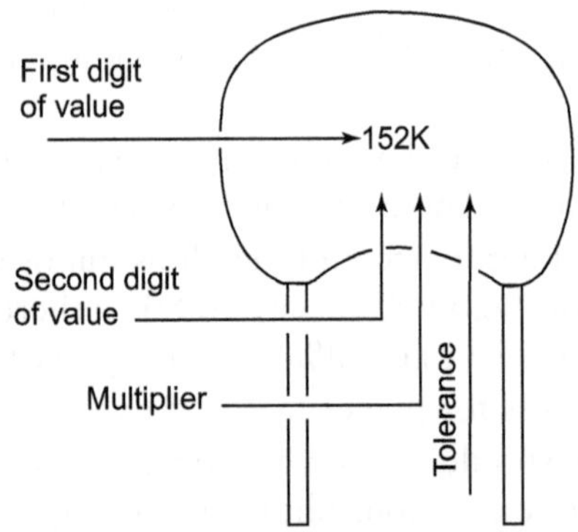

Film type capacitor

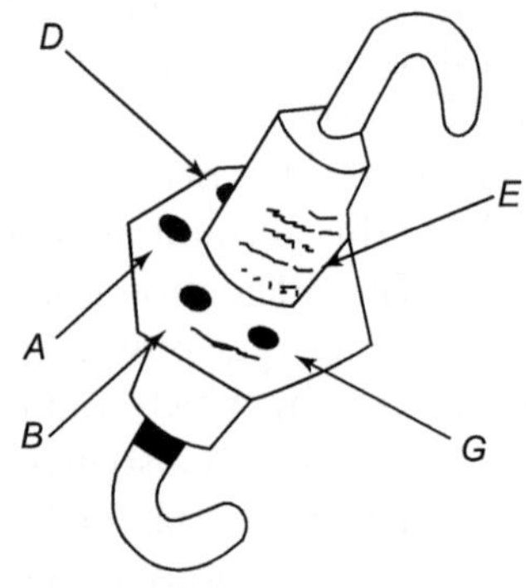

Ceramic feed through capacitors

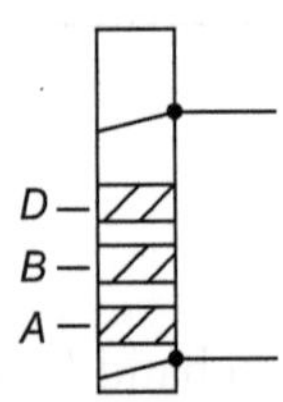

Ceramic capacitor class II

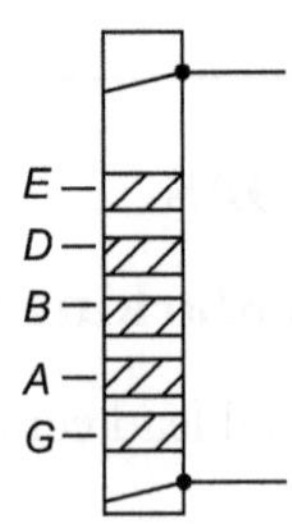

Ceramic capacitor class I

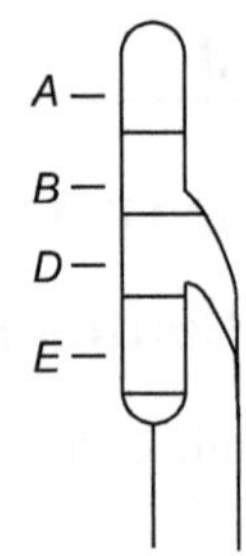

Ceramic capacitor pin-up

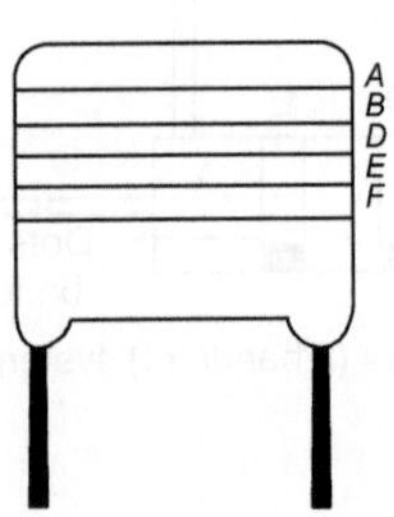

Polystyrene capacitor

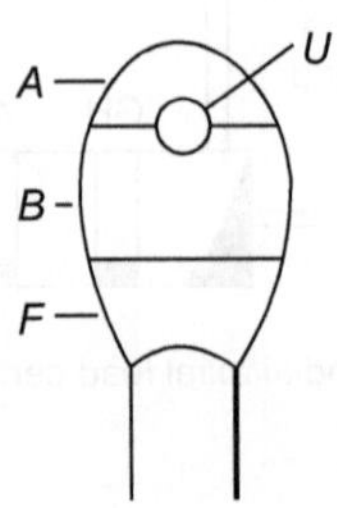

Tantalum cap.

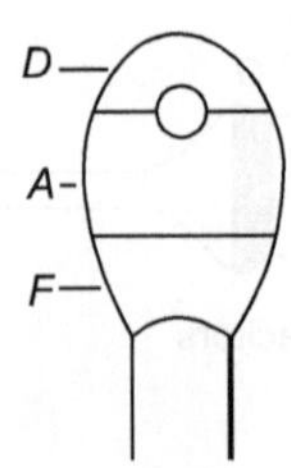

Tantalum cap.

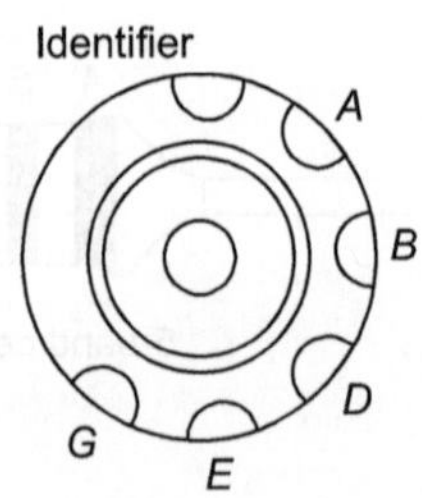

Button mica cap.

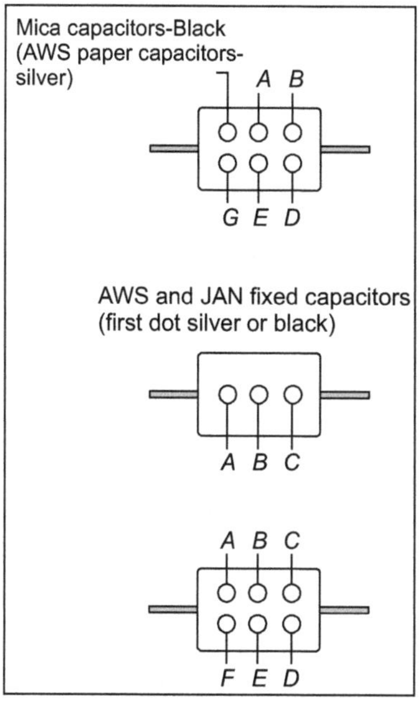

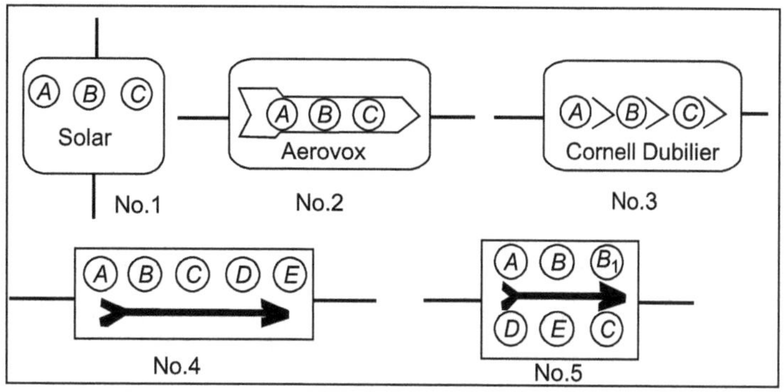

Stamp type capacitors

[2] Colour Codes

- A: first significant figure
- B: second significant figure
- C: third significant figure
- D: multiplier (factor by which the significant figures are multiplied to yield the nominal value)
- E: tolerance

Colour	*Digit A*	*Digit B*	*Multiplier D*	*Tolerance T > 10 pf*	*Tolerance T < 10 pf*	*Temp. coeff. TC*	*Working voltage V*
Black	0	0	x1	± 20%	± 2.0 pF		
Brown	1	1	x10	± 1%	± 0.1 pF	-33×10^{-6}	
Red	2	2	x100	± 2%	± 0.25 pF	-75×10^{-6}	250 V
Orange	3	3	x1000	± 3%		-150×10^{-6}	
Yellow	4	4	x10k	+100%,-0%		-220×10^{-6}	400 V
Green	5	5	x100k	± 5%	± 0.5 pF	-330×10^{-6}	100 V
Blue	6	6	x1m			-470×10^{-6}	630 V
Violet	7	7				-750×10^{-6}	
Grey	8	8	x0.01	+80%,-20%			
White	9	9	x0.1	± 10%			

3-dot to 6-dot colour codes are identical in reading from A onwards sequentially, except that the details increase with number of dots. Feed-through or button capacitors have markings as shown to read the sequence.

[3] Alpha Numeric Codes

(a) Electronic capacitor codes:

In digital notation, capacitor tolerance is indicated as per following codes:

	Letter	*B*	*C*	*D*	*F*	*G*	*J*	*K*	*M*	*Z*
Tolerance	C <10 pF ±pF	0.1	0.25	0.5	1	2				
	C >10 pF ±%			0.5	1	2	5	10	20	+80 −20

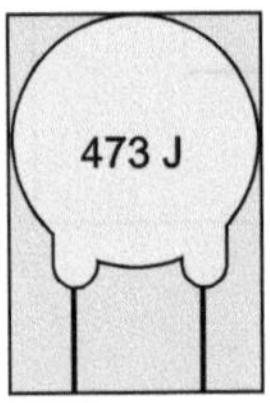

A ceramic disc type capacitor with code 473J printed.
Capacitor value is 47 pF × 1000(3 zeros) = 47,000 pF, or 0.047 μF
Hence capacitor rating is .047 μF ± 5%

(b) Detailed alpha numeric codes:

DM 5 — Style

F — Characteristic

D — Working Voltage
Y: 50VDC
A:100VDC
C: 300VDC
D: 500VDC

181 — Capacitance

J — Capacitance tolerance
D: ± 0.5 pF
K: ± 10%
J: ± 5%
G: ± 2%
F: ± 1%
E: ± 0.5%

O — Temperature range
O:–55°C to +125°C
O:–55°C to +150°C

3 — Vibration grade frequency
10 Hz to 2 kHz

Letter	Temperature coefficient [PP M^{10}C]	Capacitance drifft
C	−200 to +200	± (0.5% + 0.1 pF)
D	−100 to +100	± (0.3% + 0.1 pF)
E	−20 to +100	± (0.1% + 0.1 pF)
F	0 to +70	± (0.05% + 0.1 pF)

[4] Codes for Ceremac Capacitors

First symbol (a letter)	*Low temperature limit*	*Second symbol (a number)*	*High temperature limit*	*Third symbol (a letter)*	*Max. capacitance change over temperature*
Z	+10 deg. C	2	+45 °C	A	+1.0%
Y	–30 deg. C	4	+65 °C	B	+/–1.5%
X	–55 deg. C	5	+85 °C	C	+/–2.2%
		6	+105 °C	D	+/– 3.3%
		7	+125 °C	E	+/– 4.7%
		8	+150 °C	F	+/–7.5%
		9	+ 200 °C	P	+/– 10.0%
				R	+/– 15.0%
				S	+/– 22.0%
				T	+22%, –33%
				U	+ 22%, –56%
				V	+22%, –82%

APPENDIX K

Indian Standards for Electrical Equipment Capacitors and their IEC Equivalents

IS No.		*Year*	*Details*	*IEC Standard*
IS	590	1964	Specification for fixed paper dielectric capacitors for DC	
IS	1569	1976	Capacitors for use in tubular fluorescent high pressure mercury and low pressure sodium vapour discharge lamp circuit	IEC 61049
IS	1709	1984	Specification for capacitors for electric fan motors	IEC 60252-1
IS	1885 Part 42	1993	Electrotechnical Vocabulary: Part 42 Power capacitors	IEC 60050-436
IS	1885 Part 45	1977	Electrotechnical Vocabulary - Part XLV: Capacitors	
IS	1980	1967	Ceramic dielectric capacitors, Type 1	
IS	2001	1968	Specification for fixed silvered mica capacitors	
IS	2993	1998	AC motor capacitors	IEC 60252=1
	-	-	AC motor start capacitors	IEC 60252-2
IS	3156 Part 4	1992	Voltage transformers: Part 4: Capacitor voltage transformers	IEC 60186
IS	3723 Part 1-2-3	1978	Specification for capacitors for radio interference suppression - Part I: General Requirements and Methods of Tests Part 2: Type FCRS 1 Part 3: Type FCRS 2	IEC 60938-1
IS	4317	1983	Specification for aluminium electrolytic capacitors with non-solid electrolyte	IEC 60384-4.1

IS No.		*Year*	*Details*	*IEC Standard*
IS	4633	1968	Specification for fixed metallized-paper dielectric capacitor for direct current®	
IS	5361	1969	Polyester film dielectric capacitors for direct current	
IS	5547	1983	Application guide for capacitor voltage transformers	IEC 60044-5
IS	7305	1984	Specification for fixed capacitors used in electronic equipment	
IS	8186	1976	Marking codes for values and tolerance of resistors and capacitors	IEC 60062
IS	8507 Sec 1-2	1977-1983	Fixed hermetically sealed tantalum capacitors	
IS	9251	1979	Capacitors for inductive heat generating plants operating at frequencies between 40 and 24,000 Hz	IEC 60110 Part 1-2
IS	9256 Part 1-3	1979	Specification for fixed metallized polyester film dielectric capacitors - Part I: General Requirements and Methods of Tests, Part 2: Type FCPM 1, Pt 3-Type FCPM 2	
IS	9348	1998	Coupling capacitor and capacitor dividers	IEC 60358
IS	9402	1980	High voltage fuses for the external protection of shunt power capacitors	IEC 60549
IS	9368 Part 1	1980	Specification for fixed polyester film dielectric capacitors for direct current - Part 1: General Requirements and Methods of Tests	60384-23 60384-23.1
IS	9835	1981	Series capacitors for power systems	IEC 60143-1
IS	9835 Part 1	2001	Series capacitors for power systems - Part 1: General Performance, Testing and Rating - Safety Requirements - Guide for Installation	IEC 60143 -2
IS	9960	1981	Electrical grade castor oil for use as capacitor impregnant	
IS	10825 Part 1-3	1075-1984	Specification for ceramic dielectric capacitors type 1 - Part 1: General requirements and methods of tests, Part 2: FCCT 1, Part 3: FCCT 2	
IS	10991 Part 1,	1984	Specification for fixed metallized polypropylene film dielectric capacitors - Part 1: General Requirements and Methods of Tests	
IS	10991 Part 3	1991	Plastic films for electrical purposes: Part 3 Specifications for individual materials, Sec 1 Polypropylene films for capacitors	
IS	11298 Part 3 sec 1	1991	Plastic films for electrical purposes: Part 3 Specifications for individual materials, Sec 1 Polypropylene films for capacitors	IEC 60674.3
IS	11530	1985	Voltage grading capacitors	IEC 60358
IS	11548	1986	Capacitors for surge protection 650 V to 33 kV	IEC 61643
IS	12298 (part 3 sec 1)	1990	Polypropylene film for capacitors	IEC 60674-17-1
IS	12298 (part 3 sec 2)	1990	Metallized polypropylene films	IEC 60384
IS	12672	1989	Internal fuses and internal overpressure disconnectors for shunt capacitors	IEC 60593 BS 7631
IS	12677	1989	Internal fuses for series capacitors	IEC 60143-3 BS 7633
IS	12730	1989	Internal fuses and internal overpressure disconnectors for capacitors for inductive heat generating plants	IEC 61071-2

IS No.		*Year*	*Details*	*IEC Standard*
IS	13067	1991	Impregnants for power capacitors	
IS	13340	1993	Power capacitors of self-healing type for AC power systems having rated voltage up to 650 V - Specification	IEC 60831 Part 1
IS	13341	1992	Requirements for ageing test, self-healing test and destruction test on shunt capacitors of the self-healing type for ac power systems having a rated voltage up to and including 650 V	IEC 60831 Part 2
IS	13580	1992	Internal fuses and internal overpressure disconnectors for power electronic capacitors	IEC 61071-1
IS	13585 Part 1	1994	Shunt capacitors of non self-healing type for ac power systems having a rated voltage up to and including 650 V	IEC 60931 Part 1-2-3
IS	13648	1993	Power electronics capacitors	IEC 61071
IS	13666	1993	Energy storage capacitors	
IS	13925 Part 1	1998	Shunt capacitors for ac power systems having a rated voltage above 1000 V Part 1:General performance, testing and rating safety requirements - Guide for installation and operation	IEC 60871 Part 1
IS	13925 Part 2-3	2002	Shunt capacitors for AC power systems having a rated voltage above 1000 V - Part 2: Endurance Testing Part 3: Protection of Shunt Capacitors and Shunt Capacitor Banks	IEC 60871 Part 2-3-4
IS	15406 Part 1	2003	Capacitors for microwave ovens - Part 1: General	IEC 61270
IS	13252	2003	Information technology equipment - Safety - Part 1: General requirements	IEC 60950-1 IEC950
IS	15039	2001	Information technology equipment - Immunity characteristics - Limits and methods of measurement.	EN 55024
			Harmonized system of quality assessment for electronic components. Sectional specification. Fixed capacitors for electromagnetic interference suppression and connection to the supply mains (Assessment Level D)	EN 132400/ IEC 384-14
			Guide for surge voltages in low-voltage AC power circuits.	IEEE62-14-1980 (Formerly IEEE587)

Note: The list is not exhaustive. Only the most commonly referred standards are given. Please check the ISI Standards list for other standards.

REFERENCES

Articles

1. Analysis for EDLC Application on Electric Railway System, Hanmin Lee, Gildong Kim, Changmu Lee, Korea Railroad Research Institute, South Korea Enable IPC Subsidiary,
2. SolRayo, Unveils New Lower Cost, Higher Power Ultracapacitor Incorporating Nanotechnology Coatings, Rich Kaiser, Enable IPC, July 8, 2009
3. Low ESR Capacitors- Fact or Fiction?: Mark Gebbia, Illinois Capacitor, Inc.
4. Avoiding capacitor failures in large UPS systems, white paper no. 60, American Power Corporation,
5. Dielectrics, ASC Capacitors, Literature
6. Introduction to microelectromechanical systems MEMS):Complaint Mechanism Research group, Brigham Young University
7. Understanding dielectric adsorption :Sencore Tech Tips No. 105
8. Utility capacitor bank switching studies: Electrotek Concepts, Inc., U.S.A.
9. Wet Tantalum capacitors meet DSCC 9302. Power Electronics Technology: June1, 2009.
10. Introduction to safety certified capacitors. Johanson Dielectrics.
11. MKP Capacitors for lighting applications: Conis Company Ltd.
12. Application guide snubber capacitors. CDE Cornell Dubilier, New Bedford.
13. Power Quality Equipment: Harmonic Mitigating Transformers: Reliant Energy, Houston, Texas
14. Power system Harmonic Fundamental considerations: Larry Ray, P.E.& Louis Hapeshis, P.E.
15. Capacitor Bank Switching Transients: Northeast Power Systems Inc., N.Y.
16. High voltage capacitors suit mission critical military and aerospace applications: Joe Moxley, Custom Electronic Inc., EDN, Sept 23, 2010
17. ABC's of SAFETY (Interference Suppression) Capacitors for tube radios: Just Radios, Feb15, 2008
18. Low-cost drives can be an expensive alternative: Duan Smith, Fluke Corp., Sep1, 2004, EC&M Electrical Resource
19. Use of Capacitors for Power Factor Correction: PHD Chamber of Commerce & Industry, revised 2nd Ed.
20. Application Note- Capacitive Discharge Ignition Company Literature: ST Microelectronics. AN810
21. Summary of capacitor bank switching studies: Mitsubishi Electric Power Products, Inc.
22. A Hybrid Electrochemical Capacitor with Asymmetric Electrodes and Aqueous Electrolyte, By Jong H. Kim,, Jong H. Park, Chang S. Jin, and Kyung H. Shina, The 58th Annual Meeting International Society of Electrochemistry, 9th Sep.2007,Banff, Canada
23. Tantalum Hybrid Capacitor Life Test, David A. Evans, Evans Capacitor Company, 12th Annual Seminar on Double Layer Capacitors, Dec 9–11, 2002, Deerfield Beach, Florida

24. Gentlemen, stop your engines, Erick Schonfeld and Jeanette Borzo, Business 2.0, September 20 2006
25. RMFMEMs variable capacitors for tunable filters: Charles L. Goldsmith & others, Raytheon Systems Corpn. Texas.
26. General Descriptions of Aluminum Electrolytic Capacitors- Nichicon Corporation-Catalogue
27. Online certifications directory; UL
28. Harmonics in Polyphase power systems: Polyphase AC circuits, OpAmp Electronics
29. The Ultra Battery, Kevin Bullis, Technology Review, Feb13, 2006
30. EEstor - Blog, Greg Allen, 17th June 2008 Ultracapacitor.org
31. Power Capacitors-Series capacitor banks: Cooper Power Systems- Electrical Apparatus 230-31
32. Ultracapacitors challenge the battery, John M. Miller, World and I, Vol.19, June 2004
33. BASF Patent For Barium Titanate Ultracapacitor, Rob McMillin Rossmoor, California, Peak Oil Optimist, 12th Dec. 2007
34. Carbon aerogels and ultracapacitor, Greg Allen, 13th Apr. 2008, Ultracapacitor.org Article
35. Introduction to film Capacitors, Mark Gebbia, Illinois Capacitor, Inc
36. Maxwell and the Promise of Ultracapacitors: Greentech Media, Jan 29, 2010
37. Supercapacitor applications: Vina Technology Co., Ltd., Korea
38. Motor run capacitors: Martin Hudis and David C. Howe, Aerovox Inc. Mass., U.S.A.

Catalogues and Company Documents

1. ABB Capacitor literature
2. ABB Switzerland Ltd, advanced power electronics Literature Literature of ZEZ SILKO, s.r.o., Pod Cerným lesem 683, 564 22 Žamberk, Czech Republic
3. Literature of VEE-TECH Engineers 55, Chandralok Bldg. Mayo Hospital Road, Nagpur.-440018-
4. WIMA, U.S.A.
5. Ioxus, N.Y., U.S.A.
6. Professional Plastics, Inc, Fullerton, CA92831-USA: PEN literature
7. Steiner film Catalogue,
8. Jaivic Capacitors, Nashik :Motor Run Capacitors,
9. Martin Hudis & David C. Howe, Aerovox Capacitors
10. Yesha Electricals Pvt. Ltd. Literature
11. Magnewin, Sangli, Maharashtra :Literature.
12. Jognic's: Catalogue for capacitor machinery
13. What's SNU: Cornell Dubilier
14. Dynacomp – The top class reactive compensator: ABB Ltd.
15. Metallized Plastic Film Catalogue: EPCOS Ltd. Product Catalogue: Fomex Technology Co., ChinaKemet Electronic Corporation, CA, literatureCalculations and Tables. ESTAprop/ESTA dry Power Factor Correction Capacitor Low Voltage

16. Toray Film CatalogueLiterature, CDE Cornell Dubilier • 1605 E. Rodney French Blvd. • New Bedford
17. Electrolytic capacitors. Sam Parler, Cornell Dubilier Electronics, Inc., U.S.A
18. EMI Suppression capacitors: EPCOS
19. Noise suppression capacitors: Okaya Electric Industries Co., Ltd
20. Capacitors for power electronics. Westcode Semiconductor Inc., U.S.A. - Catalogue
21. Low Voltage Capacitor Discharge Welders: Catalog, Heron North America, Ontario, Canada.
22. KOPAFILM Elektrofolien literature.
23. Literature pseudocapacitor, NESSCAP Company Ltd., South Korea
24. Maxwell Double layer capacitors Data, Maxwell Technologies U.S.A. Catalogue
25. Boostcap ELDC Capacitor Data, Maxwell Technologies U.S.A. – Catalogue
26. Ultracapacitor Capacitor Literature, LS Cable, South Korea
27. HY-CAP Supercapacitor Literature: Vina Technology Co., Ltd., Korea
28. IGBT Snubber Capacitors for power Electronics, Aerovox Inc., MA., U.S.A.
29. Aerovox Capacitor catalog

Index

D

E

9 780071 848565

Printed by Libri Plureos GmbH in Hamburg,
Germany